transistor electronics

Basic Instruction in Electricity
and Electronics, with Major Emphasis
on Solid State Components

by

HOWARD H. GERRISH

Professor Emeritus
Humboldt State University
Arcata, California

and

WILLIAM E. DUGGER, JR.

Professor and Program Area Leader
Virginia Polytechnic Institute
and State University
Blacksburg, Virginia

South Holland, Illinois
THE GOODHEART-WILLCOX COMPANY, INC.
Publishers

Library of Congress Cataloging in Publication Data

Gerrish, Howard H.
  Transistor electronics.

  Includes index.
  1. Electronics.   2. Transistor circuits.
I. Dugger, William, joint author.   II. Title.
TK7816.G44   1981   621.3815'28   81—6740
ISBN 0—87006—394-4

INTRODUCTION

TRANSISTOR ELECTRONICS is intended to fulfill today's need for basic instruction in electricity and electronics with major emphasis on solid state components. Its depth of coverage will provide you with a comprehensive background in the field of electronics.

This revised text features new chapters on "Integrated Circuits" and "Microcomputers." It explores the use of solid state components in the fields of communication, industrial controls, automation, computer technology and space technology. It shows how solid state electronics has made possible the miniaturization of electronic equipment and improved circuit reliability to a remarkable degree.

TRANSISTOR ELECTRONICS also includes many new projects and experiments. Each completed project will aid you in understanding the theory presented and help develop your manipulative skill in electronic construction. In addition, a comprehensive course in personal safety is presented for students working with electricity and electronics. Carefully follow this safety program, both to insure maximum learning and also to assist in the development of safe working habits.

A TRANSISTOR ELECTRONICS LABORATORY MANUAL that is correlated to this textbook is available for the student or experimenter. This laboratory manual contains numerous experiments that help clarify the theory presented in TRANSISTOR ELECTRONICS.

Howard H. Gerrish
William E. Dugger

CONTENTS

SAFETY PRECAUTIONS FOR THE
ELECTRICITY-ELECTRONICS SHOP

There is always an element of danger when working with electricity. Observe all safety rules that concern each project and be particularly careful not to contact any live wire or terminal regardless of whether it is connected to either a low voltage or a high voltage. Projects do not specify dangerous voltage levels. However, keep in mind at all times that it is possible to experience a surprising electric shock under certain circumstances. Even a normal healthy person can be injured or seriously hurt by the shock or what happens as a result of it. Do not fool around. The lab is no place for horseplay.

PROJECTS AND ACTIVITIES

To develop a degree of skill and understanding, the following activities and "take home" projects are included in this text. They are listed in numerical order by page number.

Chapter 1

THEORY OF ENERGY

The fundamentals of electricity and electronics are founded in the atomic building blocks that are common to all matter. In this chapter, you will:

1. Discover the nature and atomic structure of matter.
2. Investigate the energy level of the atom.
3. Learn the meaning of electrical potential and electrostatic fields.

The early hunter probably hid on a rock ledge and hurled a stone at a wandering animal. Although this was a crude method of converting energy, it did provide food and clothing for the family. Unaware of the principles of science, the hunter bound a stone to the end of a stick to give the world its first hammer.

Since the beginning of time, people have searched for ways to convert energy into useful work. Only in recent years have we emerged from the dark ages of mystery and superstition and made intelligent investigations into the sources and uses of energy. There is a great expanse of time between the crude inventions of the past and the now sophisticated age of orbiting satellites, nuclear power plants and computerized machinery. It is said that more has been learned and recorded in the field of science during the last one hundred years than in all the previous years human beings have existed on Earth.

Astronauts who traveled to the moon have disclosed it to be a cold and barren outpost. Can you imagine the life, if any, on Earth if it were not for the sun and its continuous supply of light and heat energy? Energy is our life. It grows our food and supplies our water. It is harnessed to manufacture our homes, our clothes and many conveniences. Its conversion to transportation has provided us with ships and trains, automobiles and planes. Energy conversion has joined remote world locations within a world community by communications.

Is there a question in your mind why youth of the 20th century should study and investigate the origin and conversion of energy into useful work? Electronics is a study of energy conversion. It is a study of the discovery of the development of an electrical potential and its transfer and conversion into work.

The structure of all matter is a combination of atoms in elementary form or in combinations with others. Each atom is a source of energy.

THE STRUCTURE OF MATTER

Anything which occupies space and has mass is called "matter." Matter may be composed of any number of elementary substances found in nature. Matter may be a compound or mixture of elements or elements in their pure form. The ELEMENT is the basic building block of nature. The element was originally defined as a substance which could not be divided or decomposed into simpler substances. However, through the development, construction and use of accelerators or atom smashers, scientists have divided some of the heavier elements and found the source of atomic energy.

Familiar examples of ELEMENTS which you use daily are iron (Fe), copper (Cu), aluminum (Al), carbon (C), gold (Au) and silver (Ag). There are over one hundred of these elements which have been identified in nature and in the laboratory. It is a revealing experience to study the Periodic Table of Elements and identify those elements which are familiar.

ATOMS AND MOLECULES

Just as the element is the building block of nature, the element is composed of small particles called ATOMS. An atom is defined as the smallest particle of an element that retains all the properties of the element. DALTON'S* conception of the atom was, "all materials are composed of minute indestructible particles" and "the smallest component part of an element that enters into a chemical reaction."

Combinations of two or more atoms, either of the same kind or different, are called MOLECULES and constitute matter. The molecule may be defined as the smallest particle of a compound. If further subdivision is made, the molecule will not be the same substance. For example: a grain of salt (NaCl) is a chemical combination of sodium (Na) and chlorine (Cl). If you could divide the grain of salt an infinite number of times, you would still have salt. In spite of the infinite division, you would still have an atom of sodium and an atom of chlorine. If the smallest division or molecule were chemically divided, it would no longer be salt. It is possible, also, to have molecules made up of atoms of the same kind.

ATOMIC STRUCTURE

Atoms of all elements are composed of minute particles of electrical charges. Elements differ only in the number of particles and their arrangement. THOMSON** proposed a theory that an atom was a small ball containing an equal amount of positive and negative charges of electricity. Its overall charge was said to be neutral. Further studies by RUTHERFORD*** proved that the POSITIVE charges were con-

centrated at the center or NUCLEUS of the atom and the NEGATIVE charges rotated in specific orbits around the nucleus. These negative particles of electricity or charges are called ELECTRONS. The positive particles are called PROTONS. A third particle, which is neutral (neither positive nor negative), is called a NEUTRON. These neutrons are found in the nucleus of the atom. Thus emerges our present concept of the structure of an atom. Examples of model structures of hydrogen, carbon, silicon and germanium are shown in Fig. 1-1.

All elements are classified in the PERIODIC TABLE OF ELEMENTS by atomic number and weight. The atomic number designates the number of oribtal electrons in the atom. Refering again to Fig. 1-1, the atomic number of hydrogen is 1, carbon is 6, silicon is 14 and germanium is 32. The atomic weight of an atom is governed by the number of neutrons and protons contained in its nucleus. The weight of an atom is referenced to the weight of an OXYGEN atom, which is 16. It is important to remember that the mass of any atom is made up of the neutrons and protons. The mass of an electron is only 1/1845 that of a proton and therefore does not contribute significantly to the total mass and weight.

---

* DALTON, JOHN (1766-1844). The English chemist and physicist who is known for his ATOMIC THEORY. He first formulated a rudimentary table of atomic weights. He believed that all atoms of a given element were identical. He devised a system of classification of chemical compounds based on the number of atoms involved. His observations were crude, but his atomic theory and concept of atomic weights were fundamental. (Encyclopedia of Science)

** THOMSON, SIR JOSEPH JOHN (1856-1940). An English physicist who investigated the conductivity of gases exposed to X rays. He was able to determine the mass of cathode ray particles. In 1906 he received the Nobel Prize for his investigation on the transmission of electricity through gases. His discoveries are the foundation for the present day theory of "electron flow" as an electric current.

*** RUTHERFORD, ERNEST (1871-1937). A British physicist who is best known for his discoveries in the field of radioactive materials. He proposed the theory that an atom consisted of a compact positive nucleus surrounded by negatively charged electrons in motion. It remains valid to the present day.

ATOMIC ENERGY LEVELS

In our previous discussion, reference has been made to orbital electrons. It is hard to believe that electrons are truly in motion in an apparently solid material such as a piece of copper. Yet, it is nevertheless true. The question arises about the dimension or diameter of the orbits of electrons around the nucleus. In the quantum mechanical model of the atom,

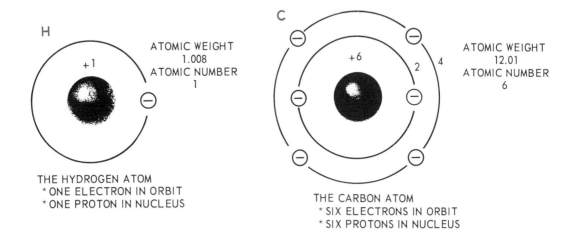

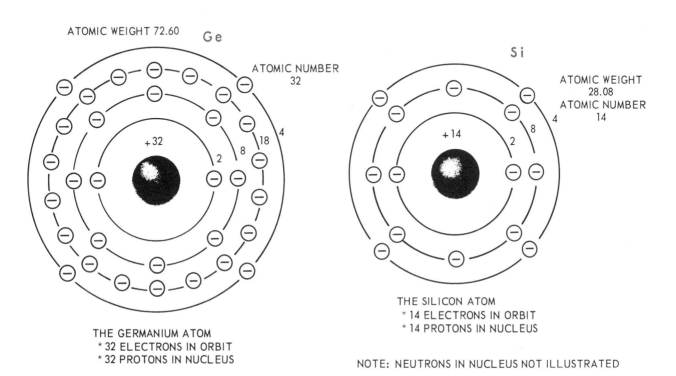

Fig. 1-1. The atomic structure of hydrogen, carbon, silicon and germanium.

the orbiting electron is held an average distance from the nucleus. At this average diameter of the orbit, these forces are in balance.

The atom contains both mass and motion, and therefore two kinds of energy:

1. POTENTIAL ENERGY, which is a result of its position. Example: There is considerable potential energy stored in a lake in the mountains because of elevation. If an outlet is provided, the water will rush down the mountainside.

2. KINETIC ENERGY, which is the energy of motion. Example: If an automobile moving at five miles per hour (mph) were driven into a wall, it would be only partially damaged. At 50 mph (80.5 km), the car and driver might be completely destroyed. Energy of motion causes objects in motion to continue in motion.

In the atom, the sum of the potential and kinetic energy determines the energy level of the atom and the radii of the orbital electrons.

ENERGY SHELLS

The concept of energy levels in an atom is most important in the study of solid state electronics. This concept suggests that electrons orbit in specified shells or layers around the nucleus. Each shell corresponds to a fixed energy level. The number of electrons in each shell may be predicted by the formula, $2N^2$, where N is the number of the shell. See Fig. 1-2.

Each major shell is then divided into sub-shells which represent a level of energy. These sub-shells are illustrated in Fig. 1-3 with letter designations of S, P, D and F. The number of electrons in each sub-shell may also be predicted since the quantity of electrons in each shell is limited to:

S shell − 2	D shell − 10
P shell − 6	F shell − 14

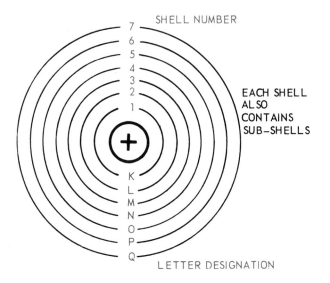

Fig. 1-2. Electrons are contained in shells of specified energy levels around the nucleus. Shell letter and number designations are shown.

PHOTONS AND PHONONS

The interesting study of energy shells of an atom would not be complete without mention of EXCITIATION. The addition of energy to an atom by means of heat, light or an electric field will cause electrons to jump outward into the next energy shell.

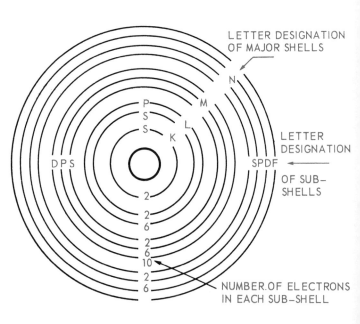

Fig. 1-3. Each shell is divided into sub-shells. Each sub-shell represents a specific energy level. Sub-shell letter designations are shown with electrons in each shell.

A discreet quantity of heat energy is called a PHONON and a discreet quantity of light energy is a PHOTON. When these discreet quantities of energy are accepted by the atom, the electrons move outward from their sub-shell to the next higher energy sub-shell. They can also move from one major shell to the next higher major shell.

Also, if electrons move to the next lower shell, energy is given off. Example: As a warm object cools, it gives off heat. Conversely, to heat an object requires the addition of heat.

Under the influence of an electric field, electrons may also be caused to move to outer orbits. This will be studied later in this chapter.

VALENCE

In semiconductors, the number of electrons in the outer shell, less than the full permissible complement for that shell, determines the VALENCE of the atom. The outer shell, therefore, is called the VALENCE SHELL. The valence of an atom is an indication of its ability to gain or lose electrons, and it determines the electrical and chemical properties of the atom. An atom which has almost its full complement of electrons in the valence shell will easily gain electrons to complete its shell. However, a relatively large amount of energy is required to free any of its electrons. Conversely, an atom which has only a small number of electrons in the valence shell compared to its permitted amount will lose these electrons quite easily.

Elements are arranged in groups on the Periodic Table, Fig. 1-4, according to their valence. Those in Group IVA have four valence electrons; Group IIIA are TRIVALENTS, meaning they have three valence electrons; Group VA are PENTAVALENTS, meaning they have five valence electrons.

IONIZATION

Up to this point in our discussion, we have assumed an equal number of protons and electrons in the atom. The electric charge is neutral. It is possible by adding external energy to the atom to upset this balance or neutrality. If electrons are driven out of an atom (by adding energy), the protons and electrons are out of balance. With the protons in a majority, the atom becomes a POSITIVE ION. On the other hand, if an atom with an incomplete shell accepts an electron from some source, the electrons are in the majority and the atom becomes a NEGATIVE ION. The process is called IONIZATION.

To remove an electron from an atom, energy levels must be raised. As stated earlier, this may be accomplished by heat, light or an electric field. The energy required to do this is called the IONIZATION POTENTIAL.

In Chapter 2 of this text, several ways of producing an electric potential will be discussed.

COVALENT BONDING

Most inorganic materials have their atoms arranged in crystalline structure. The crystals of a substance may be arranged in random fashion, or they may assume a LATTICE formation, Fig. 1-5. Every crystal will be like its adjacent neighbor and arranged in a precise manner.

For simplicity, only the valence electrons are shown in Fig. 1-5. Please note that the valence of germanium is 4 and it has four valence electrons. To complete the valence band of germanium, eight electrons are required. Since no free electrons are available, each atom shares its valence electrons with the other. The atoms are firmly attached in a crystal-lattice structure by COVALENT BONDING. If an electrical potential was applied to a crystal of this type, no electric current would flow through the crystal since there would be no "free electrons" to transfer the energy.

LAW OF CHARGES

In the discussion of ionization, it was pointed out that under certain conditions atoms

PERIODIC TABLE OF THE ELEMENTS

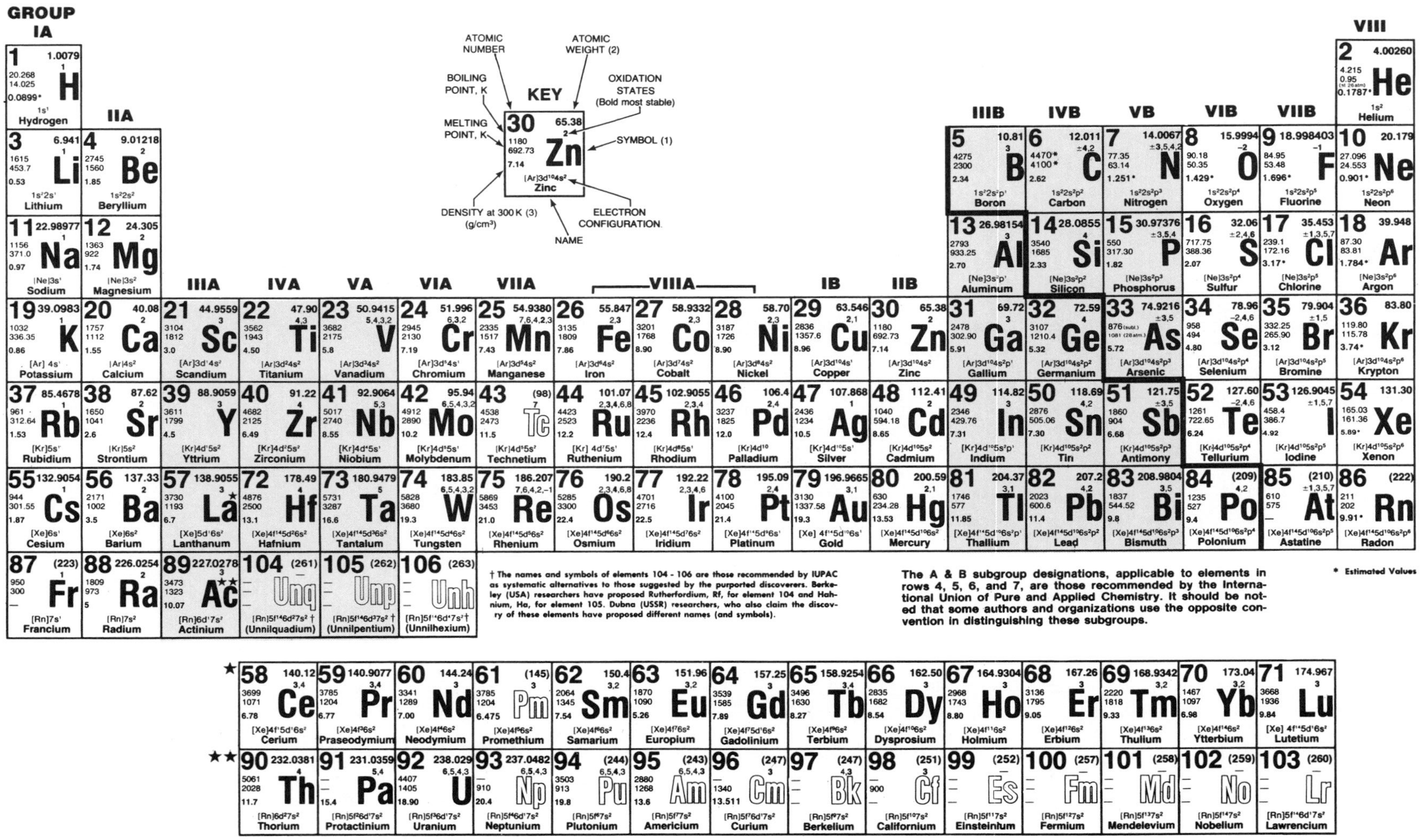

Fig. 1-4. The Periodic Table of Elements classifies all known elements by atomic weight and atomic number. See "key" at upper center. (Sargent-Welch Scientific Co.)

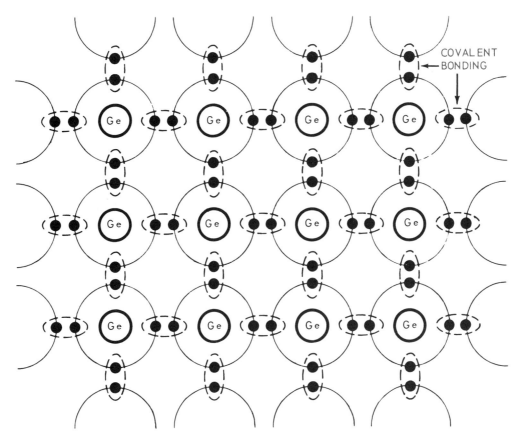

Fig. 1-5. The lattice crystalline structure of germanium. Only valence electrons are shown for simplification.

could lose or gain electrons. When such is the case, a material or body which has gained electrons is charged NEGATIVELY. Existing in space around such a charged body is an invisible ELECTROSTATIC FIELD. By convention, a negative electrostatic field is illustrated by lines and arrows pointing toward the charged mass. See Fig. 1-6.

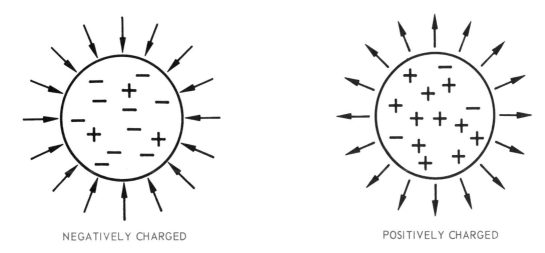

NEGATIVELY CHARGED

POSITIVELY CHARGED

Fig. 1-6. The electrostatic field of force existing around a negatively and positively charged body.

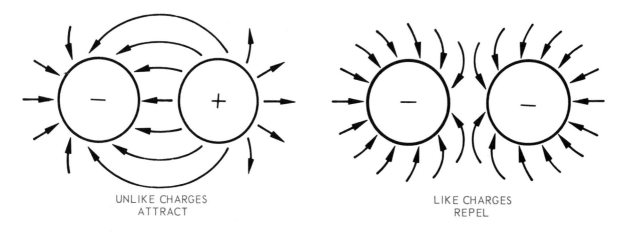

UNLIKE CHARGES
ATTRACT

LIKE CHARGES
REPEL

Fig. 1-7. The first law of electric charges. Unlike charges attract. Like charges repel.

A similar condition applies to a body or mass from which electrons have been driven. It is POSITIVELY charged and also supports an electrostatic field. Conventionally, a positively charged body is designated by lines and arrows pointing outward, Fig. 1-6.

If these two unlike charged masses or bodies are brought close together so that their fields will interact, an attractive force will appear which tends to bring the bodies together. Refer to Fig. 1-7.

If the two bodies possess the same charge, either positive or negative, a repulsive force would tend to hold the bodies apart. This introduces one of the more important laws in the study of electricity, the Law of Charges:

LIKE CHARGES REPEL EACH OTHER
UNLIKE CHARGES ATTRACT EACH OTHER

This law will be applied to explain the theory and operation of many components, circuits and devices.

THE COULOMB

The atom is an infinitesimally small particle of matter. In fact, several billion and more atoms would be required to make up the "head of a pin." The electron is even smaller. In order to have a quantity of electrons of sufficient size to use, the term COULOMB* is used as a quantitative measure of electrical charges. It represents a charge of 6.24×10^{18} electrons or

6,240,000,000,000,000,000 electrons. The letter symbol Q has been assigned to the coulomb. Coulomb's law states that:

THE FORCE BETWEEN TWO CHARGES IS DIRECTLY PROPORTIONAL TO THE PRODUCT OF THE CHARGES AND INVERSELY PROPORTIONAL TO THE SQUARE OF THE DISTANCE BETWEEN THE CHARGES.

THE ELECTROSCOPE

An instrument used in the laboratory to detect a charge and determine its polarity is called the ELECTROSCOPE, Fig. 1-8. It consists of a glass jar with a center rod and ball. Two gold leaves hang on the rod at the center of the jar. If the two sections of gold leaf become charged alike, either negatively or positively, they will expand because like charges on both leaves will repel each other.

To demonstrate this phenomenon of static electricity, vigorously rub a vulcanite rod with a piece of wool or cat's fur. This action places a negative charge on the rod. Then, bring the rod close to the ball on top of the electroscope, but do not touch the ball. The gold leaves will expand. See Fig. 1-9.

---

* COULOMB, CHARLES (1736-1806). A French physicist who conducted extensive research in the measurement of small forces. His experiments verified the laws of electrostatics. His work made possible the definition of quantity of electric charge which is named in his honor. (Encyclopedia of Science)

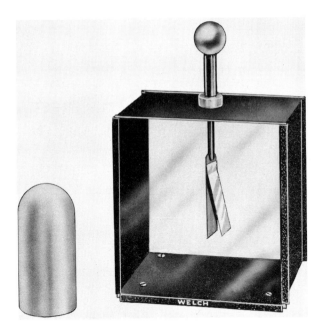

Fig. 1-8. An electroscope will detect a charge and reveal its polarity. (Sargent-Welch Scienntific Co.)

While the rod is in position and the leaves are expanded, touch the ball with your finger. Remove your finger and remove the rod. The leaves remain expanded. What happened? When you touched the ball, electrons escaped through your finger and into your body. When the finger and rod were removed, the electroscope had no chance to regain its lost electrons. It therefore became charged positively and the leaves expanded. This is charging an electroscope by INDUCTION since only the electric fields were used.

Touching the electroscope will permit it to regain its lost electrons and become neutral with its leaves collapsed.

Rub the vulcanite rod again, then touch the ball of the electroscope with the rod. Remove the rod and the electroscope remains charged NEGATIVELY. The rod shared its electrons with the electroscope. This is charging by CONTACT. Refer to Fig. 1-10.

These experiments may be repeated, using a glass rod and a piece of silk. Rubbing will place

Here is what happened: The negative electrostatic field around the rod repelled the electrons from the ball down to the gold leaves. Since both leaves would have a negative charge, they would repel each other.

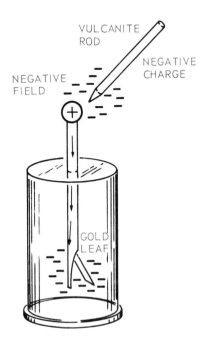

Fig. 1-9. Negative electrostatic field around the rod drives electrons down into the jar and causes gold leaves to expand.

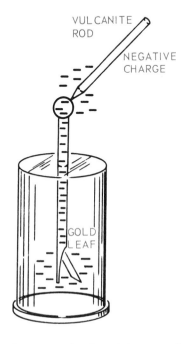

Fig. 1-10. The electroscope is charged by contact with a negatively charged rod.

a positive charge on the glass rod. A positive rod will charge the electroscope negatively by induction and positively by contact.

LESSON IN SAFETY: When working with electricity, as well as with all kinds of machinery, never permit yourself to be distracted. Equally important, do not distract your fellow worker. Do not clown around! Jokes have no place in the electrical shop. Your seemingly harmless interruption may cause a serious accident and also damage to expensive equipment.

ELECTROSCOPE PROJECT

The electroscope shown in Fig. 1-11 is made with a glass beaker found in the chemistry department. Any glass jar (such as a Mason jar) would be suitable, if you made an insulated cover for it.

Secure a piece of 3/8 in. brass rod and file one end to a flat side. Round the opposite end for a finished appearance. Attach a gold leaf with just a drop of cement at one end only. Assemble the electroscope as shown in Fig. 1-11 and try it out.

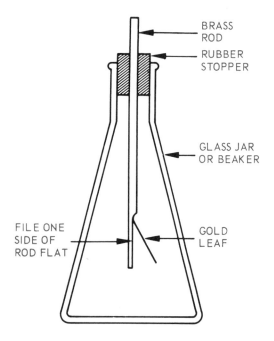

Fig. 1-11. A plan for an electroscope.

LESSON IN SAFETY: Is an electric shock fatal? The effects of an electric current can be generally predicted by reference to Fig. 1-12. The current does the damage! Notice that currents as small as 100 milliamperes (0.1 amperes) can be fatal. Currents below 100 milliamperes can be serious and fatal.

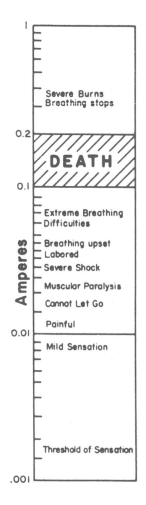

Fig. 1-12. Physiological effects of electric currents.

TEST YOUR KNOWLEDGE

Write your answers to these questions on a separate sheet of paper. Do not write in this book.

1. Anything which occupies space and has mass is called _____.
2. What is an element?
3. What is an atom?

4. What is a molecule?
5. What is a proton, an electron, a neutron?
6. The atomic weight of an atom is governed by:
   a. The number of neutrons, protons and electrons in the atom.
   b. The number of neutrons and protons contained in its nucleus.
   c. The number of orbital electrons in the atom.
7. What is a positive ion, a negative ion?
8. Considering the valence of an atom, which would be a better conductor: an atom with one valence electron; an atom with five valence electrons?
9. Potential energy of an atom is the energy of:
   a. Mass.
   b. Motion.
   c. Position.
10. Kinetic energy of an atom is the energy of:
   a. Mass.
   b. Motion.
   c. Position.
11. The weight of an atom is referenced to the weight of _____ , which is 16.
12. State the First Law of Charges.
13. Existing in space around a charged body is an invisible _____ field.
14. What is a coulomb?
15. What is the letter symbol for coulomb?
   a. C.
   b. K.
   c. Q.
16. What is an electroscope?
17. How does the distance between two charged bodies affect the forces between them?
18. Electric currents as small as _____ milliamperes can be fatal.

Chapter 2

SOURCES OF ELECTRICITY

Where does electricity come from or where can it be found as a source? We know energy cannot be created or destroyed. However, it can be converted from one source to another. In this chapter, you will:

1. Investigate common sources of electrical energy.
2. Become familiar with accepted units of measurement used in electricity.
3. Experiment and study the effects of connecting voltage sources in series and parallel.

ENERGY SOURCES

To produce the multitude of electrical and electronic phenomena, a source of energy must always be provided. Consider how static electricity was produced in Chapter 1 by rubbing a vulcanite rod with a piece of wool.

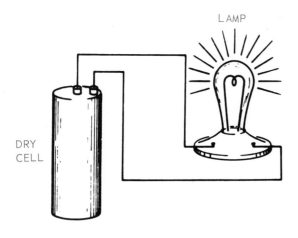

Fig. 2-1. A battery converts chemical energy to electrical energy. The lamp converts electrical energy to light and heat energy.

Remember, however, that a SOURCE is not necessarily a reservoir or storage of energy readily available for use. More correctly, a source is a means of converting some other forms of energy into electrical energy. A flashlight battery stores no electrical energy. It is a means of converting chemical energy into electrical energy. A generator is only a means of converting mechanical energy into electrical energy. In electricity-electronics, many circuits, components and devices are used to reconvert the electrical source energy into useful work and desired effects.

POTENTIAL DIFFERENCE

Through the phenomenon of ionization, a body may become charged either positively or negatively by losing or gaining electrons respectively. When two bodies or terminals have a difference of charge, they can be described as having a DIFFERENCE OF POTENTIAL (PD). A wire, circuit or device connected between these terminals will permit a movement or transfer of energy between the terminals, and the moving energy can be made to do WORK. This transfer of energy will continue as long as there is a difference of potential.

In Fig. 2-1, a battery is connected to a single lamp. Follow the energy conversion: The battery changes chemical energy to electrical energy, which is converted to heat and light energy by the lamp. The lamp will become dimmer as the chemical energy of the battery is used up and the POTENTIAL DIFFERENCE between its terminals approaches zero.

The unit of measurement of potential difference is the VOLT. The letter "V" will be used as the symbol for voltage in transistor circuits in this text. In parts dealing only with basic electricity and vacuum tubes, the letter "E" is used to represent voltage or ELECTRO-MOTIVE FORCE.

A scientific determination of the value of one volt can be understood when you realize that a certain potential difference is required to move a quantity of electricity or electric charge through a circuit. When this movement occurs, WORK IS DONE, and it is expressed as "WORK PER UNIT CHARGE" or "JOULES PER COULOMB."

NOTE: A JOULE is approximately .738 foot pounds of work. A COULOMB is a quantity of electrons. See Chapter 1.

A VOLT is equal to the potential difference required to move one coulomb of charge and accomplish one joule of work. This is expressed by the equation:

$$V = \frac{W}{Q} \quad \begin{array}{l} \text{work in joules} \\ \text{coulombs} \end{array}$$

The fixed value of one volt has been established by the International Electrical Congress and by law in the United States.

ELECTRIC CURRENT

Movement of electrical energy through a conductor is named CURRENT. It is the current which does the work. Voltage only will cause a current to flow. The actual mechanics of current flow in a conductor will be described in detail in Chapter 3 under the headings CONDUCTORS and SEMICONDUCTORS.

An AMPERE* of current is described as the current flow in a conductor at the rate of one coulomb per second at any given point in the circuit. Current is assigned the letter "I." It is measured in amperes, milliamperes and micro-amperes. In this text, current will be considered as "electron flow" and will always be in the direction of minus to plus, or negative to positive.

POTENTIAL DIFFERENCE BY CHEMICAL ACTION

At the close of the eighteenth century, an Italian scientist by the name of GALVANI** was doing some experimentation with a dead frog. He had the frog fastened in spread eagle fashion on his bench by attaching copper wires to its legs. Each time that Galvani touched the frog with his steel scalpel, the frog jumped! How could this happen? The frog was dead. Galvani reasoned that the frog contained animal electricity which caused its muscles to contract when touched with a knife. This assumption was proved to be incorrect by another Italian scientist, ALESSANDRO VOLTA.*** Volta produced electricity by immersing two dissimilar metals in a fluid which chemically reacted with one of the metals.

We must give a great deal of credit to these scientists. The cells and batteries which we use so frequently to power electronic circuits had their origin through the inventiveness and genius of the early experimenters. It will be interesting to perform the following experiments since they will lead to a better understanding of this valuable source of electrical energy.

---

* AMPERE, ANDRE (1775-1836). Professor of physics at the College DeFrance in Paris. He discovered the cause and nature of electric currents. He identified the electrochemical and electromagnetic effects as properties of current. He invented the term "galvanometer" as the name of a current measuring device. Ampere, the unit of measurement of current, is so named in his honor.

** GALVANI, LUIGI (1737-1798). An Italian physiologist and professor at the University of Bologna who investigated the theory of animal electricity known as "galvanism." His work led to the development of the voltaic cell.

*** VOLTA, COUNT ALESSANDRO (1745-1827). An Italian physicist who disproved the theory of Galvani by the discovery that "animal electricity" was, in truth, "metallic electricity." He discovered the voltaic cell which bears his name. Volt, the unit of potential difference or electromotive force is also named in his honor.

SIMPLE CELLS

Secure a dime, a penny and a small square of blotting paper. Soak the blotting paper in a strong salt solution. Now assemble your simple CELL as shown in Fig. 2-2. A voltmeter connected to this simple cell will indicate a potential difference.

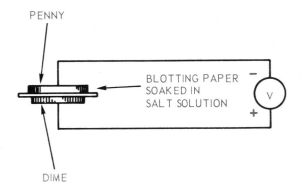

Fig. 2-2. A voltmeter connected to this cell will show a potential difference.

Further experimentation may be done by inserting a piece of copper and a piece of zinc in a lemon. Refer to Fig. 2-3. The voltmeter will indicate a potential difference, which is measured in VOLTS. Notice that in both experiments, plus (+) and minus (−) signs were placed by the terminals to indicate POSITIVE and NEGATIVE. It is extremely important to always observe the correct polarity when connecting a meter to a circuit.

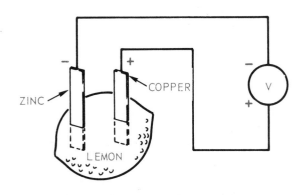

Fig. 2-3. Zinc and copper electrodes inserted in a lemon will produce a potential difference.

A brief scientific explanation of this phenomenon is in order. In Fig. 2-4, a ZINC ELECTRODE is first placed in a jar containing a mild solution of ELECTROLYTE (acid and water). Small bubbles of hydrogen will appear around the zinc, showing that the acid is reacting with the zinc. When the zinc goes into the solution, the terminal of the zinc electrode will be found to have a negative charge. Then, the carbon rod is placed in the solution. Although the acid and the carbon do not chemically react, it will be found that the carbon terminal will become positively charged.

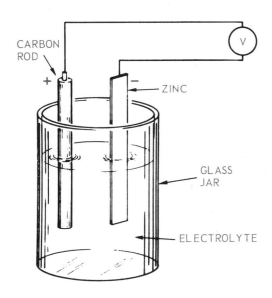

Fig. 2-4. Carbon and zinc electrodes in acid electrolyte will produce a potential difference.

In Fig. 2-5, a chemical explanation is given, using sulfuric acid and water as the electrolyte. The acid disassociates in the water to form positive hydrogen ions $2H^+$ and negative ions, SO_4^-. The sulfate ions move toward the zinc and combine with it to form $ZnSO_4$. This action causes the zinc to become negatively charged, since the neutral zinc must give up some positive ions to form the zinc sulfate. The $2H^+$ ions move toward the carbon rod and form a blanket of hydrogen bubbles or positive ions around the rod. It becomes positively charged. You may wish to ask your chemistry instructor for a more detailed explanation. Refer to Figs. 2-5 and 2-6.

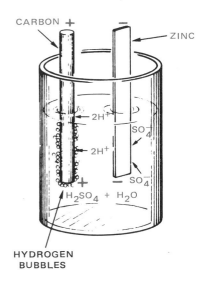

Fig. 2-5. The chemical reaction in a simple voltaic cell.

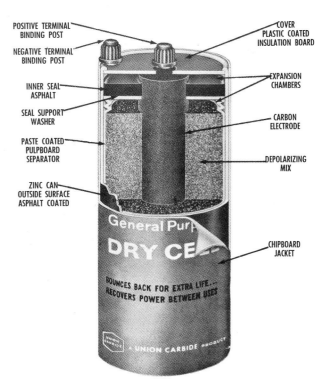

Fig. 2-7. A cutaway view of a No. 6 dry cell.
(Union Carbide Co.)

Fig. 2-6. An experimental setup of a voltaic cell.
(Lab-Volt Div., Buck Engineering Co.)

Notice in your experimental voltaic cell, that the zinc is eaten away or used up as the cell is used. This cell cannot be recharged and is classified as a PRIMARY CELL.

Another example of a primary cell is the familiar dry cell. In Fig. 2-7, it is easy to identify the zinc electrode or case of a dry cell. The case is negative. The carbon electrode through the center is the positive electrode. The electrolyte is in a paste form consisting mostly

of ammomium chloride and manganese dioxide. The chemical equation may be represented as:

$$Zn + 2NH_4Cl + 2\,MnO_2 \rightarrow Mn_2O_3 + H_2O + 2NH_3 + ZnCl_2$$

The left side of the equation shows the chemical content of a new dry cell. As current is drawn from the cell, the zinc, ammonium chloride and manganese dioxide combine to produce manganese dioxide, water, ammonia and zinc chloride. The dry cell can be recharged if certain precautions are observed:

1. Voltage of dry cell should not be below one volt when removed from service.
2. Charging rate should be low and distributed over 10 to 15 hours.
3. Recharged cells should be used at once, since they have limited shelf life. Several companies manufacturer and sell devices for recharging or rejuvenating dry cells.

In the chemical equation given, note that maganese dioxide, MnO_2, appears. What is its purpose? It is a DEPOLARIZER. During the discharge of a cell, hydrogen bubbles collect around the carbon electrode and form an effective insulating blanket. The depolarizer combines with this hydrogen and other products of the action to form water and Mn_2O_3. The depolarizer adds substantially to the life and output of the cell.

MERCURY CELLS

A widely used type of dry cell is the mercury cell. Much of the newly developed miniaturized electronic gear would not have been possible without the development of the mercury cell as an energy source. Although a detailed technical explanation of the cell is beyond the scope of this text, the general construction is illustrated in Fig. 2-8.

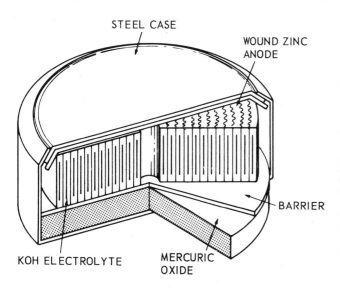

Fig. 2-8. A cutaway view of a mercury cell.

The negative electrode is made of highly purified compressed zinc powder. The positive plate is made of compressed mercuric oxide and graphite. The electrolyte is a solution of potassium hydroxide saturated with zinc ions. The plates are separated by a porous material. The distinct advantages of the mercury cell over the zinc-carbon cell are high density of electrical energy in a very small package and a reliable terminal voltage. Mercury cells also enjoy a long shelf life without deterioration.

ALKALINE CELLS

Many of the cells which use a caustic electrolyte may be classified as alkaline cells. However, we will limit our discussion to the RECHARGEABLE NICKEL-CADMIUM CELL, which is popular in cordless types of home appliances, such as an electric toothbrush or a cordless electric carving knife. These cells have contributed to the development and availability of a whole new family of devices for home and industrial use. These cells have a very low internal resistance and can deliver high currents with very little loss of terminal voltage. They can be recharged many times and will hold their charge for relatively long periods of time. You will read about them under the name of "ENERGY CELLS" or "POWERPACKS."

The nickel-cadmium cell using sintered plates is described in Fig. 2-9. It consists of a positive plate, a separator, a negative plate, the electrolyte and a cell container. The plates are made by sintering powdered nickel into a nickel wire screen, which makes a quite strong and flexible plate. The plate is impregnated with nickel salt solutions for a positive plate and with cadmium salt solutions for a negative plate. The separator is made of some absorbent insulating material. The electrolyte is a solution of potassium hydroxide. The container is a steel can, which is sealed after the cell is placed in it. The open circuit terminal voltage of this type of cell is 1.33 volts at room temperature.

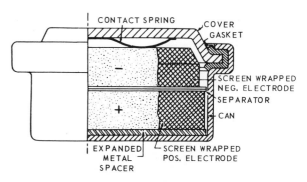

Fig. 2-9. A cutaway view of a rechargeable nickel-cadmium cell. (Union Carbide Co.)

BATTERIES

Up to this point in your studies, only the CELL has been discussed. When two or more cells are connected either in series or parallel, the combination is rightly called a BATTERY. Do not be confused by this term, since many people call a flashlight cell a flashlight battery. This is technically incorrect.

Cells connected in SERIES means that they are connected end to end, or plus to minus, as shown in Fig. 2-10. Note that both the pictorial views and the schematic symbols are compared. The series method is used to increase the voltage of the battery above a single cell. The output voltage may be computed by the following formula:

$$E_{out} = N \times E_c$$

where,

E_{out} = voltage output of battery

N = the number of cells

E_c = voltage of one cell

In Fig. 2-11, the four cells are connected in parallel. With this method, the voltage output does not increase, but the total supply of energy and the capacity to produce a given current over a given time is substantially increased.

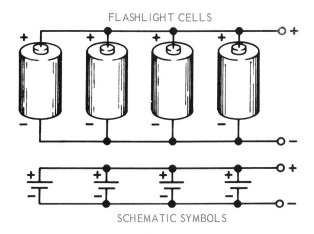

Fig. 2-11. A battery of four parallel cells.

Cells may also be grouped as series groups with groups connected in parallel; or parallel groups connected in series. These groupings are shown schematically in Fig. 2-12.

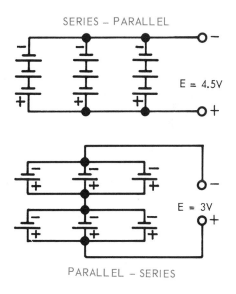

Fig. 2-12. Combination groups of 1.5 volt cells.

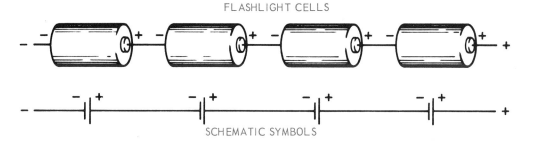

Fig. 2-10. A battery of four series cells.

Which grouping is best for a given condition? It depends upon the internal resistance of an individual cell. Generally, if the resistance of the load connected to the battery is high, then use the series connection. If the resistance of the load is low, then the parallel grouping will produce a larger current.

SECONDARY CELLS

A SECONDARY CELL is a cell in which the chemical action may be reversed, and the cell can be RECHARGED by applying a reverse voltage. The classic example of this type is the LEAD-ACID storage battery used in the automobile. It is made up of three (6 volts) or six (12 volts) secondary cells in series. The "storage" battery is incorrectly named, since it does not "store" electrical energy. It converts chemical energy to electrical energy.

Fig. 2-13 is an excellent view of a typical 12 volt battery used in an automobile. Appropriate numbers indicate the various components, which should be a part of your basic knowledge.

LESSON IN SAFETY: During the battery charging process, highly explosive hydrogen gas may be present. Do not light matches near charging batteries. Charge only in a well ventilated room. Batteries should be connected to the charger before the power is applied. Otherwise, sparks made during connection might ignite the hydrogen gas and cause an explosion.

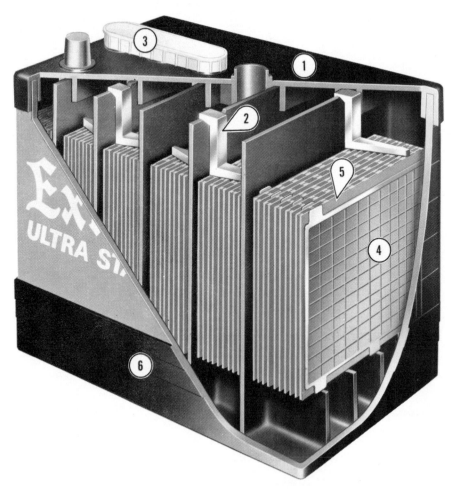

Fig. 2-13. A typical 12 volt automotive battery. 1—One piece battery cover. 2—Intercell connector (six cells in series). 3—Gang vent plugs and filler caps. 4—Negative plate. 5—Separator. 6—Battery container.

CONSTRUCTION OF PLATES FOR LEAD-ACID CELL

Both the positive and negative plates of the storage battery cells are made by pressing a paste of very finely ground lead on a lead alloy grid. The grid lends mechanical strength to the plates. Next, the paste plates are dried to form a strong thin plate of active material. Then, the plates are assembled in groups of alternate negative and positive plates. The polarity of the plates is formed by charging and reaction with the acid electrolyte. In a fully charged state, the positive plates are lead peroxide (PbO_2) and the negative plates are spongy lead (Pb).

Each group of plates (cell) has one additional negative plate. Example: A 13 plate cell has 7 negative and 6 positive plates. Each plate is separated from its adjoining plates by separators of porous plastic or rubber. These separators electrically insulate one plate from another, yet are sufficiently porous to provide free passage of the electrolyte.

The groups of cells are then set in the compartments of the battery case and connected in series by lead interconnecting links. The cover is sealed in place. Electrolyte is poured through the cell vents and the vents are capped. The electrolyte used in an automotive storage battery is sulfuric acid (H_2SO_4) and distilled water.

SPECIFIC GRAVITY

In discussing battery electrolyte, SPECIFIC GRAVITY becomes an important consideration. Specific gravity is a classification of liquids according to their weight in respect to an equal volume of water. Water is assigned the number 1.000. Sulfuric acid is almost twice as heavy as water. In fact, it has a specific gravity of 1.835. The electrolyte used in the lead-acid battery is a mixture of acid and water with a mixed weight of approximately 1.300 times heavier than water. During the discharge of a lead-acid cell, the electrolyte changes to almost pure water. During charging of the cell, the

Fig. 2-14. A hydrometer is used to measure the state of charge of a lead-acid cell.

electrolyte becomes richer in acid and approaches 1.300. This information becomes very useful since the STATE OF CHARGE of a lead-acid cell can be determined by measuring the specific gravity of the electrolyte. This is done by means of a HYDROMETER, Fig. 2-14. The state of charge broken down by specific gravity readings is as follows:

State of Charge	Specific Gravity
100%	1.260
75%	1.230
50%	1.200
25%	1.170

The hydrometer operates on the physical law that:

A FLOATING OBJECT WILL SINK INTO A LIQUID UNTIL THE WEIGHT OF THE LIQUID IT DISPLACES IS EQUAL TO THE WEIGHT OF THE FLOATING OBJECT.

The float in the hydrometer sinks into the liquid drawn from the battery. A scale on the hydrometer float reads in specific gravity.

CAUTION: When checking the state of charge of a battery, you are handling acid. Be very careful that drops of acid clinging to the end of the hydrometer do not drop off on your hands or clothes. It will burn your hands. It will eat holes in your clothes. It is always wise to have some BAKING SODA nearby to neutralize any acid accidently spilled.

LEAD-ACID CELL CHEMISTRY

The chemical action within the lead-acid cell is graphically described in Fig. 2-15. Note on

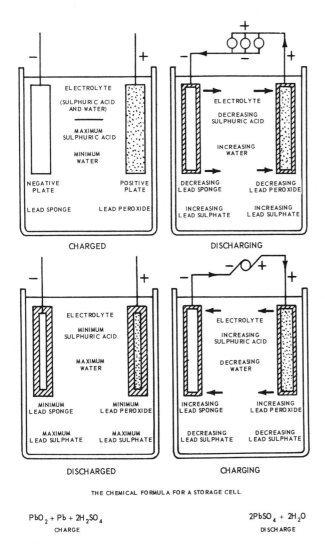

THE CHEMICAL FORMULA FOR A STORAGE CELL

$$PbO_2 + Pb + 2H_2SO_4 \qquad\qquad 2PbSO_4 + 2H_2O$$

CHARGE   DISCHARGE

Fig. 2-15. A diagram of the chemical action within a lead-acid cell. (ESB, Inc., Automotive Div.)

these diagrams some of the facts previously discussed.

1. In a charged state, negative plates are spongy lead and positive plates are lead peroxide.
2. During discharge, electrolyte increases in its water content. This is why a battery would freeze quite easily when discharged.

RATINGS OF BATTERIES

The ability of a battery to deliver power continuously over a given time is an indication of the quality of a battery. It is measured in AMPERE-HOURS. Example: A battery rated as 100 AMP-HOURS would continuously supply 5

amperes for 20 hours (5 x 20 = 100 amp hours). Roughly speaking, the same battery might deliver 10 amperes for 10 hours (10 x 10 = 100 amp hours) or one ampere for 100 hours (1 x 100 = 100 amp hours). This amp hour rating is not exactly true since RATE OF DISCHARGE also affects the capability of the battery. If rapidly discharged, it will not live up to its amp hour rating.

If you were buying a new battery for your car, you should compare the ampere-hour ratings of the batteries on sale. Also, the number of plates and the kind of separators used will indicate the quality of the battery. A well-known manufacturer with a good reputation is one of your best assurances of getting the most for your money.

DRY CHARGED BATTERIES

A dry charged battery is a convenient type to ship and store. The battery is shipped with charged plates, but without acid. The battery is "ready for use" when filled with the correct grade of acid electrolyte. However, the dry battery should be activated with a boost charge prior to installation, especially during cold weather.

LIGHT ENERGY AND ELECTRICAL ENERGY

The action of light energy, called PHOTONS, on certain atoms is a means of converting light energy to electrical energy. This leads to many interesting electrical phenomena. The beginning student frequently and mistakenly groups all photoelectric devices and components into a single classification. Actually there are three general classes of these devices.

PHOTOVOLTAIC CELL. Since we are studying sources of an electrical potential, the photovoltaic cell, Fig. 2-16, will be considered first. When a light shines on certain solid materials, electric charges will move across a barrier and establish a potential difference.

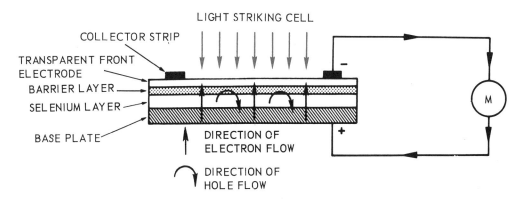

Fig. 2-16. Typical cell structure of a photovoltaic cell.
(International Rectifier Corp.)

In Fig. 2-16 showing the cell structure of a photovoltaic cell, light energy causes electrons from the selenium to move across the barrier layer and collect on the transparent front electrode. This action will make the front electrode negative and leave the selenium with a deficiency of electrons, or positive. If an external circuit is now connected to the terminals of the cell, a small current will flow.

Here is one of the more challenging fields of science. Already some inroads have been made to convert energy from the sun to useful electrical energy. This may be one of our great sources of power in future years. Many of our orbiting satellites now have banks of many photovoltaic cells which supply the energy to operate communication equipment and other instruments.

A typical sun cell is shown in Fig. 2-17. You will perform experiments and build projects which utilize these cells.

An application of the sun cell may be found in the common light meter used by the photographer. Since the voltage developed depends upon the light, it is a simple matter to have the meter indicate the intensity of the light.

PHOTOCONDUCTIVE CELL. Light energy falling on a photoconductive cell will cause a change in the resistance of the cell and will consequently change the current flow in the circuit. Many of these photoresistive cells are also made of selenium. Light energy causes the formation of electron-hole pairs and frees electrons for conduction purposes. This phenomenon will be studied in detail in Chapter 3.

Another type of conductive cell which is widely used is the cadmium sulfide cell. Refer to Fig. 2-18. This device depends upon a small

Fig. 2-17. A typical photovoltaic cell or sun cell.
(International Rectifier Corp.)

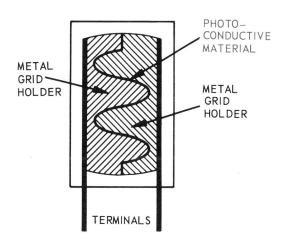

Fig. 2-18. A cadmium-sulfide photoconductive cell.

cadmium sulfide crystal which is doped with an impurity called an "activator." Silver, antimony and indium are used as impurities. The dark resistance of the CdS cell in its manufactured form can be in the order of 600 to 1, when compared to resistance of the cell under a bright light. The cadmium sulfide cell is a very practical device and can be used to directly operate switching relays without further amplification. It likewise is very sensitive to minute changes of light intensities. A simple circuit using a CdS cell and relay is illustrated in Fig. 2-19.

An example of a photocell circuit is shown and explained in Fig. 2-20.

Still another type of photoconductive cell is the PHOTODIODE. It will be necessary to delay the understanding of this diode until you study Chapter 3. At this time, it is only necessary to realize that light energy directed on the diode will increase its conduction.

PHOTOEMISSION CELL. The third photo device is called a photoemission device. When light energy is directed on such materials as

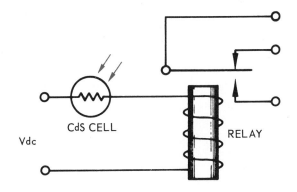

Fig. 2-19. Light directed toward the CdS cell will decrease its resistance and permit current to flow in circuit. The current will activate the relay.

sodium, cesium, potassium and alkali earths like strontium and barium, electrons will be ejected from the surface of these materials. Laws regarding photoemission state that emission is directly proportional to light intensity. This device is also frequency sensitive, which makes it useful in the measurement of electromagnetic radiation. In the study of vacuum tubes, a photoemissive cathode is enclosed in a tube with a plate which is held at a positive potential. Emitted electrons are collected by the plate, which means that the tube

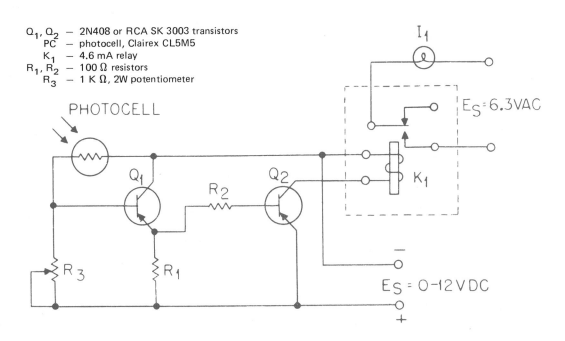

Fig. 2-20. Photocell circuit operation. Light shining on photocell drives Q_1 into conduction. Charge is amplified by Q_2 which activates relay K_1. The light, I_1 or an electrical device, can be turned on or off. R_3 is a sensitivity control.

is conducting. Incident light on the emitter surface will vary the conduction of the tube.

> LESSON IN SAFETY: Always use the proper eye protection when grinding, chipping or working with metals. Eyes cannot be replaced!

ELECTRIC ENERGY FROM HEAT ENERGY

In Fig. 2-21, two dissimilar wires are twisted together. One is a copper wire, the other is iron. When heat is applied to the twisted connection, a small potential difference is developed between the two wires. This voltage is directly proportional to the heat intensity. Such a device is called a THERMOCOUPLE and is widely used as a temperature sensing device and as a temperature control device. Many combinations of metals are used, including copper-constantan, iron-constantan and platinum-platinum rhodium.

Several thermocouples may be connected in series to increase the output voltage. A device composed of several thermocouples is called a THERMOPILE.

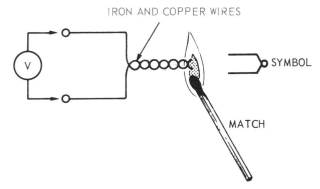

Fig. 2-21. Dissimilar metals in contact with each other produce a small voltage when heated.

ELECTRICAL ENERGY FROM MECHANICAL STRESS

When certain crystalline substances such as quartz, tourmaline, rochelle salts and barium titanate are subject to stress or pressure, an electrical potential will develop across certain points of their crystal structure. This phenomenon is called PIEZOELECTRIC EFFECT. The voltage developed across the crystal surfaces is proportional to the applied distortion or stress on the crystal.

Quartz crystals are, of course, natural. They are minded and cut. The other crystals can be manufactured by growing processes from a liquid solution.

Piezoelectricity is a very useful effect, and many of our modern conveniences depend upon it. A good example is the PICKUP used on the record player. Refer to Fig. 2-22.

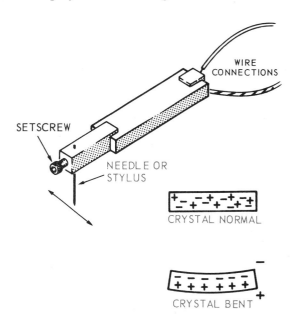

Fig. 2-22. As needle rides in the groove of a record, it moves back and forth according to the cut groove. This movement causes crystal to flex and produce a voltage. This is a CRYSTAL PICKUP.

The phonograph needle is attached to one end of the thin crystal. As the needle rides in the grooves of a record, it vibrates back and forth. This vibration will flex and bend the crystal to produce a varying voltage, which is amplified and reproduced as the music or sound in the SPEAKER.

Another familiar application is the typical crystal microphone. This action is illustrated in

Fig. 2-23. The alternate condensations and rarefactions of the sound waves cause the diaphragm to move in and out. A mechanical linkage between the diaphragm and the crystal causes the crystal to also flex inward and outward. This develops a voltage in proportion to the magnitude and frequency of the sound wave that strikes the diaphragm. The voltage is

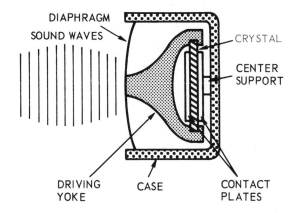

Fig. 2-23. A sketch shows the construction of a crystal microphone.

then amplified and its output may be heard from a speaker. In Fig. 2-24, several crystal microphones are shown.

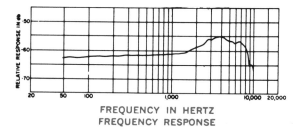

FREQUENCY IN HERTZ
FREQUENCY RESPONSE

Fig. 2-24. Crystal microphones used in communications. (Shure Bros. Inc.)

Frequently, crystals are joined together either in series, parallel or series parallel, to produce a desired output. The construction of a crystal BIMORPH CELL of two crystals in parallel is shown in Fig. 2-25.

Another very important application of crystals in electronics is the ability of a crystal to precisely control frequency. This characteristic will be discussed in Chapter 16 concerning oscillators.

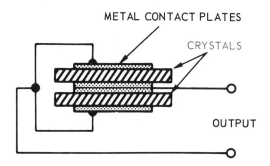

Fig. 2-25. The construction of a crystal BIMORPH CELL.

LESSON IN SAFETY: Many people are killed by supposedly "unloaded guns." Many technicians are injured by supposedly "dead circuits." Be sure your equipment is turned off and disconnected from the main power source before you make circuit changes and replace components.

TO REVIEW

Several sources or methods of producing an electrical potential difference have been discussed. They include:

1. Static electricity by friction.
2. Batteries by chemical action.
3. Electrical energy from light energy.
4. Electrical energy from heat energy.
5. Electrical energy from distortion and pressure.
6. Electrical energy from magnetic field and motion (mechanical energy).

The most important source is the generation of an electrical potential by the movement of conductors in a magnetic field. The principles of the generator and alternator will be discussed in Chapter 6.

TEST YOUR KNOWLEDGE

1. What is the principle of the voltaic cell?
2. What is polarization?
3. Can your experimental zinc and carbon electrode voltaic cell be recharged?
4. Chargers are available that will rejuvenate a dry cell to some extent. True or False?
5. A _____ cell can be recharged by applying reverse voltage.
6. Lead-acid batteries should be connected to the charger _____ the power is applied.
7. What is used as a depolarizer in a flashlight cell?
8. Give three advantages that a mercury cell has over a typical flashlight cell.
9. What is the major advantage of a nickel-cadmium cell?
10. _____ _____ is a classification of liquids according to their weight in respect to an equal volume of water.
11. Which combination of cells produces the highest terminal voltage?
    a. Series.
    b. Parallel.
12. Which combination of cells should be used for a heavy load?
    a. Series.
    b. Parallel.
13. Name five sources of electrical energy.
14. What is meant by piezoelectric effect?
15. How could you use a CdS cell to turn on your house lights when darkness arrives?

Chapter 3

CONDUCTORS, SEMICONDUCTORS, INSULATORS

Three of the most important materials in electricity and electronics are conductors, semiconductors and insulators. To gain an understanding of these materials, this chapter will discuss:

1. The theory of transferring electrical energy.
2. The characteristics of insulators, semiconductors and conductors.
3. A study of the properties of resistance and resistance performance in a simple dc circuit.

The several methods devised by science to produce an electrical potential have been discussed in the previous chapter. This potential difference, you will recall, is measured in VOLTAGE. Voltage has been established as a source of energy "at rest," but ready to perform useful work if connected to a circuit containing wires and components. In other words, we should investigate the properties of certain materials with respect to their ability to conduct electricity.

INSULATORS

The ability of a material to conduct electricity can be understood by reviewing the atomic theory in Chapter 1. In some materials, however, the electrons are held securely, close to the nucleus, and require large amounts of energy to break loose for conduction. These materials are classified as INSULATORS. An example of this type of material is GLASS.

You will remember that GERMANIUM, in its pure form, has all of its valence electrons securely locked in covalent bonds. None were free to circulate. In Fig. 3-1, note that very little conduction takes place in a crystal of germanium.

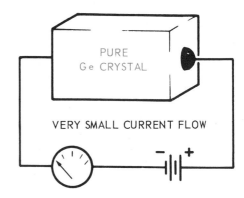

Fig. 3-1. A pure germanium crystal is not a good conductor.

By adding large amounts of energy to an insulating material, it can be made to conduct. A very high potential could force the electrons to move, but this could destroy the material.

ENERGY DIAGRAMS

A simple ENERGY DIAGRAM in Fig. 3-2 shows how some materials conduct and others do not. In the diagram of the INSULATOR, the electrons are held in the VALENCE BAND, and a large amount of energy is needed to cause electrons to jump the FORBIDDEN GAP and

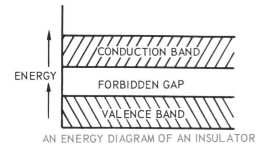

AN ENERGY DIAGRAM OF AN INSULATOR

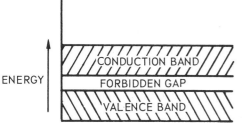

AN ENERGY DIAGRAM OF A SEMICONDUCTOR

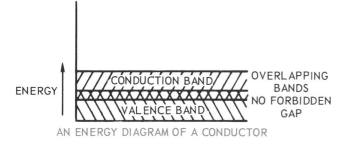

AN ENERGY DIAGRAM OF A CONDUCTOR

Fig. 3-2. These energy diagrams illustrate the difference between a conductor, semiconductor and an insulator.

find freedom of movement in the CONDUCTION BAND.

Some materials are neither good insulators nor conductors. These are classed as SEMI-CONDUCTORS. Note in the diagram that energy is required to make them conduct, but the forbidden gap is narrow when compared to the insulator.

The diagram of a CONDUCTOR requires no energy at all to free electrons for conduction. A material which is a good conductor has many free electrons ready to transfer energy by conduction.

No current carriers can exist in the forbidden gap region of the energy diagram. A discrete quantity of energy must be supplied to make the carriers "jump the gap."

INSULATORS serve many useful purposes. They have an extremely high resistance to the flow of electricity. Insulating materials are used to cover and support wires and components, to protect personnel from "shocks" and to prevent short circuits between wires and parts. Examples of insulators include rubber, glass, bakelite, plastics, air, asbestos, and many other special materials sold under manufacturer's trade names.

INTRINSIC CONDUCTORS

An INTRINSIC CONDUCTOR is made of atoms of all the same kind. Pure GERMANIUM or SILICON are examples of this type. These materials are classified as SEMICONDUCTORS. Refer again to Fig. 3-2. Note that energy is required to make semiconductors conduct. Actually, when energy is added to the crystal (with either heat, light or an electric potential), electrons jump from the valence band to the conduction band. This leaves a HOLE or POSITIVE SITE in the valence band. There should be an electron in the hole, but it has departed to the conduction band. Under the influence of heat and light energy, there is always the formation of electron-hole pairs and movement back and forth across the forbidden band. So the intrinsic crystal will conduct to some extent and is classified as a SEMICONDUCTOR.

A resistance chart in Fig. 3-3 shows the relationships between conductors, semiconductors and insulators.

CONDUCTION BY FREE ELECTRONS

In Fig. 3-4, a germanium crystal is connected to a source of potential. Remember, there are some free electrons which have moved up to the conduction level. Electrons leave the negative power source and enter the crystal. For every

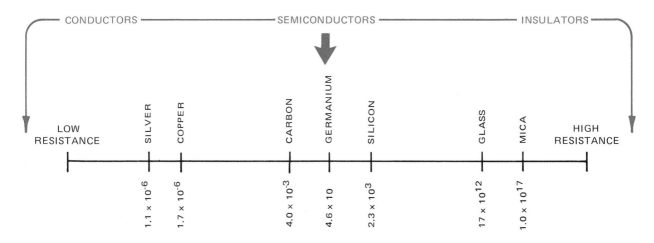

Fig. 3-3. Resistance chart of certain conductors, semiconductors and insulators (in ohms/cm).

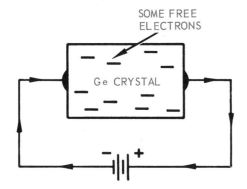

Fig. 3-4. Conduction takes place by means of electrons.

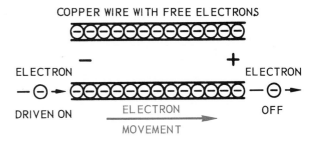

Fig. 3-5. Energy exchanged between electrons constitutes a current.

Fig. 3-6. Transfer of energy in a drift pattern.

electron entering the crystal, one must leave the positive end and is attracted to the positive source. THE CRYSTAL ALWAYS MAINTAINS AN OVERALL NET CHARGE OF ZERO.

Further explanation is shown in Fig. 3-5. In this sketch, a copper wire with many free electrons is used. An electron forced on one end of the wire causes an electron to leave the other end. This illustration oversimplifies the conduction phenomenon as the electrons appear as a string of balls in a tube. Actually, current flow is a drift pattern illustrated in an enlarged view in Fig. 3-6 as an exchange or transfer of energy between atoms.

The speed of energy transfer is interesting to note. It approximates the speed of light, which is 186,000 miles/sec. or 3×10^8 metres per

second. However, this is not the electron movement speed. It is relatively slow.

HOLE CONDUCTION

As mentioned, when carriers become free in the intrinsic crystal, HOLES are created in the valence band. Conduction also can be accomplished by these holes. In Fig. 3-7, an electron is attracted from the crystal by the positive source leaving a hole needing to be filled. This is named HOLE INJECTION.

The electron from the next atom jumps over to fill the hole and leaves a hole behind. Holes

ELECTRON OUT
LEAVES HOLE

HOLE

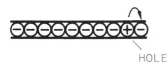

HOLE

ELECTRON FROM ADJACENT
ATOM MOVES TO HOLE

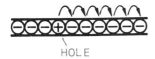

HOLE

HOLES ARE FILLED BY
ELECTRONS, LEAVING HOLES

HOLE
ELECTRON

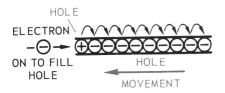

ON TO FILL   HOLE
HOLE
          MOVEMENT

HOLES HAVE TRANSFERRED
ENERGY

Fig. 3-7. Conduction is by "hole" movement. It is opposite to electron conduction.

are deserted and occupied in random fashion until the hole arrives at the negative power source where it is finally filled by an electron from the source. The crystal always maintains a net charge of zero.

Remember:

1. HOLE movement is in the opposite direction to ELECTRON movement.

2. The current in the external circuit is always electron flow and moves from negative to positive.

A crystal containing both electrons and holes is connected in a circuit in Fig. 3-8. Note the direction of movement in each case.

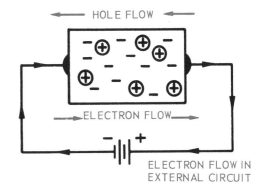

HOLE FLOW

ELECTRON FLOW

ELECTRON FLOW IN
EXTERNAL CIRCUIT

Fig. 3-8. Conduction in a crystal containing both electrons and holes.

N TYPE CRYSTAL FORMATION

In order to make the pure intrinsic crystal a slightly better conductor, certain impurities are added to the crystal during crystallization. Remember that both germanium and silicon have FOUR valence electrons and, in their crystalline-lattice structure, these electrons are held in COVALENT BOND. Impurities are selected which have FIVE valence electrons. These include phosphorus (P), arsenic (As), bismuth (Bi) and antimony (Sb). They are classified as PENTAVALENTS.

The process of adding impurities to crystal is called "doping" and the impurities are "dopants." Only a small amount of dopant is used, such as one atom of impurity to 10 million atoms of germanium. The crystal now becomes EXTRINSIC and its conduction characteristics will change.

N type crystal formation is shown in Fig. 3-9. An arsenic atom with five valence electrons is added to the pure germanium. Four of the valence electrons of the arsenic join in covalent bonds with the germanium. Since the outer ring of the germanium is now satisfied, the EXTRA ELECTRON FROM THE ARSENIC is free to move about the crystal. This negative electron will conduct an electric current through the crystal. A crystal doped in this manner is an N type crystal. Its MAJORITY CARRIERS ARE ELECTRONS. A few holes do exist in the crystal due to the formation of

Fig. 3-9. Doping the crystal with a pentavalent will produce free electrons.

electron-hole pairs, and these are MINORITY CARRIERS. The minority carrier current is very small indeed. Conduction in the N type crystal is shown in Fig. 3-10. Impurities that add free electrons to a crystal are called DONOR IMPURITIES.

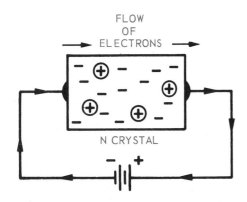

Fig. 3-10. Conduction by electrons in a N type crystal.

P TYPE CRYSTAL FORMATION

Beginning again with a pure crystal of germanium or silicon, a minute quantity of a TRIVALENT impurity is added during crystallization. Common trivalents which have three valence electrons are indium (In), gallium (Ga) and boron (B).

P type crystal formation is shown in Fig. 3-11. The three valence electrons of the indium join in covalent bond with the germanium atoms. But the germanium atom is not satisfied; another electron is needed to complete the covalent bonding structure. There is, therefore, a POSITIVE SITE or HOLE which has a strong attraction for an electron, if one happened to be in the vicinity. The trivalent dopant creates many of these positive holes and conduction through this crystal is by HOLES.

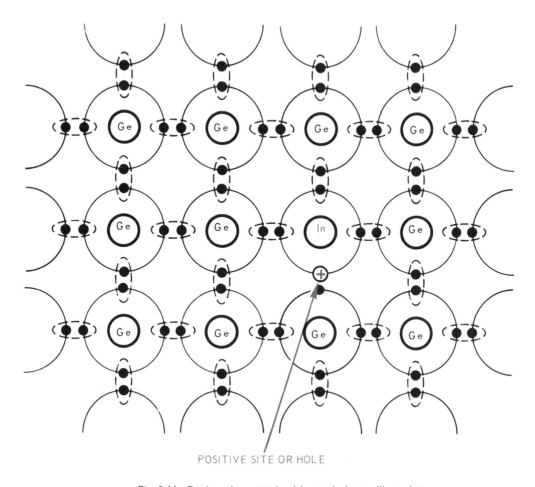

Fig. 3-11. Doping the crystal with a trivalent will produce
"holes" or "positive sites."

A crystal formed in this manner becomes a P type crystal, and its MAJORITY CARRIERS ARE HOLES. A few electrons freed by electron-hole pairs will permit a very small conduction, and the MINORITY CARRIERS in a P crystal are electrons. Since this dopant creates holes which will accept electrons, it is called an ACCEPTOR IMPURITY. Conduction through the P type crystal is illustrated in Fig. 3-12.

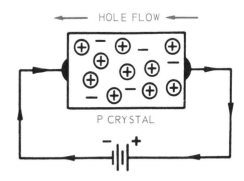

Fig. 3-12. Conduction in a P type crystal by holes.

ENERGY LEVELS IN DOPED CRYSTALS

Why does a crystal with an impurity conduct better than a pure crystal? From the mechanical point of view, the answer would include the introduction of carriers such as electrons or holes. Take a closer look at the energy diagrams in Fig. 3-13. In the case of the DONOR impurity which supplies free electrons, the free electron cannot fit into the valence band since that band is already full. You might assume that the electron should be found in the conduction band but, energy-wise, it cannot quite make it. It is customary to indicate a new discrete energy level for these electrons slightly below the conduction band.

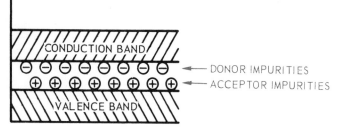

Fig. 3-13. Energy-wise, electrons from donor impurities are located below the conduction band. Acceptor impurities form a new energy level just above the valence band.

In the case of the acceptor impurity which creates positive sites or holes, a new discrete energy level for acceptor impurities or holes will exist slightly above the valence band. This is due to a natural tendency to form covalent bonding with the valence electrons.

With donor impurities, ·the materials will conduct since only a small amount of energy is required to move the electrons into the conduction band. With the acceptor impurities or holes, only a small amount of energy is required to move an electron from the valence band to a positive site. This leaves a hole in the valence band. Conduction in a P type crystal is by holes in the valence band. The conditions of added energy to the doped crystals are illustrated in Fig. 3-14.

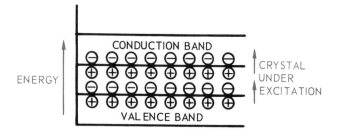

Fig. 3-14. Excitation causes electrons to move to the conduction band other electrons to move from the valence band to positive holes.

CONDUCTORS

Electrons in the valence band of an atom are associated with the nucleus of the atom. In order to move these electrons into the conduction band to become current carriers, it is necessary to add sufficient energy to cause the electrons to jump the forbidden gap.

Look at Fig. 3-15, which is an energy diagram for copper. It has no forbidden gap. In fact, the valence band and conduction band overlap. Large numbers of electrons are immediately available for conduction. The quality of a

Fig. 3-15. In a conductor, large numbers of electrons are free to conduct.

conductor is determined by the number of free carriers available. To express it another way, the resistance to current flow in copper is very low. In the table, Fig. 3-16, the SPECIFIC RESIS-

MATERIAL	ρ (rho) *
ALUMINUM	17.
CARBON	2500 – 7500 x copper
CONSTANTAN	295.
COPPER	10.4
GOLD	14.
IRON	58.
NICHROME	676.
SILVER	9.8
STEEL	100.
TUNGSTEN	33.8

Fig. 3-16. Table showing specific resistance of several materials. *Specific resistance is the circular-mil ohms per foot at 68 deg. F (20 C) temperature.

TANCE of several METALS is tabulated. The Greek symbol for SPECIFIC RESISTANCE is "ρ" (rho).

RESISTANCE

Materials vary in their ability to conduct electricity. To say it another way: a material resists the flow of electricity. The reasons for this have already been explored.

Resistance has been assigned the letter symbol R. It is measured in OHMS in memory of GEORGE SIMON OHM.* The Greek symbol for OHM is Ω (omega). One ohm of resistance will permit a current of one ampere to flow at an applied potential of one volt.

Materials with greater resistance require greater energy to raise electrons to the conduction band. Consequently, energy is consumed by resistance and appears as HEAT. In fact, one of the primary uses of resistance is to convert electrical energy into heat energy. Such an application is found in your electric range or heater at home.

Resistance finds another primary use in producing LIGHT. The resistance filament in a light bulb becomes white hot and converts electrical energy to light energy. Resistance units find an infinite number of uses in electrical and electronic circuits. Resistance is represented in circuit drawings by the symbol shown in Fig. 3-17.

Resistance in a conductor is influenced by four factors:

1. Material of conductor.
2. Size of conductor.
3. Length of conductor.
4. Temperature of conductor.

RESISTANCE AND MATERIAL

As we have established, the ability of a conductor to conduct depends upon the material from which it is made. By referring back to the table in Fig. 3-16, it is apparent that copper has approximately one-tenth the resistance of a steel wire. Aluminum offers only a small amount more resistance than copper (17 to 10.4). Silver offers the least resistance of all materials, but it is expensive. Conducting wires generally are made of copper, except in cases

Fig. 3-17. The schematic symbol for resistance.

where mechanical strength also must be considered. Particular note should be made of carbon. It is a highly resistive material. Compounds of carbon and other materials are used in small resistors found in electronic circuits.

RESISTANCE AND SIZE OF CONDUCTOR

It is just common sense to expect that a larger wire will conduct electricity more freely and in greater quantities than a small wire. There are many more free carriers to do the work. That is: the resistance of a wire is inversely proportional to its size. The larger the wire, the less resistance; the smaller the wire, the greater the resistance.

The size of a wire is specified by a number. Fig. 3-18 pictures an American Standard Wire Gauge. The larger the number, the smaller the diameter of the wire. Numbers 20 and 22 are commonly used for hookup wire in electronic equipment. Numbers 12 and 14 will be found in light circuits in your home. Number 6 wire

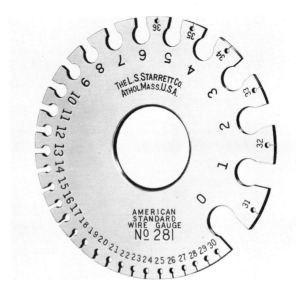

Fig. 3-18. A gauge used to determine wire size. (L. S. Starrett Co.)

---

*OHM, GEORGE SIMON (1787-1854). A German physicist who developed the mathematical relationship between potential difference, electric current and resistance.

would be used to connect an electric range. The size of a wire is determined by the amount of current the wire is required to carry. If a wire is too small, it will heat up due to the loss of energy as it overcomes the resistance.

The inch English system is not used in the measurement of wire size. Wire is sized by CIRCULAR-MILS, which is its cross-sectional area. One circular-mil is the area of a circle one mil or .001 inch in diameter. Work this problem to illustrate. A wire has a diameter of .050 in. or a radius of .025 in. Its cross-sectional area in square inches would be:

$$A = \pi R^2 = \pi \left(\frac{D}{2}\right)^2 = 3.1416 \times \frac{D^2}{4} = .7854\, D^2$$

If the diameter were in mils then:

$$A \text{ (in square mils)} = .7854\ D^2$$

By definition, a circular-mil is the area of a wire having a diameter of .001 in. or one mil. This wire then would contain:

$$.7854 \times (1)^2 = .7854 \text{ square mils}$$

To find the circular-mil area of any wire, divide its square mil area by .7854.

$$A \text{ (in circular mils)} = \frac{A \text{ in square mils}}{A \text{ of one circular mil}}$$

$$= \frac{.7854\ D^2}{.7854} = D^2$$

The circular-mil area of a wire is equal to its diameter in mils squared. In the table in Fig. 3-19, the number size of wires and circular-mil area is given.

A common method of designating the resistance of a wire is by ohms per circular-mil foot. This term was used in Fig. 3-16 for the values of resistivity. A circular-mil foot is a wire having one circular-mil cross sectional area and is one foot long.

PROBLEM: What is the resistance of 100 ft. of No. 22 copper wire?

$$R \text{ (in ohms)} = \rho \times \frac{\text{length}}{\text{area}}$$

$$= 10.4 \times \frac{100}{642.4}$$

$$= 1.62 \text{ ohms}$$

ρ is found in Fig. 3-16.

Circular-mil area is found in the table in Fig. 3-19.

Conductors are made in a wide variety of sizes, both bare and insulated, to fit the various needs.

Some wires are stranded with either 7, 19 or 37 separate wires twisted together. This provides the necessary cross-sectional area to carry the current, and also provides flexibility and ease of handling.

RESISTANCE AND CONDUCTOR LENGTH

A wire increases in resistance as its length is increased. Resistance and length are directly in proportion. If a given wire has a resistance of one ohm per foot, the resistance of 10 ft. of wire would be 10 ohms. The length of wire was a factor in computing the previous problem involving circular-mil area and resistivity.

RESISTANCE AND HEAT

Most conductors increase in resistance at increased temperatures. This is not a conflict with the energy theory. Under increased temperatures, there are so many free electrons for conduction, the drift path becomes complicated and there are many more collisions. Consequently the wire does not conduct as well.

This is not the case with semiconductors. Increased temperature will bring many electrons into the conduction band, and the resistance of

GAGE NO.	DIAM. MILS	CIRCULAR MIL AREA	OHMS PER 1,000 FT. OF COPPER WIRE AT 25°C	GAGE NO.	DIAM. MILS	CIRCULAR MIL AREA	OHMS PER 1,000 FT. OF COPPER WIRE AT 25°C
1	289.3	83,690	0.1264	21	28.46	810.1	13.05
2	257.6	66,370	0.1593	22	25.35	642.4	16.46
3	229.4	52,640	0.2009	23	25.57	509.5	20.76
4	204.3	41,740	0.2533	24	20.10	404.0	26.17
5	181.9	33,100	0.3195	25	17.90	320.4	33.00
6	162.0	26,250	0.4028	26	15.94	254.1	41.62
7	144.3	20,820	0.5080	27	14.20	201.5	52.48
8	128.5	16,510	0.6405	28	12.64	159.8	66.17
9	114.4	13,090	0.8077	29	11.26	126.7	83.44
10	101.9	10,380	1.018	30	10.03	100.5	105.2
11	90.74	8,234	1.284	31	8.928	79.70	132.7
12	80.81	6,530	1.619	32	7.950	63.21	167.3
13	71 96	5,178	2.042	33	7.080	50.13	211.0
14	64.08	4,107	2.575	34	6.305	39.75	266.0
15	57.07	3,257	3.247	35	5.615	31.52	335.0
16	50.82	2,583	4.094	36	5.000	25.00	423.0
17	45.26	2,048	5.163	37	4.453	19.83	533.4
18	40.30	1,624	6.510	38	3.965	15.72	672.6
19	35.89	1,288	8.210	39	3.531	12.47	848.1
20	31.96	1,022	10.35	40	3.145	9.88	1,069.

Fig. 3-19. Copper wire table.

the crystal decreases to become a straight conductor.

In summary:

Resistance depends upon the <u>kind of material</u>.
Resistance is inversely proportional to the <u>size</u> of the wire.
Resistance is directly proportional to the <u>length</u> of the wire.
Resistance is directly proportional to <u>temperature</u>.

TYPES OF RESISTORS

Resistors are manufactured in various types, values and sizes:

1. CARBON RESISTORS are the most common of all resistors used in electronic circuits. They are manufactured in a wide range of values from a fraction of an ohm to millions of ohms. Also, resistors of a specified value are made in various sizes such as 1/4 watt, 1/2 watt, 1 watt and 2 watts. The larger sizes give a greater surface area for dissipation of heat and will carry higher currents without self-destruction. An assortment of these resistors is shown in Fig. 3-20.

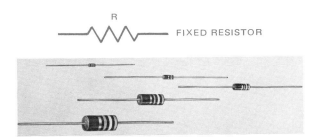

Fig. 3-20. Various types and sizes of carbon resistors. (Ohmite Mfg. Co.)

The construction of a carbon resistor is illustrated in the cutaway view in Fig. 3-21.

2. POWER RESISTORS, Fig. 3-22, usually are resistive wire wound on a ceramic form. Their large size predicts their use in high power circuits with considerable current. A special type of power resistor is shown in Fig. 3-23 with circuit symbol. They are

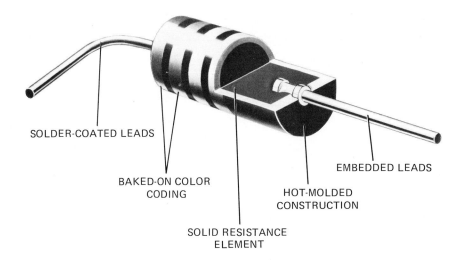

SOLDER-COATED LEADS

BAKED-ON COLOR CODING

HOT-MOLDED CONSTRUCTION

EMBEDDED LEADS

SOLID RESISTANCE ELEMENT

Fig. 3-21. A cutaway view showing the construction of a carbon resistor. (Allen-Bradley Co.)

manufactured with either fixed taps or an adjustable slider. They are useful in voltage divider circuits, which will be discussed later in this text.

TAPPED RESISTOR

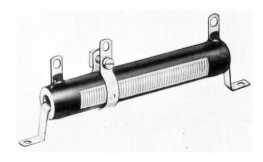

Fig. 3-23. An adjustable power resistor. (Ohmite Mfg. Co.)

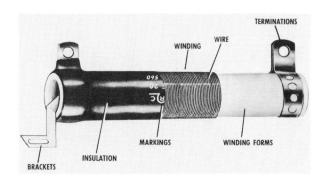

Fig. 3-22. A cutaway view of a power resistor. (IRC, Inc.)

3. VARIABLE RESISTORS have wide application in the field. One popular type, the POTENTIOMETER, is shown in Fig. 3-24. with its circuit symbol. The resistance unit of this device may be a molded carbon ring for low power or a wire-wound circular form for high power applications. A contact arm, controlled by a shaft and knob, rides in contact with the resistance unit. A desired amount of resistance may be selected by turning the knob. There are three connections to this component; one at each end of

VARIABLE RESISTOR

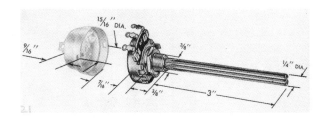

Fig. 3-24. A typical potentiometer and its schematic symbol. (Centralab)

Fig. 3-25. A wire-wound potentiometer. (Ohmite Mfg. Co.)

Fig. 3-26. A precision screwdriver-adjust trimmer potentiometer. (IRC, Inc.)

the total resistance and a third attached to the moving arm. These also are made in a wide selection of ohmic values and sizes to meet circuit needs. See Fig. 3-25.

A precision type of potentiometer with an end screwdriver adjustment is shown in Fig. 3-26. These are made in an exceptionally wide variety of values, terminals and mountings to fit most any application.

4. FILM RESISTORS are sometimes called THIN FILM RESISTORS. They usually are made by depositing a resistive film on a ceramic tube. The leads are connected to the ceramic tube by caps that slip over the end of the tube. These resistors are very accurate,

and they are becoming more popular in the electronic field. See Fig. 3-27.

RESISTANCE COLOR CODING

Most low wattage resistors are identified by color codes. These colors are standard, and they have been adopted by the Electronics Industries Association (EIA) and The United States Armed Forces (MIL).

Refer to Fig. 3-28 and note that color bands are painted around the body of the resistor.

Fig. 3-27. Thin film resistor construction. (Allen-Bradley Co.)

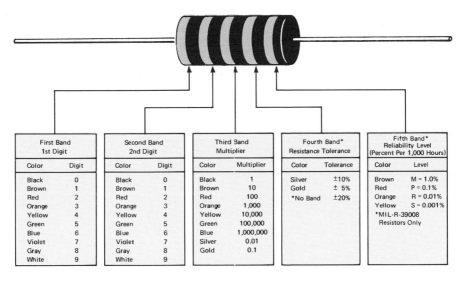

First Band 1st Digit		Second Band 2nd Digit		Third Band Multiplier		Fourth Band* Resistance Tolerance		Fifth Band* Reliability Level (Percent Per 1,000 Hours)	
Color	Digit	Color	Digit	Color	Multiplier	Color	Tolerance	Color	Level
Black	0	Black	0	Black	1	Silver	±10%	Brown	M = 1.0%
Brown	1	Brown	1	Brown	10	Gold	± 5%	Red	P = 0.1%
Red	2	Red	2	Red	100	*No Band	±20%	Orange	R = 0.01%
Orange	3	Orange	3	Orange	1,000			Yellow	S = 0.001%
Yellow	4	Yellow	4	Yellow	10,000			*MIL-R-39008	
Green	5	Green	5	Green	100,000			Resistors Only	
Blue	6	Blue	6	Blue	1,000,000				
Violet	7	Violet	7	Silver	0.01				
Gray	8	Gray	8	Gold	0.1				
White	9	White	9						

Fig. 3-28. Standard color code for resistors. (Allen-Bradley Co.)

Each color stands for a certain digit. Normally, resistors only have four color bands. However, a fifth color band may be used in some resistors to indicate the reliability level.

A couple of examples of reading resistor values, using the color code, are shown in Fig. 3-29.

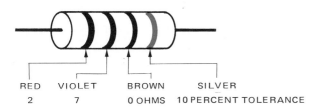

RED   VIOLET   BROWN   SILVER
2   7   0 OHMS 10 PERCENT TOLERANCE

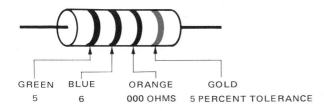

GREEN BLUE   ORANGE   GOLD
5   6   000 OHMS 5 PERCENT TOLERANCE

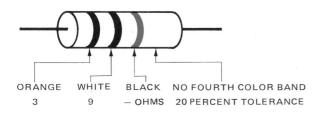

ORANGE WHITE BLACK NO FOURTH COLOR BAND
3   9   — OHMS 20 PERCENT TOLERANCE

Fig. 3-29. Examples of standard color-coded resistors.

TEST YOUR KNOWLEDGE

1. An _____ conductor is one made of atoms all of the same kind.
2. An _____ conductor is pure crystal with certain impurities added during crystallation.
3. Impurities that add free electrons to a crystal are called _____ impurities, while those which create holes that will accept electrons are called _____ impurities.
4. What causes the formation of electron-hole pairs?
5. Explain what is meant by the term "current."
6. Hole movement is in the opposite direction to electron movement. True or False?
7. Name four pentavalents.
8. What are the majority and minority carriers in an N type crystal?
9. What are the majority and minority carriers in a P type crystal?
10. Name three trivalent or acceptor impurities added to a pure crystal.
    - a. Phosphorus.   d. Gallium.
    - b. Arsenic.   e. Bismuth.
    - c. Indium.   f. Boron.
11. What is specific resistance?
12. The kind of a material used in a conductor affects its resistance. Name three other factors that affect resistance.
13. What is the resistance of 50 ft. of No. 24 copper wire?
14. Why is a 100 ohm fixed carbon resistor made in several sizes?
    What are the color codes for the following resistors:
15. 82 ohms, 10 percent tolerance.
16. 1.6 Meg ohms, 20 percent tolerance.
17. 36,000 ohms, 5 percent tolerance.
18. 470 ohms, 10 percent tolerance.
19. 7,500 ohms, 5 percent tolerance.
20. .62 Meg ohms, 20 percent tolerance.

Chapter 4

RESISTIVE CIRCUITS

Resistance is that property in a circuit that opposes electric current. In some circuits, resistance is wanted. In other circuits, it is unwanted. In this chapter, you will learn about resistance circuits by:

1. Studying the relationship between resistance, volts and amperes in Ohm's Law.
2. Investigating the properties of a series circuit.
3. Investigating the properties of a parallel circuit.
4. Understanding the conversion of electrical energy into power in resistive circuits.

The fundamental relationship between VOLTS, AMPERES and OHMS is basic to all studies of electricity and electronics. There is no lesson in this text which is more important to understand and apply. Ohm's Law is the foundation of circuit design and service.

These relationships were investigated and formulated into a mathematical equation by George Simon Ohm, the German scientist, during the 19th century.

1. The current in any circuit is directly PROPORTIONAL to the applied voltage. That is: if the voltage is increased, the current will increase. If the voltage is decreased, the current will decrease. See the simple circuit shown in Fig. 4-1. Note the symbol (A) used for AMMETER in this illustration.

2. The current in a circuit is INVERSELY PROPORTIONAL to its resistance. In Fig. 4-2, for example, the circuit has 10 ohms of resistance and a current of 1.0 amperes flowing. If the resistance is increased to 20 ohms, the current DECREASES to .5 amperes. If the resistance is reduced to 5 ohms, the current INCREASES to 2 amperes.

These relationships may be expressed in an equation:

$$I = \frac{E}{R}$$

where,

I = intensity of the current in amperes
E = electromotive force in volts
R = resistance in ohms

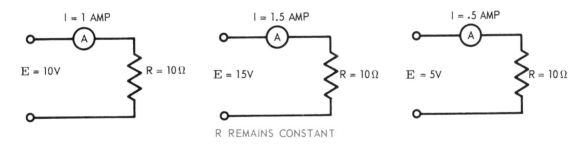

R REMAINS CONSTANT

Fig. 4-1. Current in a circuit is directly proportional to applied voltage.

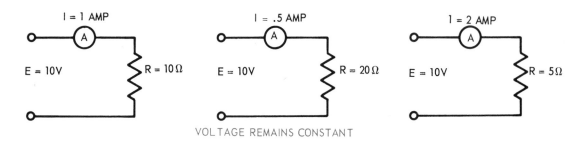

VOLTAGE REMAINS CONSTANT

Fig. 4-2. The current in a circuit is inversely proportional to the resistance.

This equation can be arranged in three different ways:

$$I = \frac{E}{R} \qquad E = I \times R \qquad R = \frac{E}{I}$$

Then knowing any two of these quantities of a circuit are known, the third may easily be computed.

PROBLEM: A current of .1 ampere flows in a circuit that has a resistance of 1000 ohms. What is the applied voltage?

$$E = IR = .1A \times 1000\,\Omega = 100\ \text{volts}$$

PROBLEM: A circuit has a current of .1 ampere when the applied voltage is 100 volts. What is the circuit resistance?

$$R = \frac{E}{I} = \frac{100V}{.1A} = 1000\ \text{ohms}$$

IMPORTANT: WHEN USING OHM'S LAW, ALL QUANTITIES MUST BE IN THE SAME BASIC UNITS: E IN VOLTS, I IN AMPERES, R IN OHMS.

If a quantity such as ohms is given in kilohms, it must be changed to ohms. If current were given in milliamperes, it must be changed to amperes. See APPENDIX A for CONVERSIONS.

EXAMPLE: A circuit with a resistance of 1.5 KΩ has a current of 50 milliamperes. What is the applied voltage?

1. $1.5\ \text{K}\Omega = 1.5 \times 10^3\,\Omega$

    $= 1500\ \text{ohms}$

2. $50\ \text{mA} = 5 \times 10^{-2}\,A$

    $= .05\ \text{amp}$

3. $E = I \times R$

    $= 1.5 \times 10^3 \times 5 \times 10^{-2}$

    $= 7.5 \times 10^1$

    $= 75\ \text{volts}$

The use of scientific notation or powers of ten is explained in APPENDIX A.

SERIES CIRCUITS

When components are connected end-to-end so that all the circuit current flows through each component, the circuit is a SERIES CIRCUIT. In Fig. 4-3, a series circuit of three resistors is drawn schematically. The number subscripts by each resistor are used to identify a particular resistor.

A summary of the previous discussion is a modified statement of Kirchoff's Laws for series circuits.

KIRCHOFF'S VOLTAGE LAW FOR SERIES CIRCUITS: The sum of the voltage drops around a series circuit is equal to the applied voltage.

KIRCHOFF'S CURRENT LAW FOR SERIES CIRCUITS: The current is the same at all points in a series circuit.

Examples of various applications of these laws are shown in Figs. 4-3 to 4-7.

VOLTAGE DROP POLARITIES

The study of voltage polarities, whether negative or positive, is of extreme importance in transistor and semiconductor circuits. The situation is severely aggravated by disagreements in theory and practice about the direction of current flow. Engineering and technical texts are about evenly divided between the concept that current flows from negative to positive and that current flows from positive to negative. Frequently, current from positive to negative is

called conventional current flow, whereas current from negative to positive is labeled as "electron flow."

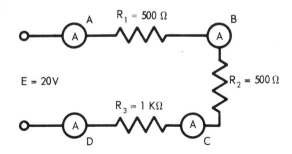

Fig. 4-5. In a SERIES circuit, the current is the SAME value, regardless of the point at which it is measured. Four meters have been connected in the circuit. Meters A, B, C and D all read exactly the same value of current.

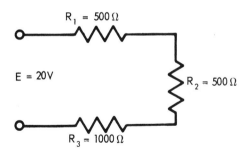

Fig. 4-3. The total resistance, R_T of a series circuit is equal to the SUM of the individual resistances:
$$R_T = R_1 + R_2 + R_3 \ldots$$
$$R_T = 500\,\Omega + 500\,\Omega + 1000\,\Omega = 2000\,\Omega$$

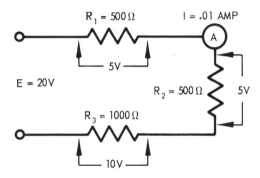

Fig. 4-6. When a current flows through each resistor in the series circuit, a definite amount of electrical energy is consumed. This is called VOLTAGE DROP. It may be found for each resistor by Ohm's Law, E = I x R. In this circuit voltage drops are computed.

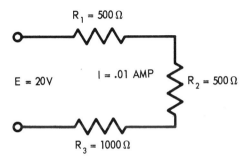

Fig. 4-4. The current flowing in a series circuit is equal to the applied voltage, E, divided by the total resistance, R_T. If 20 volts were applied to the circuit, the current can be computed:
$$I = \frac{20\ VOLTS}{2000\ \Omega} = .01A$$

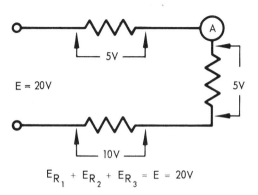

Fig. 4-7. The sum of the VOLTAGE DROPS around a series circuit is equal to the applied voltage.
$$E_S = E_{R_1} + E_{R_2} + E_{R_3} \quad \text{or}$$
$$E_S = 5V + 5V + 10V = 20V$$

For the purposes of this text, an electric current will always be considered the movement of electrons from NEGATIVE TO POSITIVE. Since most commercial test instruments follow this theory, it will save a lot of grief and maintenance.

In Fig. 4-8, a comparison is made between potential or voltage levels and a staircase.

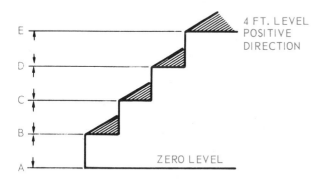

Fig. 4-8. Using a staircase to show the meaning of voltage level.

On the staircase:

Point B is one foot higher or more positive than zero level.

Point C is two feet higher or more positive than zero level.

Point D is three feet higher or more positive than zero level.

Point E is four feet higher or more positive than zero level.

In this illustration, the reference level has been taken as ZERO. In Fig. 4-9, the same

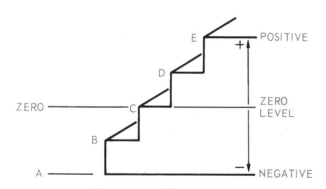

Fig. 4-9. A change in reference level gives both positive and negative values.

staircase is used but the ZERO reference level is taken at point C.

As a result:

Point A is two feet lower or more negative than zero level C.

Point B is one foot lower or more negative than zero level C.

Point C is at ZERO REFERENCE LEVEL.

Point D is one foot more positive than level C.

Point E is two feet more positive than C.

In both of these illustrations, the TOTAL DIFFERENCE in height has been four feet. Compare this with the circuit of four resistors in Fig. 4-10. Each resistor equals one step or level. Polarities are indicated.

Point B is 1 volt more positive than the zero level or point A.

Point C is 2 volts more positive than A.

Point D is 3 volts more positive than A.

Point E is 4 volts more positive than A.

The circuit is redrawn in Fig. 4-11, using the EXACT SAME VOLTAGES. However, ALL MEASUREMENT WILL BE MADE FROM POINT C, THE ZERO REFERENCE LEVEL.

The circuit of Fig. 4-6 is redrawn in Fig. 4-12 to show the polarity of voltage drops. The heavy arrows indicate the negative connection or black test lead of a meter used to measure the voltage.

NOTE: The polarity of each voltage across each resistor is opposing the source voltage.

Since they are opposite in polarity, we can write KIRCHOFF'S voltage law as:

$$+E_c + (-E_{R_1}) + (-E_{R_2}) + (-E_{R_3}) = 0 \text{ volts}$$

or, using our previous values:

$$+20V - 5V - 5V - 10V = 0 \text{ volts}$$

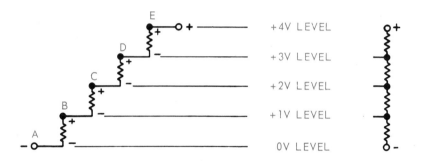

Fig. 4-10. With four equal resistors in sereis, voltage levels compare with equal steps.

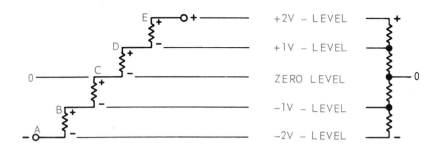

Fig. 4-11. ZERO REFERENCE is changed to Point C to provide a negative voltage.

This is no different than previously explained. It tells us that the sum of all voltages around a circuit including the source voltage equals ZERO.

Note also in Fig. 4-12 that all the current flowing to POINT A is equal to the current leaving Point A. This is another way of stating Kirchoff's current law. The algebraic sum of the currents INTO a point or junction in a circuit is equal to the current OUT of that junction.

> LESSON IN SAFETY: Do not work on wet floors. Your contact resistance is substantially reduced and an otherwise harmless shock may become serious. Work on a rubber mat or insulated platform if you are testing high voltage circuits.

RULE OF TEN TO ONE

Often, practical adjustments are made to scientific laws if the error created is insignificant. For example, look at circuit in Fig. 4-13.

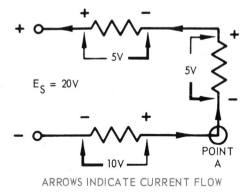

ARROWS INDICATE CURRENT FLOW

Fig. 4-12. This circuit illustrates the polarity of voltage drops. Heavy arrows indicate negative or black test lead of the voltmeter.

Total resistance is 1100 ohms; circuit current is .01 amp. In the second circuit, the 100 ohm resistor is disregarded. Total resistance is considered as 1000 ohms only. The current only changed one milliampere to .011 amp. The error created is hardly worth the time and effort to compute the values exactly. After all, the resistors may not be precisely made. A 10 percent tolerance resistor for 1000 ohms could be anywhere between 900 and 1100 ohms.

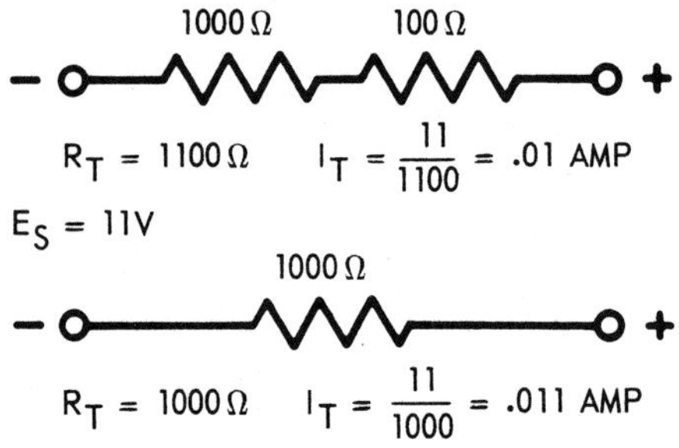

1000 Ω
100 Ω
R_T = 1100 Ω
$I_T = \frac{11}{1100}$ = .01 AMP
E_S = 11V
1000 Ω
R_T = 1000 Ω
$I_T = \frac{11}{1000}$ = .011 AMP

Fig. 4-13. When the ratio of one resistor to a second resistor in series is ten to one or more, the smaller resistor may be ignored for practical purposes.

Simplify your mathematics unless you are working with extremely accurate circuits.

PROPORTIONAL VOLTAGE IN SERIES CIRCUIT

Since voltage varies directly with resistance, a useful proportion may be set up to determine the voltage drop across any resistor in the series string.

$$E_R = \frac{R}{R_T} \times E_S$$

In Fig. 4-14, the total resistance of the circuit is 100 kilohms (KΩ) and E_S is 50 volts. To find voltage drop across the resistors:

$$E_{R_1} = \frac{R_1}{R_T} \times E_S$$

$$= \frac{10\ K\Omega}{100\ K\Omega} \times 50V = 5 \text{ volts}$$

For R_2,

$$E_{R_2} = \frac{R_2}{R_T} \times E_S$$

$$= \frac{40\ K\Omega}{100\ K\Omega} \times 50V = 20 \text{ volts}$$

$$E_{R_3} = \frac{R_3}{R_T} \times E_S$$

$$= \frac{50\ K\Omega}{100\ K\Omega} \times 50V = 25 \text{ volts}$$

R_1 = 10 KΩ, E_{R_1} = 5V
E_S = 50V
R_2 = 40 KΩ, E_{R_2} = 20V
R_3 = 50 KΩ, E_{R_3} = 25V

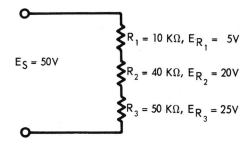

Fig. 4-14. Voltage drops are proportional to resistance.

LESSON IN SAFETY: Never remove the grounding prong of a three-wire appliance plug. By its removal, you have eliminated the grounding feature of the equipment or instrument. This could create a serious shock hazard. Some types of instruments will not work correctly if the ground is removed.

POWER

ENERGY IS THE ABILITY TO DO WORK, and POWER IS THE RATE OF DOING WORK. But what is work? In physics, work is the product of force and distance:

$$W = F \times D$$

Work is measured in foot-pounds.

If you lifted a one pound weight one foot, you would accomplish one foot-pound of work. No mention has been made of the time required to lift the weight. If you lifted the weight once each second, your rate of doing work or POWER would be "one foot-pound per second." Therefore, power is work divided by time or:

$$P = \frac{W}{t}$$

where,

W equals work in foot-pounds
t equals time

In mechanics, an engine may be rated in HORSEPOWER. If you have a one horsepower engine, it can work at a rate of:

$$\frac{33{,}000 \text{ ft. lb.}}{\text{minute}} \quad \text{or} \quad \frac{550 \text{ ft. lb.}}{\text{second}}$$

In electricity, we have learned that voltage is a force and current is the movement of electrons per second. One ampere equals one coulomb per second. Therefore, we can write the equation that:

$$P = I \times E$$

Power is measured in WATTS, named after JAMES WATT,* the inventor of the steam engine. To equate mechanical power with electrical power:

$$1 \text{ HP} = 746 \text{ watts}$$

Rearrangement of the power law will give us three formulas.

$$P = I \times E \qquad E = \frac{P}{I} \qquad I = \frac{P}{E}$$

Then, knowing any two values of current, voltage or power, it is possible to find the third unknown value.

Problem: A device operates on 110 volts and draws 5 amperes of current. What is its power?

$$P = I \times E$$

$$= 5 \text{ amps } \times 110 \text{ volts} = 550 \text{ watts}$$

Problem: How much current is drawn by a 100 watt light bulb at 110 volts?

$$I = \frac{P}{E} \qquad I = \frac{110W}{110V} = .91 \text{ amp}$$

When a current meets a resistor or load in a circuit, it has work to do to overcome that resistance. Energy is expended which has been

computed as voltage drop across the resistor. This energy is converted to heat energy and is radiated into space. The resistor consumes power at a certain rate in WATTS. So the POWER dissipated by a RESISTANCE is a function of VOLTAGE and CURRENT.

Consider the series circuit in Fig. 4-15.

$$R_T = 500\,\Omega + 1000\,\Omega = 1500 \text{ ohms}$$

$$I_T = \frac{150 \text{ volts}}{1500\,\Omega} = .1 \text{ amp}$$

$$P_T = .1A \times 150 \text{ volts} = 15 \text{ watts}$$

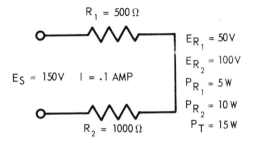

Fig. 4-15. Power is computed by finding the product of the voltage E and the current I.

The power expended by the individual resistors can be computed by finding the voltages across these resistors.

$$E_{R_1} = .1A \times 500\,\Omega = 50 \text{ volts}$$

$$E_{R_2} = .1A \times 1000\,\Omega = 100 \text{ volts}$$

and

$$P_{R_1} = .1A \times 50V = 5 \text{ watts}$$

$$P_{R_2} = .1A \times 100V = 10 \text{ watts}$$

and

$$P_T = P_{R_1} + P_{R_2}$$

$$= 5W + 10W = 15 \text{ watts}$$

---

*WATT, JAMES (1736-1819). A Scottish mechanical engineer and inventor. He is credited with the invention of the condensing steam engine. He also invented the centrifugal governor for controlling the speed of an engine. The unit of electrical power is named "watt" in his honor.

OHM'S LAW AND THE POWER LAW

By simple algebraic manipulation, it is possible to combine Ohm's Law and the Power Law so that an unknown may be found directly if any two of the other variables are known.

The original laws and methods used in combining laws are explained in Fig. 4-16.

FORMULA	SOURCE
1. $E = IR$	Ohm's Law
2. $E = \dfrac{P}{I}$	Power Law
3. $E = \sqrt{PR}$	By transposing Equation 12 and taking the square root
4. $I = \dfrac{E}{R}$	Ohm's Law
5. $I = \dfrac{P}{E}$	Power Law
6. $I = \sqrt{\dfrac{P}{R}}$	By transposing Equation 9 and taking the square root
7. $R = \dfrac{E}{I}$	Ohm's Law
8. $R = \dfrac{E^2}{P}$	By transposing Equation 12
9. $R = \dfrac{P}{I^2}$	By transposing Equation 11
10. $P = IE$	Power Law
11. $P = I^2R$	By substituting IR from equation 1 for E
12. $P = \dfrac{E^2}{R}$	By substituting $\dfrac{E}{R}$ from equation 4 for I

Fig. 4-16. To use this table, select the formula which contains any two known quantities. By completing the problems at the end of this chapter, you will become familiar with these important laws.

PARALLEL CIRCUITS

In a parallel circuit, components are connected side by side and provide multiple paths for current flow. Consider Fig. 4-17.

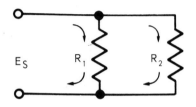

Fig. 4-17. Parallel circuit provides multiple paths for current flow.

Resistance R_1 and R_2 are connected in parallel to a common voltage source. The voltage across R_1 is equal to E_s. The voltage across R_2 is also equal to E_s. This must be true since both resistors are connected to a common source.

KIRCHOFF'S VOLTAGE LAW FOR PARALLEL CIRCUITS: The voltages across all branches of a parallel circuit are equal.

KIRCHOFF'S CURRENT LAW FOR PARALLEL CIRCUITS: The current in a parallel circuit is equal to the sum of its branch currents.

In Fig. 4-17:

$$I_{R_1} = \frac{E_s}{R_1} \text{ and } I_{R_2} = \frac{E_s}{R_2}$$

The total circuit current is the sum of these currents.

$$I_T = I_{R_1} + I_{R_2} = \frac{E_s}{R_1} + \frac{E_s}{R_2}$$

RESISTANCE IN A PARALLEL CIRCUIT

It is sometimes difficult to realize that each resistor added in parallel to a circuit DECREASES THE TOTAL RESISTANCE of the circuit. Actually, each resistor provides an additional current path and increases the total current. With a constant voltage, an increase in current can only mean a decrease in resistance.

In Fig. 4-18, two EQUAL resistors are connected in parallel. When two or more equal resistances are connected in parallel, the total circuit resistance may be found by this formula:

$$R_T = \frac{R}{N}$$

where,

R is the value of any single resistor
N is the number of resistors in parallel.

In the circuit in Fig. 4-18:

$$R_T = \frac{100 \, \Omega}{2} = 50 \text{ ohms}$$

Current in the circuit may be found by:

$$I_T = \frac{E_s}{R_T} = \frac{100V}{50 \, \Omega} = 2 \text{ amps}$$

To prove that the total current is equal to the sum of the branch current:

$$I_{R_1} = \frac{E_s}{R_1} = \frac{100V}{100 \, \Omega} = 1A$$

$$I_{R_2} = \frac{E_s}{R_2} = \frac{100V}{100 \, \Omega} = 1A$$

$$I_T = I_{R_1} + I_{R_2}$$

$$= 1A + 1A = 2 \text{ amps}$$

Power may be computed by using the formula:

$$P = IE \quad \text{or} \quad P = I^2R$$

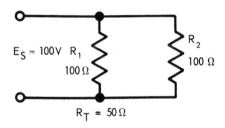

Fig. 4-18. Two equal resistors in parallel.

TWO UNEQUAL RESISTORS IN PARALLEL

A simplified formula may be used to find the total resistance of a parallel circuit having two unequal resistors.

$$R_T = \frac{R_1 R_2}{R_1 + R_2}$$

In the circuit shown in Fig. 4-19:

$$R_T = \frac{20 \times 30}{20 + 30} = \frac{600}{50} = 12 \text{ ohms}$$

IMPORTANT: The total resistance of any parallel circuit MUST BE LESS THAN the value of any single resistor in the parallel circuit. Use this statement to check your mathematics when working problems.

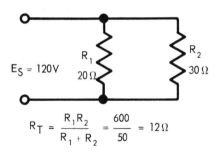

Fig. 4-19. Two unequal resistors in parallel.

To prove that the computed resistance in Fig. 4-19 is correct, the individual currents are found.

$$I_{R_1} = \frac{E_s}{R_1} = \frac{120V}{20 \, \Omega} = 6A$$

$$I_{R_2} = \frac{E_s}{R_2} = \frac{120V}{30 \, \Omega} = 4A$$

$$I_T = I_{R_1} + I_{R_2} = \frac{E_s}{R_1} + \frac{E_s}{R_2}$$

$$= 6 \text{ amps} + 4 \text{ amps} = 10 \text{ amps}$$

The resistance must be:

$$R = \frac{E_s}{I_T} = \frac{120V}{10A} = 12 \text{ ohms}$$

This agrees with our previous computation.

Using $P = I^2 R$ formula:

$$P_{R_1} = I_{R_1}{}^2 R_1 = (6)^2 \times 20\,\Omega = 36 \times 20 = 720W$$

$$P_{R_2} = I_{R_2}{}^2 R_2 = (4)^2 \times 30\,\Omega = 16 \times 30 = 480W$$

$$P_T = P_{R_1} + P_{R_2} = 720W + 480W = 1200 \text{ watts}$$

To prove this:

$$P_T = I_T E_s = 10A \times 120V = 1200 \text{ watts}$$

CONDUCTANCE

The ability to conduct is opposite to the ability to resist. Either may be used in the computation of circuit values.

Conductance of a circuit, symbol G, is the reciprocal of its resistance and is measured in MHOS or SIEMENS. The conductance of a 10 ohm resistor is:

$$G = \frac{1}{R} \qquad G = \frac{1}{10} \qquad = .1 \text{ SIEMENS}$$

If Ohm's Law says that $R = \frac{E}{I}$, then $G = \frac{I}{E}$

Conductance should be remembered. It is used in later studies of transistor circuit parameters (controlling elements of a circuit such as voltage, current, resistance, inductance and capacitance).

UNEQUAL RESISTORS IN PARALLEL

When a circuit contains two or more unequal resistors in parallel, it is more convenient to find the conductance first. Then, the conductance is converted to ohms.

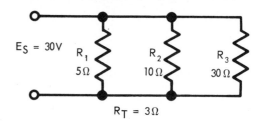

Fig. 4-20. Two or more unequal resistors in parallel.

In Fig. 4-20, three unequal resistors are connected in parallel. The total conductance of the circuit equals the SUM of each conductance.

$$G_T = \frac{1}{R_1} + \frac{1}{R_2} + \frac{1}{R_3} \cdots$$

$$G_T = \frac{1}{5} + \frac{1}{10} + \frac{1}{30}$$

The least common denominator of the fractions is 30.

$$G_T = \frac{6}{30} + \frac{3}{30} + \frac{1}{30} = \frac{10}{30}$$

If $G_T = \frac{1}{R_T}$, then $R_T = \frac{1}{G_T}$

$$R_T = \frac{\frac{1}{10}}{30} = \frac{30}{10} = 3 \text{ ohms}$$

To prove this total resistance, we will compute the individual currents with $E_s = 30$ volts.

$$I_{R_1} = \frac{30V}{5\,\Omega} = 6A$$

$$I_{R_2} = \frac{30V}{10\,\Omega} = 3A$$

$$I_{R_3} = \frac{30V}{30\,\Omega} = 1A$$

Total current equals:

$$I_T = I_{R_1} + I_{R_2} + I_{R_3}$$

$$= 6 + 3 + 1 = 10 \text{ amps}$$

$$R_T = \frac{E_c}{I_T} = \frac{30V}{10A} = 3 \text{ ohms}$$

Power may be computed using $P = IE$

$$P_{R_1} = I_{R_1} \times E = 6 \times 30 = 180W$$

$$P_{R_2} = I_{R_2} \times E = 3 \times 30 = 90W$$

$$P_{R_3} = I_{R_3} \times E = 1 \times 30 = 30W$$

$$P_T = P_{R_1} + P_{R_2} + P_{R_3}$$

$$= 180 + 90 + 30 = 300 \text{ watts}$$

PROOF

$$P_T = I_T \times E = 10 \times 30 = 300 \text{ watts}$$

$$P_T = I_T^2 \times R_T = 10^2 \times 3 = 300 \text{ watts}$$

PROPORTIONAL CURRENT DIVISION

Current is inversely proportional to resistance. Therefore, current through a branch of a parallel circuit may be found by setting up an inverse proportion. Refer to Fig. 4-21. The total resistance of this circuit is 12 ohms and, at $E_s = 12$ volts, the total current is one ampere.

$$I_{R_1} = \frac{R_2}{R_1 + R_2} \times I_T$$

$$= \frac{30}{50} \times 1A = .6 \text{ amp}$$

$$I_{R_2} = \frac{R_1}{R_1 + R_2} \times I_T$$

$$= \frac{20}{50} \times 1A = .4 \text{ amp}$$

RULE OF TEN TO ONE

Once again the ten-to-one ratio may be used to simplify mathematical computation and produce an approximate answer. Consider the

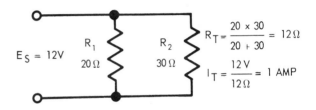

$$R_T = \frac{20 \times 30}{20 + 30} = 12\,\Omega$$

$$I_T = \frac{12V}{12\,\Omega} = 1 \text{ AMP}$$

Fig. 4-21. This circuit illustrates the proportional method of finding branch currents. Refer to text.

circuit in Fig. 4-22. The total resistance of this circuit is:

$$R_T = \frac{R_1 R_2}{R_1 + R_2} = \frac{100 \times 1000}{100 + 1000}$$

$$= \frac{10^5}{1100} = 90.9 \text{ ohms}$$

This is only nine ohms less than a circuit using R_1 only. This error is acceptable in the practical design of a circuit and is actually less than might be caused by using resistors with a 10 percent tolerance. It is interesting to note that if the ratio is greater than ten to one, the approximate answer becomes more accurate.

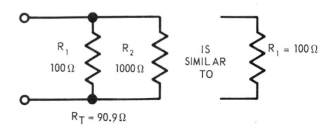

$$R_T = 90.9\,\Omega$$

Fig. 4-22. If the ratio of R_2 to R_1 is 10 to 1 or greater, then for practical purposes R_2 may be disregarded.

EQUIVALENT CIRCUITS

Remember that when a power source is connected to a complicated circuit of series and parallel resistors, it "sees" only the total resistance of the circuit or its equivalent resistance. The power source cannot determine how the circuit is connected. The equivalent resistance is represented by a single resistor which is the "load" the power source "sees."

In Fig. 4-23, a combination circuit is schematically drawn. Follow each step as its total resistance is computed.

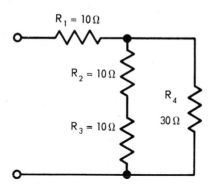

Fig. 4-23. This complex circuit is to be reduced to R_E.

1. Combine series resistors R_2 and R_3.

$$R_2 + R_3 = 10\,\Omega + 10\,\Omega = 20 \text{ ohms}$$

Redraw the circuit with R_2 and R_3 combined. See Fig. 4-24.

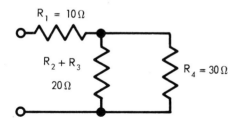

Fig. 4-24. Series R in parallel branches combined.

2. Combine $R_2 + R_3$ and R_4 in parallel.

$$R_1, R_2, R_4, = \frac{R_2, R_3 \times R_4}{R_2, R_3 + R_4} = \frac{20 \times 30}{20 + 30}$$

$$= \frac{600}{50} = 12 \text{ ohms}$$

Redraw circuit as in Fig. 4-25.

3. Combine series resistances, R_3, R_2, R_4 and R_1.

$$R_2, R_3, R_4 + R_1 = 12\,\Omega + 10\,\Omega = 22 \text{ ohms}$$

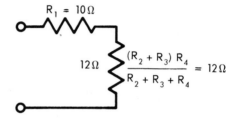

Fig. 4-25. Parallel branch resistances combine.

4. The circuit is redrawn in Fig. 4-26 with only one resistor of 22 Ω, which is the equivalent of the circuit in Fig. 4-23. The power source "sees" a load of 22 ohms.

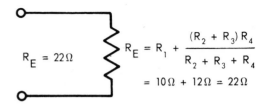

Fig. 4-26. Remaining series resistances are added to find R_E.

WHEATSTONE BRIDGE CIRCUITS

The bridge circuit using resistors, as well as other components, finds wide usage in electronic circuits. The basic Wheatstone circuit is illustrated in Fig. 4-27. The two input terminals are connected to a voltage source. The meter between points A and B is a sensitive galvanometer which measures magnitude and direction of current. R_1 and R_2 represent the ratio arm of the bridge. R_s is a variable standard resistor and R_X is unknown.

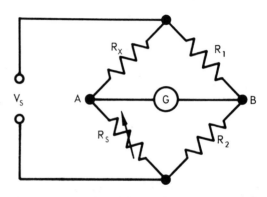

Fig. 4-27. The WHEATSTONE BRIDGE for measuring unknown resistance.

The bridge is "in balance" when the voltage drops across R_s and R_2 are equal and there is a proportional division of voltage across the bridge. In this "null" condition, the voltage at point A is equal to voltage at point B. The meter reads ZERO since there is no difference in potential. A slight change in resistance R_x would cause an unbalanced condition, and the proportional voltage division would be upset. The differences in voltages between A and B would cause the meter to deflect. A small adjustment of R_s can bring the bridge back to balance.

The proportion of voltages set up by the bridge is represented mathematically as:

$$\frac{E_{R_X}}{E_{R_S}} = \frac{E_{R_1}}{E_{R_2}}$$

$$\text{or} \quad \frac{IR_X}{IR_S} = \frac{IR_1}{IR_2}$$

$$\text{or} \quad \frac{R_X}{R_S} = \frac{R_1}{R_2}$$

If the bridge is used to find the value of an unknown resistor, the equation may be transposed to:

$$R_X = R_S \times \frac{R_1}{R_2}$$

If R_x is substituted with some resistive component that changes its resistance slightly because of temperature, light, pressure, humidity and other physical or mechanical effects, the slight change will be indicated on the meter. The meter can be calibrated and the dial marked to read directly in the physical quantity being measured.

SWITCHING

A switch is used to open and close a circuit. You will find this device anywhere it is neces-sary to disconnect power from a circuit or a piece of equipment. Switches are manufactured in hundred of sizes and types. We will discuss the more common ones shown in Figs. 4-28 through 4-35.

Push button switches, Rigs. 4-32 and 4-33, also come in assemblies. Fig. 4-34 shows one of these.

Fig. 4-28. Single-Pole Single-Throw. Switch disconnects one side of a line or a single wire circuit.

Fig. 4-29. Single-Pole Double-Throw. This switch actually has two "ON" positions. In the center it is "OFF." This switch finds wide usage in the switching from one circuit to another. It operates in a single wire circuit.

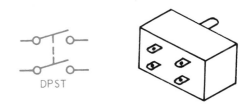

Fig. 4-30. Double-Pole Single-Throw. This is similar to SPST except both sides of a two wire line may be switched at once. It may also be used as two single-pole switches acting together.

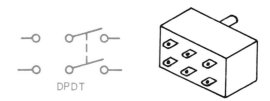

Fig. 4-31. Double-Pole Double-Throw. This switch, two wires to the center, may be switched to either of two other circuits. One to the right; the other to the left. It may also be used as two DPST switches acting together.

Fig. 4-32. Push Button, NO (normally open). Used to switch bells, alarms or momentarily close a circuit.

Fig. 4-33. Push Button, NC (normally closed). Used to momentarily open a circuit.

Fig. 4-34. Push button switch assemblies. (Centralab)

Fig. 4-35. Typical rotary switch. (Centralab)

ROTARY SWITCHES

Today, many electronic devices use rotary switches. Probably the best known rotary switch is the channel selector switch on your television set. Fig. 4-35 shows a front view of a typical rotary switch. There are many different types of rotary switch circuits. A single pole rotary switch symbol and standard section circuit diagram are shown in Fig. 4-36.

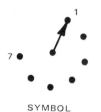

SYMBOL

ONE POLE, 7 POSITION

Fig. 4-36. Rotary switch symbol and standard section circuit diagram. (Centralab)

LESSON IN SAFETY: Would you believe that a soldering iron can be a frequent cause of injury? Never leave a hot iron on the bench where an unsuspecting friend can pick it up. Do not place your soldering iron or gun where you can accidentally touch the hot end with your hands, elbows or clothing.

RESISTOR SUBSTITUTION BOX

The resistor substitution box shown in Fig. 4-37 utilizes information covered in this chapter on resistive circuits and switches. A schematic and parts list is given in Fig. 4-38. However, the specific resistors may vary, depending on what is available and what is needed. Also, a rotary switch may be used that has more than 12 contacts if it is available.

The resistor substitution box project can be "dressed up" by using rub-on decals to identify the resistor values and other information. Chart tape or printed circuit tape can be used to separate certain portions of the front panel from other areas.

Fig. 4-37. Resistor substitution box.

58

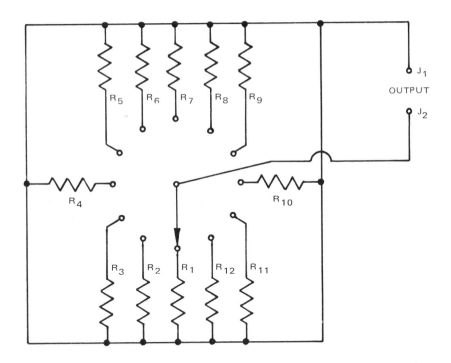

PARTS LIST FOR RESISTOR SUBSTITUTION BOX

*R_1 — 10 Ω, 1W resistor
R_2 — 47 Ω, 1W resistor
R_3 — 100 Ω, 1W resistor
R_4 — 470 Ω, 1W resistor
R_5 — 1K Ω, 1W resistor
R_6 — 4.7K Ω, 1W resistor
R_7 — 10K Ω, 1W resistor
R_8 — 47K Ω, 1W resistor
R_9 — 100K Ω, 1W resistor
R_{10} — 470K Ω, 1W resistor
R_{11} — 2.2 meg Ω, 1W resistor
R_{12} — 6.8 meg Ω, 1W resistor

S_1 — 12 position, 1 pole
    rotary switch, Calectro
    E2-162 or equivalent
J_1, J_2 — banana jacks or 5-way
    binding posts
Misc. — 6 1/4 x 3 3/4 x 2 in.
    utility box, decals,
    chart tape
* — Other resistor values may
    be used instead of those
    listed. All resistors should
    be 5 percent tolerance if possible.

Fig. 4-38. Schematic and parts list for Resistor Substitution Box.

TEST YOUR KNOWLEDGE

1. Four lamps of the same kind are connected in a series across a 6V dc power source. (See accompanying illustration.) Current draw of the lamps is .6 amp. Are they all equal in brightness?

2. Are they as bright as a single lamp connected across the 6 volts? Explain.

3. What is the resistance of one lamp?

4. What is the voltage drop across each lamp?

5. What power is being used by all four

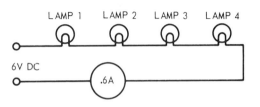

lamps?

6. How much power is used by lamp 1?

7. Connect a jumper wire across lamps 3 and 4. Do lamps 1 and 2 burn brighter or dimmer?

8. Four lamps of the same kind are connected in parallel across a 6 volt dc power source. Do all lamps burn at equal brightness?

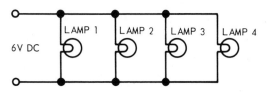

9. Do all lamps burn brighter or dimmer than a single lamp connected to the source?
10. What is the voltage across each lamp?
11. If each lamp has a resistance of 4 ohms, what is the total circuit resistance?
12. What is the current through one lamp?
13. What is the total current?
14. What power is used by one lamp? By all lamps?

Compute the following problems. Make conversions when necessary.

15. E = 100V   I = 2A   R = _____
16. I = 100 mA   R = 100 Ω   P = _____
17. I = 10 mA   E = 50V   P = _____
18. P = 500W   E = 250V   I = _____
19. P = 50W   R = 2 Ω   I = _____
20. E = 50V   R = 10 K Ω   I = _____
21. I = .01A   R = 100 Ω   E = _____
22. P = 100W   I = 2A   R = _____
23. I = 20 mA   E = 100V   R = _____
24. P = 10W   I = 1A   R = _____
25. What is the total circuit resistance with 1000 ohms, 1.2 kilohms, 5.6 kilohms and 10 kilohms in series?
26. What is the total circuit resistance with 1200 ohms, 5 kilohms and 10 kilohms in parallel?

Chapter 5

MAGNETISM

The mysteries of magnetism are legendary Folklore passed from generation to generation. A Chinese emperor, for example, used a natural magnet as a direction finding device on a chariot over five thousand years ago. Shepherds before the Christian era were mystified by small pieces of stone found in their pastures which possessed an invisible attractive and repulsive force. Early navigators found that a small piece of this weird stone, when attached to a floating block of wood in a vessel of water, would always turn in a northerly direction. This was the world's first compass. The stone was called a LODESTONE or leading stone.

In the Dark Ages of superstition, the magnet was believed to contain the devil. This natural magnet is now a curiosity, since science has found ways of making artificial magnets stronger and more permanent.

In this chapter, you will:

1. Study the nature of and properties of permanent and electromagnetic fields.
2. Become familiar with the terms used in magnetic circuits and the units of measurement.
3. Explore the conversion of electrical energy to mechanical energy by electromagnetic fields.

THE EARTH IS A MAGNET

Why does a compass point in a northerly direction? Science has discovered that the core of the earth is just an enormous magnet with its ends close to the earth's surface at magnetic north and south. From each end of the earth's great magnet are INVISIBLE LINES OF MAGNETIC FORCE. These lines seem to connect north and south in one great continuous magnetic circuit. Fig. 5-1 shows the earth and its magnetic field.

A compass anywhere on the earth's surface tends to line up parallel to these force lines and points toward magnetic north. From the scientific point of view, magnetic north is, in truth, the south end of the earth's magnetic field. Magnetic lines of force tend to exert a force from north to south. Also, magnetic north is not true North, and the angular difference between magnetic north and true North is called the ANGLE OF DECLINATION or VARIATION.

PERMANENT MAGNETS

Magnets are now manufactured from certain ferrous materials by placing a bar of this material in a strong magnetic field. The magnetic field or FLUX may be observed by placing the magnet under a sheet of cardboard. While tapping the cardboard gently, sprinkle the surface with iron filings. The iron particles will align themselves according to the individual lines of force around the magnet. See Fig. 5-2.

Notice the concentration of force lines at each end of the magnet. These are called POLES. One end is the NORTH POLE; the opposite end is the SOUTH POLE. Close examination of these force lines will indicate

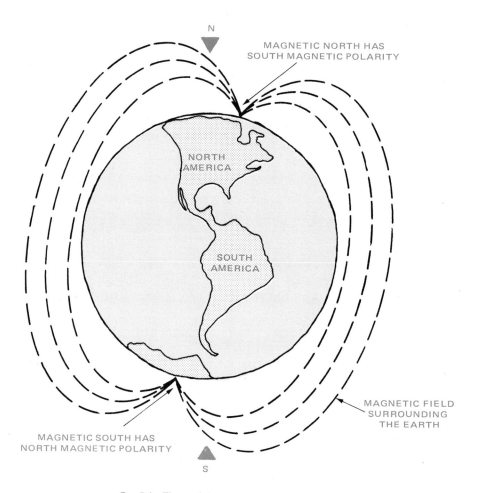

Fig. 5-1. The earth is one great magnet.

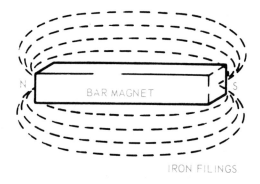

Fig. 5-2. A magnet is surrounded by a field of many magnetic lines of force.

that each line is an individual continuous line and no two lines will cross each other. At each pole, they are very dense. In the space between the poles, they tend to expand and separate. Each line repels its adjacent lines and tends to

keep them apart. The number of lines in the field will depend upon the strength of each magnet.

LAWS OF MAGNETIC FORCE

The first two laws of magnetism state that:

UNLIKE POLES ATTRACT EACH OTHER

LIKE POLES REPEL EACH OTHER

These actions are demonstrated by working with two magnets suspended from wire stands as shown in Fig. 5-3. When hung with UNLIKE poles adjacent to each other, the magnets are attracted together. When hung with either like north poles or like south poles adjacent to each other, they repel each other.

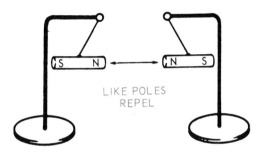 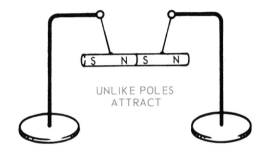

Fig. 5-3. Demonstrations of the first two laws of magnetism.

Iron filings and cardboard can again be used to further prove these laws of repulsion and attraction. Place the cardboard over two magnets with unlike poles in close proximity but fixed in place so they cannot move. Sprinkle the surface with iron filings, and they should form the pattern shown in Fig. 5-4.

Note the dense concentration of force lines between N and S at the center and also the force lines in the space between extreme ends of the pair of magnets. The pair tends to appear as one larger magnet. In fact, if not fixed in position, they would snap together and form a single magnet.

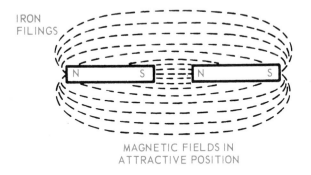

Fig. 5-4. Magnetic fields between bar magnets are shown in an attractive relationship.

The fields of magnets in the repelling positions may be studied in the same manner. Place two poles close together in a fixed position beneath the cardboard. See Fig. 5-5.

Note that the gap between the LIKE POLES of the magnets is almost empty. The lines from each magnet repel each other and do not get close unless forced to do so. If the magnets were released from their fixed positions, they would jump apart.

Hold two magnets in an attractive position but separated by several inches so that they will not come together. By moving one magnet slowly toward the other magnet, a point will be reached when the magnets will move together. You will correctly conclude that the attractive,

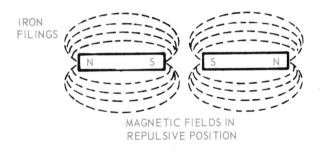

Fig. 5-5. Magnetic fields of magnets in repulsion relationship.

and also the repulsive force, between magnets varies according to the distance between the magnets. In fact, the attractive or repulsive force varies inversely as the distance squared. This is expressed in the equation:

$$F = \frac{M_1 M_2}{\mu d^2}$$

Where,

M = strength of magnet
d = distance between magnets
μ (mu) = permeability

This equation also shows that the force is directly proportional to the product of the strength of each magnet.

PERMEABILITY (μ)

PERMEABILITY is the ability of a material to conduct or carry magnetic lines of force. It is the property of a material which permits easy distribution of magnetic force lines within the material. In magnetic circuits, the Greek letter μ (mu) is assigned to permeability. The permeability of materials is signified by a number. It has no units of measurement. The number expresses the permeability of the material in respect to air, or more accurately, a vacuum, which has a permeability of 1. As a result, permeability is included in the equation for magnetic force.

RETENTIVITY

RETENTIVITY is the ability of a material to hold its magnetism after the material has been removed from a magnetizing force. In similar fashion, magnetism which remains in a material after a magnetizing force is removed is called RESIDUAL MAGNETISM. For permanent magnets, a material of high retentivity is required. On the other hand, a core of a relay or magnetic switch should retain little residual magnetism when not activated.

MAGNETIC MATERIALS

The most important magnetic materials used in electricity and electronics are FERROMAGNETIC materials. These materials are relatively easy to magnetize and include iron, steel, cobalt, permalloy and alnico. Alnico is a trade name for an alloy of iron, nickel and aluminum. It is most satisfactory for permanent magnets and holds its magnetism for long periods of time.

A PARAMAGNETIC material will become only slightly magnetized in a strong magnetic field. These materials include aluminum, chromium and platinum. Any slight magnetism of a paramagnetic material will be in the same direction as the magnetizing force.

A DIAMAGNETIC material can also be slightly magnetized, but it will assume a polarity opposite to the polarity of its magnetizing force. These materials include copper, silver, gold and mercury.

THEORY OF MAGNETISM

The exact nature of magnetism in a magnetic field has been investigated by scientists. Two acceptable theories are in general usage. The more common theory is WEBER'S MOLECULAR THEORY. It will be found in most textbooks up to the middle of this century. This theory suggests that magnetic materials consist of particles called molecular magnets. Each particle will have a N and a S pole. In an unmagnetized state, the minute molecular magnets are arranged in random fashion throughout the material. See Fig. 5-6.

MOLECULAR MAGNETS IN RANDOM ARRANGEMENT

Fig. 5-6. Molecular arrangement of steel bar in demagnetized state.

When the material is placed under the influence of a magnetic field, the molecular magnets tend to form a straight line. In a material with high retentivity, the lined-up positions remain after the magnetizing force is removed, Fig. 5-7. The polarity of the total material is the same as the polarity of the molecular magnets. In a low retentivity

MOLECULAR MAGNETS IN LINE

Fig. 5-7. Molecular arrangement of steel bar in magnetized state.

material, the molecular magnets will return to their original random arrangement when the force is removed.

The WEBER theory is substantiated by the fact that if a bar magnet is broken into pieces, each piece will be a magnet of the same polarity, Fig. 5-8.

Fig. 5-8. Each piece of a broken magnet retains its original polarity.

Further evidence that this theory is correct may be found in the care of magnets. Heat will cause molecular movement. Shock or jarring also will cause some molecular movement. Both of these influences will help the lined-up molecular magnets return to their random demagnetized state.

Magnets should not be heated nor should they be dropped on the floor.

The more modern theory of magnetism is the DOMAIN THEORY, which relates to the atomic structure of elements discussed in Chapter 1. You will remember that an atom consists of a nucleus and a definite number of orbiting electrons. The Domain Theory suggests that each electron in an atom is also spinning on its own axis, as well as orbiting around its nucleus. If an equal number of electrons are spinning in opposite directions, the atom is in an unmagnetized state. However, if more electrons are spinning in one direction than the other, the atom is in a magnetized state and is surrounded by a magnetic field.

When a number of these magnetized atoms exist in a material, they interact with adjacent atoms and form domains of atoms having the same magnetic polarity. These domains exist in random patterns throughout the material. Under the influence of a strong external magnetic

field, the domains become aligned and the total material is magnetized. The strength of the retained magnetic field depends upon the number of domains which have been lined up.

Magnetic materials contain many domains. About ten million tiny domains may be contained in a cubic centimetre of magnetic material.

MAGNETIC INDUCTION

Why does a magnet pick up a nail? See Fig. 5-9. When the nail is brought into the magnetic field, the lines of force will pass through the nail because the steel nail has a higher permeability than air. The nail becomes a part of the total magnetic circuit. The lines of force will make a magnet of the nail with the same polarity as the attracting magnet. The nail, therefore, is attracted to the magnet. Magnetization of the nail while it is in the flux field of the magnet is called MAGNETIZATION BY INDUCTION.

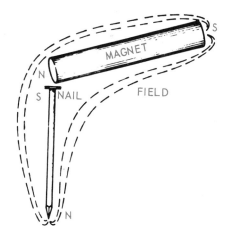

Fig. 5-9. A nail is attracted to a magnet because a field of the same polarity is induced in the nail.

MAGNETIC SHIELDING

There is no known shield against magnetism. It will pass through any material including glass, water and insulations. However, the fact that some materials have greater permeability permits their use as a means of conducting a magnetic field around a circuit or delicate instrument. This action is called SHIELDING.

In Fig. 5-10, for example, a circular shield of magnetic material surrounds a meter. Since the disturbing magnetic field will take the easiest path from N to S, the shield redirects the magnetic lines around the meter.

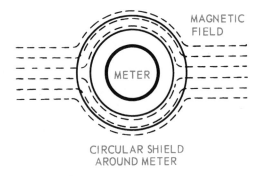

Fig. 5-10. A shield of high permeability conducts magnetic flux around meter.

MAGNETIC FIELDS AND ELECTRICITY

The relationship between electricity and magnetism was first discovered in 1819 by Danish scientist, Hans Christian Oersted. The discovery was made while doing some experiments with electrical circuits in the laboratory. A compass lying on the bench near the circuits showed very peculiar behavior when the circuit was turned on or off. This led Oersted to the conclusion that an electric current in a conductor also produced a magnetic field.

This relationship between electricity and magnetism is demonstrated in Fig. 5-11. Cur-

rent flow up through the wire produces a clockwise circular magnetic field. A reverse current produces a counterclockwise field. The LEFT HAND RULE is applied to determine the direction of the magnetic field. Grasp the conductor in your left hand with your thumb extended and pointing in the direction of the current flow. Your fingers, encircling the conductor, point in the direction of the magnetic force.

Fig. 5-12 shows cross sections of current carrying conductors. At left, the black dot in the center is the point of an arrow showing that the current is flowing toward you. The direction of the field is shown. The cross on the right sketch represents the feathers at the butt end of an arrow, indicating that current is flowing away from you. Use the left hand rule to prove that the magnetic fields are correctly drawn.

In Fig. 5-13, two conductors are illustrated. One drawing shows the magnetic fields when

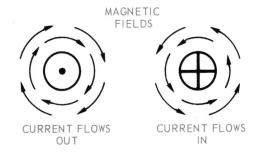

Fig. 5-12. These conventions are used to illustrate current direction and associated magnetic fields.

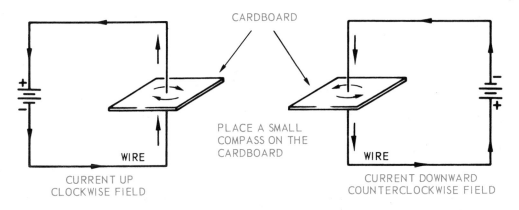

Fig. 5-11. Compasses placed on cardboard will show direction of magnetic field.

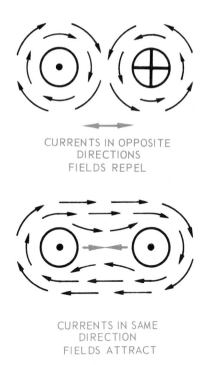

CURRENTS IN OPPOSITE
DIRECTIONS
FIELDS REPEL

CURRENTS IN SAME
DIRECTION
FIELDS ATTRACT

Fig. 5-13. The attractive or repulsive force between magnetic fields depends upon current direction in the conductors.

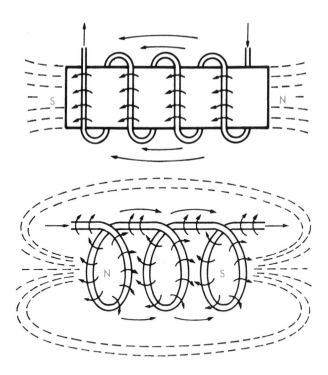

Fig. 5-14. The magnetic fields around the wires of a coil reinforce each other and give the coil magnetic polarity.

currents are flowing in opposite directions. In the second drawing, currents in the conductors flow in the same direction. Pay particular attention to the magnetic field patterns.

THE SOLENOID

When current flows through a wire wound into a coil, Fig. 5-14, the magnetic fields around the wires join and reinforce each other.

The coil assumes a POLARITY just like a magnet. One end becomes north, the other end is south. Using the LEFT HAND RULE, grasp the coil with your left hand with your fingers pointing in the direction of current flow. Your extended thumb will point toward the north pole of the coil. A coil wound in this fashion without a core is called a SOLENOID.

THE ELECTROMAGNET

A coil of wire wound on a core of magnetic material is an ELECTROMAGNET. The polarity is discovered in the same manner as the solenoid, using the left hand rule. The electromagnet is shown in Fig. 5-15.

The strength of an electromagnet is governed by:

1. The number (N) of turns of wire in coil.
2. The current (I) flowing through coil.
3. The kind of core material.
4. The ratio of coil length to its diameter.

The magnetic fields around each wire in a coil are additive. By increasing the number of turns, the number of flux lines increase and the intensity of the total magnetic field is increased.

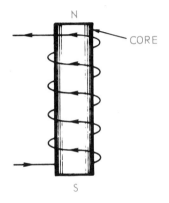

Fig. 5-15. A coil wound on a core is an electromagnet.

Since the magnetic fields around a wire are a function of current, a larger current produces stronger individual fields and the total field intensity is increased.

By winding the coil on a core of magnetic material, the overall field strength is increased. Magnetic materials have a greater permeability than air, so the passage of flux lines is greatly improved. NOTE: The field intensity of a coil can be increased only by increasing the number of turns or increasing the current. The core becomes magnetized and adds to the intensity of the overall field.

The field intensity of a coil will remain uniform throughout the cross section of the coil if the length of the coil is ten times or more greater than its diameter or twenty times the coil radius.

ELECTROMAGNETIC TERMS AND DEFINITIONS

A most confusing system of units of measurement in magnetic circuits is now in use. Some terms have not yet been universally accepted. To make it more difficult, at least three systems of terms and measurement will be found in texts and engineering reports. The three systems are: the English system units, CGS units and MKS units. For this text, we will confine ourselves to the English and Centimetre-Gram-Second (CGS) units.

FLUX is the description of the total number of lines of magnetic force. Its symbol is the Greek letter Φ (phi). In English units, flux is measured in Kilolines (1000 lines). In the CGS units, a single line of magnetic force is called a MAXWELL.

FLUX DENSITY is a term used to describe the number of flux lines in a specified cross-sectional area. A cross section view of a magnetic field is shown in Fig. 5-16.

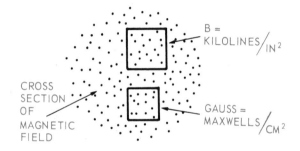

Fig. 5-16. A comparison of English and CGS systems of measuring flux density.

In the English system, density is expressed as KILOLINES PER SQUARE INCH. In CGS units, a GAUSS represents a density of ONE MAXWELL PER SQUARE CENTIMETRE. The symbol assigned to flux density is the letter B. Since density is an indication of flux lines per unit area, then:

$$\text{(density) } B = \frac{\Phi \text{ (flux lines)}}{A \text{ (unit area)}}$$

MAGNETOMOTIVE FORCE (mmf) is the total force producing a magnetic field or flux. The use of this term can be compared to Ohm's Law in electricity:

$$I = \frac{E}{R}$$

where current (I) is analgous to flux (Φ), electromotive force (E) compares to mmf, and resistance (R) is similar to reluctance (rels or R). Magnetomotive force is measured in IN (ampere-turns) in the English system, and in GILBERTS in the CGS units.

ROWLAND'S LAW states that:

$$\Phi \text{ (flux)} = \frac{\text{mmf}}{\mathcal{R}}$$

RELUCTANCE is the opposition to the establishment of flux lines in a material. It is, in effect, similar to resistance in electrical circuits. In English units, reluctance represents AMPERE TURNS PER KILOLINE. In CGS units, it represents GILBERTS PER MAXWELL.

FIELD INTENSITY, symbol H, is the magnetic force per unit length of the flux path. In English units, field intensity is expressed in AMPERE TURNS PER INCH. In CGS units, it is given in GILBERTS PER CENTIMETRE.

PERMEABILITY, as previously stated, is the ability to concentrate flux lines. Permeability may be expressed by the equation:

$$\text{(Permeability)} \quad \mu = \frac{B \text{ (flux density)}}{H \text{ (magnetizing force)}}$$

GILBERT is a unit of magnetomotive force in the CGS system. It may be converted to the English system by:

ONE GILBERT = 1.256 AMPERE-TURNS

APPLICATION OF MAGNETIC CIRCUIT TERMS

The previously discussed units of measurement take on additional meaning when applied to magnetic circuits.

The coil shown in Fig. 5-17 is assumed to have 400 turns of wire. The coils is connected to a voltage source to cause a current of 2 amperes. The magnetomotive force produced is:

I x N = 2 x 400 = 800 (ampere-turns) IN

The same force might be produced by:

4 amperes and 200 turns
8 amperes and 100 turns

Any combination of current and number of turns would give this magnetomotive force. However, most coils with which you will work have a fixed number of turns. Therefore, you will VARY THE MAGNETOMOTIVE FORCE BY VARYING THE CURRENT THROUGH THE COIL. This is not isolated theory. You will be required to understand this before you can work on generators, motors, relays and many other devices.

Field intensity differs only from force in that it is distributed over the total length of the magnetic circuit, including the core. In the previous example, if the magnetic field is 10 in. long, then field intensity (H) would be:

$$H = \frac{IN}{\text{inches}} = \frac{800}{10} = 80 \text{ IN/inch}$$

Flux density is also a function of field intensity. According to Rowland's Law, any increase in mmf by an increase in current will increase the magnetic flux (Φ).

HYSTERESIS LOOP

For closer study, the characteristics of a magnetic material within a coil may be plotted on a curve. In Fig. 5-18, the X axis represents the field intensity measured in amper-turns per inch.

The Y axis represents flux density in kilolines per square inch. As H is increased, the flux

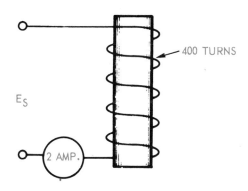

Fig. 5-17. Magnetomotive force is 800 IN (ampere-turns).

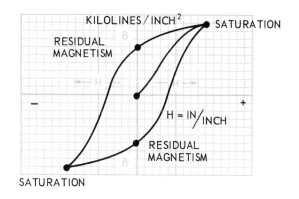

Fig. 5-18. The hysteresis loop of a magnetic material.

density increases to a point of SATURATION. This means that the magnetic core material is so full of flux lines that no more can be produced by any further increase in force.

Now, the force H is returned to ZERO. Note however, that the curve did not return to zero. This point marks the residual magnetism in the material after the magnetizing force is removed. A certain amount of force H in the opposite polarity is required to demagnetize the material. Continuation of the curve is shown as H is applied in the opposite polarity to saturation. The curve may be returned to saturation. The curve may be returned to saturation at opposite polarity by again reversing force H.

The hysteresis loop graphically shows the characteristics of a magnetic material with a magnetizing force alternating in polarity. You will need to understand this theory in order to intelligently work with chokes, transformers, memory cores of computers and many other devices and circuits. Why not prepare yourself for advanced study in electronics?

LESSON IN SAFETY: Electrical energy dissipated in resistance produces heat. Burns can be painfully serious. Resistors, vacuum tubes, transistors and many other components can become very hot. Do not put your fingers on hot components. Be cautious.

PRACTICAL APPLICATIONS OF ELECTROMAGNETS AND SOLENOIDS

In view A in Fig. 5-19, a movable core is partially placed in the hollow center of a solenoid. When the coil is activated, a magnetic field appears as illustrated. Now, magnetic lines seek the shortest possible path between poles, and the lines seem to be elastic. They exert a SUCKING FORCE on the movable core and pull it into the center of the coil. See view B.

When the coil current is turned off, a spring returns the core to its original position. This is a very popular method of converting electrical energy to electromagnetism to mechanical

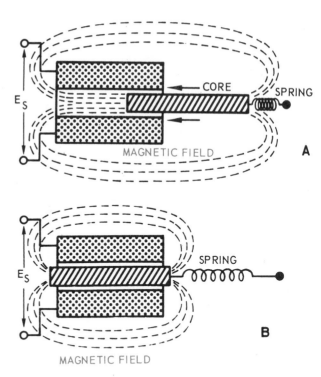

Fig. 5-19. A—The magnetic field starts to suck the core into the coil. B—The core becomes "at rest" in the center of the coil.

movement. A lever or rod attached to the core may be linked to levers, gears and switches to perform all kinds of operations. Such a device is used in many automobile starters to move the pinion gear on the starter into mesh with the flywheel gear.

Solenoids are used at home in dishwashers and washing machines to turn the hot and cold water OFF or ON and to change the cycles of operation. In Fig. 5-20, the solenoid is used in an electric door chime. The core is the plunger which strikes the chime when the coil is activated by completing the door chime circuit at the push button.

RELAYS

A relay is magnetic switch, Fig. 5-21. By design, a movable spring armature is mounted above the core of an elctromagnet. When the coil is energized, the armature is attracted and the contact points open or close (depending on the arrangement). A relay in a normally closed

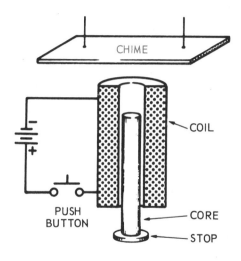

Fig. 5-20. Sketch shows the construction of a door chime.

(N.C.) position opens when activated. A normally open (N.O.) relay closes when energized. When the energizing potential is removed, spring action returns the armature to its original state. Switch points may be set up for single-pole switching or there may be several contacts for double-pole and more complicated switching operations.

The advantages of a relay include the rapid and positive switching control of machinery and devices from a remote location. This provides safety for operating personnel, since the relay operating voltages and currents can be relatively small when compared to levels required for running the machine.

The relay also permits the use of small wires and low currents for switching a machine. The contact points only have to be heavy enough to carry the line current to run the machine. Relays eliminate the use of heavy wires. Also, electrical magnetic switching can be more rapid than hand switching.

When a relay must be selected to do a specific job, consideration should be given to coil specifications. The dc resistance of the coil is the resistance of the wire from which the coil is wound. The operating current or closing voltages are indicative of the sensitivity of the relay and must conform to the particular need. The contact points must be sufficiently large and heavy to carry the required currents.

Relays may be connected in either SHUNT (parallel) or SERIES. See Fig. 5-22.

In the shunt connection, the coil is placed across the voltage source. Since current through the shunt coil connection is limited only by the resistance of the coil, it usually is made with many turns of relatively fine wire to limit the operating current. It is called a VOLTAGE COIL.

In the series connection, all the current to a specified load will flow through the coil. The coil is considered to be current operated. The windings must be heavy enough to carry the load current. In projects you will construct, relays that employ both of these methods of connection will be used.

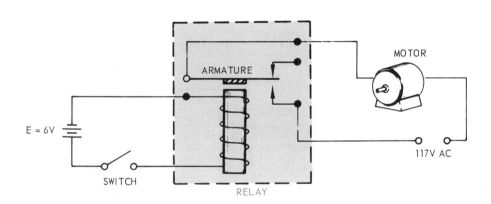

Fig. 5-21. A diagram showing the application of a relay.

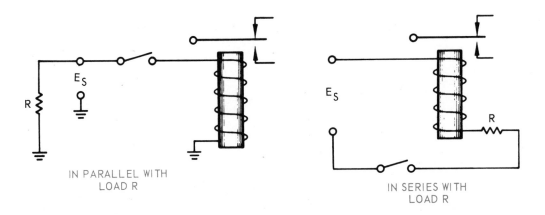

Fig. 5-22. Relays are made to be voltage or current sensitive.
They may be connected in parallel or series.

LOCKED-OUT RELAY

The locked-out method of relay connection is used on most of the machinery around your shop. Red and black push buttons in a small box are located near the operating position.

This allows the machine to be turned off rapidly. Follow the circuit diagrams for the relay in both OFF and ON positions in Figs. 5-23 and 5-24. Note that the ON button energizes the coil and turns on the machine. It remains on until the OFF button is pushed.

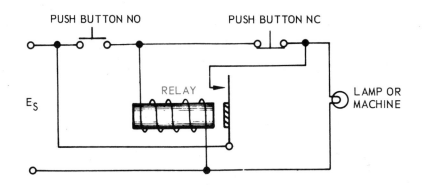

Fig. 5-23. Circuit for a LOCKED-OUT RELAY, shown in the OFF position.

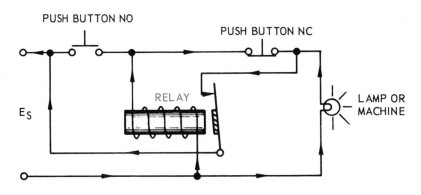

Fig. 5-24. LOCKED-OUT RELAY after pushing (NO) button.

OVER CURRENT RELAYS

A special current relay may be used as a magnetic circuit breaker or protective device. In Figs. 5-25 and 5-26, this type of circuit is illustrated. It uses a SERIES coil, and all the current to the load flows through the coil and the points. Normal load currents will NOT activate the relay. A larger-than-normal load current, due to a short circuit or overloading, will activate the relay and OPEN the circuit to the load. A STOP of some kind falls in place, which prevents the contact points from re-closing.

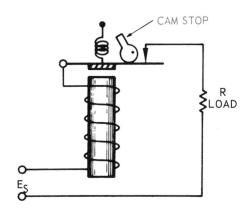

Fig. 5-25. The circuit of a simple circuit breaker.

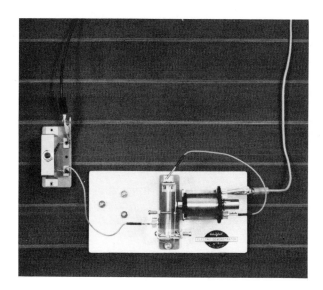

Fig. 5-26. A breadboard demonstration of a magnetic circuit breaker. (Lab—Volt)

The breaker is reset manually, after the reasons for the excessive load current have been determined. The breaker prevents destruction of motors and devices by excessive currents, and it eliminates fire hazards from conductors and devices overheated by overload currents. A breaker can be used again, merely by resetting.

REED RELAYS

The reed relay is an interesting type of electromagnetic switch. These relays are made up of two reeds (switch contacts) that are hermetically sealed in a glass capsule filled with an inert gas. Fig. 5-27 shows a cross-sectional view of a single pole, single throw (SPST) reed relay. The glass capsule is inserted in a bobbin on which a dc coil is wound. When a dc current is fed through the coil, the switch contacts open or close, depending on whether it is a normally open (N.O.) or normally closed (N.C.) arrangement.

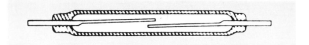

Fig. 5-27. A SPST reed relay. (Magnecraft)

Advantages of the reed relay over standard other types of relays are:

1. Absence of moving parts (except contact blades).
2. Sealed capsule which eliminates oxidation and contamination.
3. Low operating power (usually in milliwatt range).
4. Long contact life (usually 20,000,000 minimum).
5. Speed (100 operations per second is common).
6. May be used in printed circuits (see Fig. 5-28).

FUSES

Fusing is another, older method of circuit protection. FUSES will be found in many

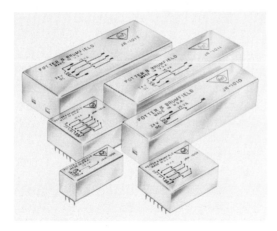

Fig. 5-28. Reed relays in printed circuit cases.
(Potter & Brumfield)

Fig. 5-30. Cartridge fuses are used in many electronic
devices. (Littelfuse)

circuits. A fuse is a link of low melting point metal. Under normal conditions, the current through the fuse will not produce excessive heat. If the circuit is overloaded, the excessive current will produce sufficient heat to melt the fusible link. It must be replaced before the circuit is again operative. You must investigate the cause of the high current before replacing a blown fuse. Fig. 5-29 shows a simple circuit that includes the symbol for a fuse.

Fig. 5-29. A simple circuit with fuse protection.

Fuses are manufactured in a wide variety of shapes and current ratings. A special type of fuse is the SLO-BLOW which will withstand surges of high current without blowing out. It is a delayed fuse. See Fig. 5-30 for some typical cartridge fuses.

BIMETAL THERMOSTATS

BIMETAL DEVICES are another type of circuit protector which are also used in thermostats. This component takes advantage of the unequal expansion rates of dissimilar metals. A simple bimetal device is illustrated in Fig. 5-31.

The construction of a bimetal device features two different kinds of metal fastened together. A rise in temperature will cause metal A in Fig. 5-31 to expand more than metal B. As a result, the metal sandwich starts to curl up and opens the switch contacts. A decrease in temperature will again close the points and reactivate the

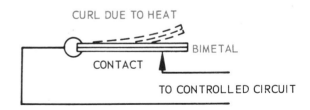

Fig. 5-31. The bimetal curls up (dotted lines) when there is an
increase in temperature.

circuit. Such a device is widely used in temperature control. You may have one in your home to control your furnace.

To use the bimetal as a circuit breaker, it is necessary to provide a stop which will fall in place when a rise in temperature opens the points. The stop must be removed or "reset" to operate the circuit again.

BUZZERS AND CHOPPERS

The circuit of an electrical buzzer is drawn in Fig. 5-32.

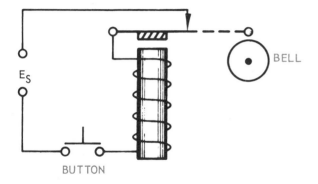

Fig. 5-32. The circuit of a buzzer or doorbell.

When the buzzer is energized, the magnetic field of the coil opens the armature contacts. Since these contacts are in series with the coil, the coil circuit also opens. This causes the magnetic field to collapse and close the points which, in turn, energizes the coil.

The operation repeats itself and the device produces the familiar "buzz" of the vibrating points. If desired, an extension with a striker can be attached to the vibrating armature to ring a bell. The buzzer then becomes a door bell.

In later chapters, you will learn that an ac or a pulsating dc current is required when transformers are used. Since a buzzer cuts the current on and off intermittently at a specified frequency, it may be used as a "chopper" and its output approaches a square wave.

The buzzer or vibrator was once used in the old fashioned automobile radio supply. Transistors are now used in auto radios and a vibrator-transformer supply for high voltage is no longer needed. The chopper does find wide usage when it is necessary to sample a voltage or signal at prescribed intervals. Even so, better and faster choppers are made by using switching transistors and eliminating all mechanical vibration and wear.

TEST YOUR KNOWLEDGE

1. The invisible lines of magnetic force are called the magnetic field or _____.

2. The angular difference between magnetic north and true North is called the _____.

3. State the first two laws of magnetism.

4. Define permeability.

5. If a huge crane used an electromagnet to pick up junk, should the core of this magnet have a high retentivity?

6. A meter can be protected from a magnetic field by installing _____ .

7. Demonstrate to your instructor that you can use and understand the left hand rule for magnetic fields around a current carrying conductor.

8. Indicate the north and south poles of these electromagnets.

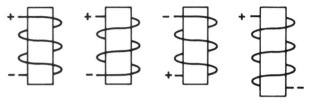

9. _____ is the ability of a material to hold its magnetism after the material has been removed from a magnetizing force.

10. Which of the following metals are paramagnetic materials?
    a. Aluminum.   d. Mercury.
    b. Silver.   e. Copper.
    c. Chromium.   f. Platinum.

11. Which of the following metals are diamagnetic materials?
    a. Aluminum.   d. Mercury.
    b. Silver.   e. Copper.
    c. Chromium.   f. Platinum.

12. Which theory of the exact nature of magnetism suggests that each electron in an atom is spinning on its own axis, as well as orbiting around its nucleus?
    a. Domain Theory.
    b. Weber's Molecular Theory.

13. Magnetization of a nail while it is in the flux field of a magnet is called magnetization by _____ .

14. A coil of wire without a core and connected to an energy source is called a solenoid. True or False?

15. A coil of wire wound on a core of magnetic material is an _____ .

16. _____ is a term used to describe the number of flux lines in a specified cross-sectional area.

17. A coil of 600 turns has a current of .5 amp. What is the magnetomotive (mmf) force in gilberts?

18. Draw a magnetic relay circuit to remotely operate an alarm system.

19. State several advantages of magnetic relays.

20. A reed relay is an electromagnetic _____ .

Chapter 6

GENERATORS

Your studies about generators will help you understand how mechanical energy is converted to electrical energy. In your earlier experiments, you produced a potential source of energy from a conductor moving in a magnetic field. This activity will now serve to open the door to many more fascinating studies of electrical phenomena and useful applications.

In this chapter, you will learn:

1. The nature and properties of an alternating current.

2. How a voltage is induced in a conductor moving through a magnetic field.

3. How the alternator and dynamo produce a useful output of electrical power.

4. The methods employed to control the voltage and current output of the alternator.

5. How the alternator may be changed to produce a dc output.

MICHAEL FARADAY

During the early part of the nineteenth century, English scientist Sir Humphrey Davies was asked, "What do you consider your greatest discovery?" His simple and forthright reply was, "Michael Faraday." FARADAY* is remembered as the "father of the dynamo."

In those days, scientists wondered about electromagnetic fields. (Electromagnets were made by wrapping turns of wire around an iron core and connecting the coil to a battery.) The scientists reasoned that there must be a relationship between electricity and magnetism. If electricity could make magnetism, then there must be a way to utilize magnetism to make electricity.

Michael Faraday devoted a major period of his life to solving this relationship between electricity and magnetism. The success of his experimentation has contributed to the present conveniences and comforts available to us from electric power. Most certainly, all nations depend upon electric power to turn the wheels of industry and production.

Faraday discovered magnetic induction. He found that if a conductor was moved through a magnetic field, a voltage was induced in the conductor. A more correct statement would be: A VOLTAGE IS INDUCED IN A CONDUCTOR IN A MAGNETIC FIELD IF THERE IS RELATIVE MOTION BETWEEN THE FIELD AND THE CONDUCTOR.

The conductor may be made to move or the magnetic field may be made to move. In Fig. 6-1, a coil is connected to a galvanometer, which will indicate a current flow in either direction. A permanent magnet pushed into the

---

* FARADAY, MICHAEL, (1791-1867). The distinguished English scientist who is remembered as the "father of the dynamo." His experimentation with and discovery of electromagnetic induction has made possible the electric power which we use daily at home and in industry. The unit of capacitance, the FARAD, is named in his honor.

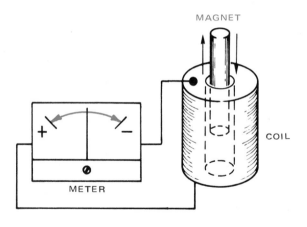

Fig. 6-1. A magnet moving in and out of the coil induces a voltage and current which will cause the meter to deflect.

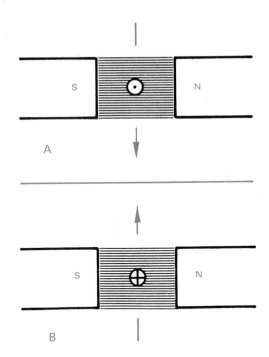

Fig. 6-2. A—Conductor moving downward through field causes current to flow out. B—Conductor moving upward through field causes current to flow inward.

hollow coil will cause the meter to DEFLECT IN ONE DIRECTION. When the magnet is pulled out, the meter will deflect in the OPPOSITE DIRECTION. When there is no motion and the magnet is held in a fixed position, the meter WILL NOT INDICATE CURRENT. Also, if you hold the magnet in a fixed position and move the coil up and down, THIS WILL PRODUCE A CURRENT, FIRST IN ONE DIRECTION, THEN IN THE OTHER. Relative motion must always be present.

POLARITY OF INDUCED VOLTAGE

We have learned that the direction of current flow in a conductor moving in a magnetic field depends upon the direction of movement. View A in Fig. 6-2 illustrates a single conductor moving DOWN through a magnetic field at a right angle. The DOT in the center of the conductor means that electron flow in the conductor is toward you. In view B, the conductor is moving up through the field and the induced current flows away from you, as indicated by the cross on the end of the conductor.

A convenient LEFT HAND RULE may be used to determine the direction of ELECTRON flow. Extend the fingers of your left hand in a flat plane above the conductor. With your fingers pointing in the direction of the magnetic field or pointing toward the South Pole, extend

your thumb to form a right angle to your four fingers. When the conductor is moving toward the palm of your hand, your thumb will point in the direction of electron flow. Of course, if an external circuit is not connected to the moving conductor, a voltage of like polarity will be induced in the conductor. How much of a voltage would be induced? This will depend on four factors:

1. The SPEED at which the conductor cuts the magnetic field.

2. The STRENGTH of the magnetic field.

3. The LENGTH or number of turns of the conductor cutting the field.

4. The ANGLE at which the conductor cuts across the field.

SPEED

If the conductor cuts the magnetic field at a greater rate of speed, the induced force on the electrons within the conductor is greater and, consequently, a greater voltage is developed.

STRENGTH OF FIELD

A stronger field or a more dense field means more magnetic lines for the conductor to cut. Remember that generated output voltage is in direct proportion to the field strength. If a conductor cuts through 10^8 flux lines per second, a voltage of one volt is induced.

ANGLE OF CUTTING

No voltage is induced in a conductor as it moves parallel to the magnetic field. Maximum voltage is induced when the conductor cuts

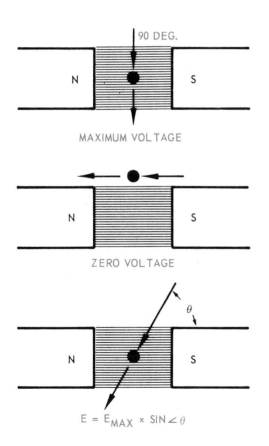

Fig. 6-3. The voltage induced in a conductor as it moves across a magnetic field depends upon the cutting angle.

directly across at right angles to the flux lines. At cutting angles between 0 and 90 deg., a voltage between zero and maximum will be

induced, Fig. 6-3. Any instantaneous voltage may be found by the formula:

$$E = \frac{\Phi \times 10^{-8} \times \sin \theta}{t}$$

where:

$\Phi \times 10^{-8}$ = number of magnetic flux lines for one volt.

θ = the angle at which the conductor cuts the field.

t = time in seconds.

E = volts induced in conductor.

NUMBER OF CONDUCTORS

If a certain voltage is developed in one conductor as it cuts across a field, then more conductors would add substantially to the total generated voltage. Ten conductors would produce ten times more and so on.

PRODUCING ELECTRICAL ENERGY

The beginning student sometimes feels that a generator produces electricity. This is only partly true. A generator CONVERTS mechanical energy into electrical energy. Consider that a current induced in a conductor as it cuts a magnetic field also produces a magnetic field around the current carrying conductor. The polarity of this field is such that it opposes the movement of the conductor and actual mechanical energy must be used to cause the conductor to move. This is the reason that water, steam and atomic power are necessary to turn the enormous alternators to produce our electricity for consumer use.

LENZ'S LAW

LENZ'S LAW provides a scientific explanation of our previous discussion:

"The polarity of an induced EMF is such that it sets up a current, the magnetic field of which always opposes the change in the existing magnetic field."

More simply, Lenz's Law states that the induced field around the moving conductor is opposed by the existing field.

THE SIMPLE ALTERNATOR

Now that you have acquired some background information, we are ready to construct a simple alternator and discuss the ways to improve it. In Fig. 6-4, the single moving conductor is replaced by a rotating single loop of wire called an ARMATURE. Note that as one side of the coil moves upward through the field, the opposite coil side moves downward through the field. External connections to the rotating armature are made through SLIP RINGS and BRUSHES.

In the position shown in Fig. 6-4, the wire loop is cutting the field at right angles. Therefore, a maximum voltage is being induced. In one quarter of a revolution, the wire loop will be momentarily moving parallel to the field and no voltage will be induced. In another quarter turn, the loop will be cutting the field at 90 deg. again. Now, however, side **A** will be moving up and side B downward through the field. As a result, THE INDUCED VOLTAGE WILL BE OPPOSITE IN POLARITY AND THE CURRENT WILL FLOW IN THE OPPOSITE DIRECTION. In Fig. 6-5, the armature is illustrated in four positions with output waves showing magnitude and polarity of the induced voltage.

DEVELOPMENT OF A SINE WAVE

Generally, the rise and fall and polarity of an induced voltage is represented by means of a ROTATING VECTOR. A vector is a line which represents the direction and magnitude of a force. In Fig. 6-6, the vector represents ten volts. Using a scale of 1/4 in. equals one volt, the vector would be 2 1/2 in. long. It is conventional to always have the vector rotate in a counterclockwise direction. To the right of the rotating vector is a graph plotted against a time interval required for one revolution. The instantaneous voltage at any point is a function of the sine of the angle of rotation. Stated in an equation:

$$E = E_{max} \times \sin \angle \theta$$

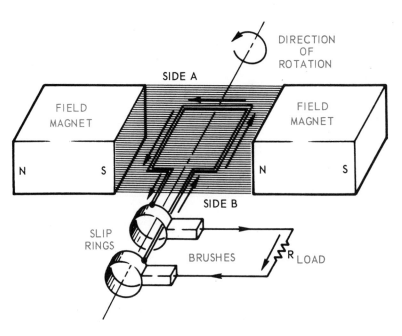

Fig. 6-4. A sketch of a simple alternator or ac generator.
Note current direction as indicated by arrows.

where:

E = maximum or peak voltage

$\angle \theta$ = angle of rotation

For example: At 30 deg. rotation, the voltage in Fig. 6-6 would be:

E = 10 volts x sin 30 deg.

= 10 volts x .5

= 5 volts

We have now constructed a graphical picture of an ALTERNATING CURRENT SINE WAVE. AC differs from DIRECT CURRENT in that AC is continually changing amplitude and polarity. Some appropriate definitions will help to describe the ac wave:

CYCLE. Starting at zero, the wave rises to maximum at one polarity, drops to zero, rises to maximum in the opposite polarity, then returns to zero. One cycle of events has occurred. These cycles may continue to repeat themselves and would represent the output of the simple ac generator.

FREQUENCY. The number of cycles occurring per second is termed the frequency of the ac wave. Originally, it was measured in cycles per second (cps). In 1967, a new and shorter unit of measurement called the HERTZ (Hz) was adopted. The term honors the memory of HEINRICH HERTZ,* the German scientist who first discovered radiating waves.

One hertz (Hz) = one cycle per second

One kilohertz (KHz) = 1000 cps

One megahertz (MHz) = 1,000,000 cps

---

*   HERTZ, HEINRICH (1857-1894). The German scientist who demonstrated the existence of electromagnetic waves. The investigations into electromagnetic nature of light remain as the outstanding 19th century achievement in science. Radio waves are called "Hertzian waves" in his honor. The term, "cycles per second" is being replaced by "HERTZ" as the standard unit of frequency measurement.

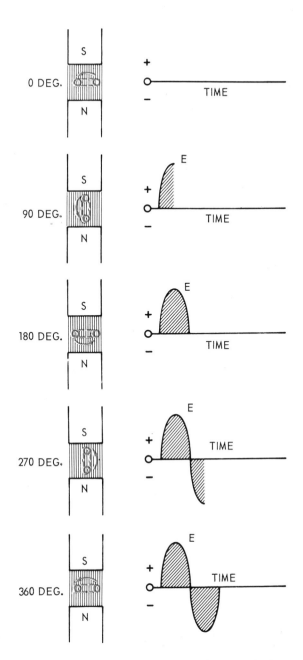

Fig. 6-5. The step-by-step development of the induced voltage during one revolution of an armature in a magnetic field.

PERIOD. The time in seconds for the duration of one cycle is said to be the PERIOD of the ac wave. Mathematically stated, the period of the wave is the reciprocal of its frequency.

$$T \text{ (period)} = \frac{1}{f \text{ in Hz}}$$

Examples:

A 100 Hz wave has a period of $\frac{1}{100}$ sec. or .01 sec.

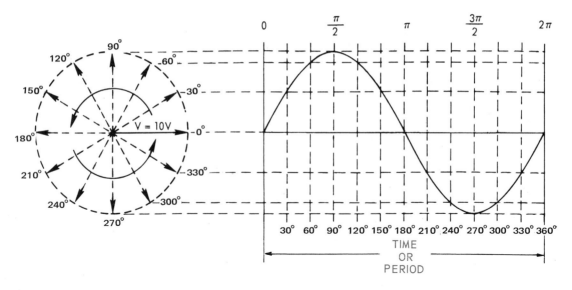

Fig. 6-6. Development of a sine wave.

A 1 KHz wave has a period of $\frac{1}{1000}$ or .001 sec. or one millisecond.

PEAK VALUE. The peak value of the ac wave is measured from the zero line and represents the maximum amplitude of the wave in either positive or negative direction.

PEAK TO PEAK VALUE. This is the value of the wave from the maximum positive peak to the maximum negative peak. It usually represents two time the peak value.

WAVELENGTH. The frequency of a wave may also be designated by its length in metres. The length of the wave is measured from the

peak of one cycle to the corresponding peak of the next cycle. See Fig. 6-7.

The Greek letter λ (lambda) has been assigned to wavelength. The actual length of a wave is determined by the distance a wave travels in space at the speed of light. Since the speed of light is 3×10^8 metres per second or 186,000 miles per second, λ may be equated as:

$$\lambda \text{ in metres} = \frac{3 \times 10^8}{f \text{ in Hz}}$$

$$\text{or} \quad \frac{3 \times 10^5}{f \text{ in KHz}}$$

$$\text{or} \quad \frac{3 \times 10^2}{f \text{ in MHz}}$$

$$\lambda \text{ in feet} = \frac{984}{f \text{ in MHz}}$$

If the wavelength is known, the frequency may be found by transposing the equation to read:

$$f \text{ in Hz} = \frac{3 \times 10^8}{\lambda \text{ in metres}}$$

Examples: What is the wavelength of a 3 MHz wave?

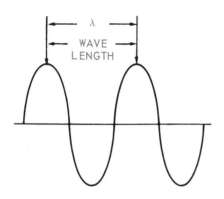

Fig. 6-7. The wavelength is measured from one peak to the corresponding peak of the next cycle.

$$\lambda = \frac{3 \times 10^2}{3\ MHz} = 100\ metres$$

What is the frequency of a 15 metre wave?

$$f = \frac{3 \times 10^2}{15} = 20\ MHz$$

AVERAGE VALUE OF AN AC WAVE

An ac wave is made up of positive and negative ALTERNATIONS. The average value of the wave is considered the average value of ONE ALTERNATION. This value is approximately 63.7 percent of the peak value, as shown in Fig. 6-8, when the area above the average line equals the area in the valley between the alternations.

The following equations may be used to find unknown values:

$$E_{av} = .637 \times E_{peak}$$

$$E_{peak} = 1.57 \times E_{average}$$

These same formulas are used for current or power sine waves as:

$$I_{av} = .637 \times I_{peak}$$

Before leaving this average value, consider that the VALUE of an ac wave, when considering both negative and positive alternations, is zero.

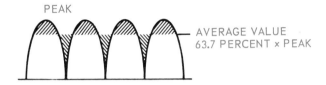

Fig. 6-8. Computation of the average value of a sine wave.

EFFECTIVE VALUE

The effective value, by far, is the most useful value to know about the ac wave. It is a direct comparison of ac to dc and represents the equivalent value of the ac wave to produce the same power in a resistive load as a steady dc value. Most meters you will use indicate the effective value of an ac voltage or current, but you should know how the effective value is computed.

A review of the power laws in Chapter 4 will show that power dissipated in a resistance is $P = I^2 R$. Now, look at the simple circuit in Fig. 6-9. An ac voltage applied to this circuit will produce a current in the circuit which will

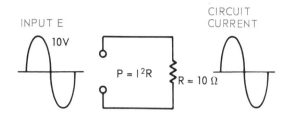

Fig. 6-9. Circuit explains the RMS value of a wave.

vary in step with the voltage as indicated. The resistance, on the other hand, is not affected by the direction of the current. It uses the same power in either direction.

By more advanced mathematics, it may be found that a value of voltage equal to .707 x the peak value of the wave will produce a current and power equal to a dc voltage of that value. We can apply this to the problem in Fig. 6-9:

$$E_{eff} = .707 \times 10V = 7.07\ volts$$

$$Power = (.707)^2\ R$$

$$= .5 \times 10 = 5\ watts$$

To produce the same power with a dc voltage, it would have to be:

$$P = \frac{E^2}{R} \quad or \quad E = \sqrt{PR}$$

$$= \sqrt{5 \times 10\ \Omega} \quad\quad = 7.07\ volts$$

Many times the effective value will be referred to as the rms value. The letters rms stand for ROOT MEAN SQUARE. This is saying that the effective value of an ac sine wave is equal to the square root of the mean value of many instantaneous currents squared.

The following equations are used to determine unknown values. They apply to current and power as well as voltage.

$$E_{eff} = .707 \times E_{peak}$$

$$I_{eff} = .707 \times I_{peak}$$

$$E_{peak} = 1.414 \times E_{eff}$$

$$I_{peak} = 1.414 \times I_{eff}$$

Note that $1.414 = \sqrt{2}$ and $\dfrac{1}{1.414} = .707$

By substitution, these equations may be written as:

$$E_{eff} = \frac{E_{peak}}{\sqrt{2}} \text{ and } I_{eff} = \frac{I_{peak}}{\sqrt{2}}$$

$$E_{peak} = \sqrt{2} \times E_{eff}$$

$$I_{peak} = \sqrt{2} \times I_{eff}$$

We can go a step further and change the formulas, using a peak-to-peak value for the sine wave.

$$\text{Now, } E_p = \frac{E_{p\text{-}p}}{2} \text{ and } I_p = \frac{I_{p\text{-}p}}{2}$$

$$\text{so, } E_{rms} = \frac{E_{p\text{-}p}}{2\sqrt{2}} \text{ and } I_{rms} = \frac{I_{p\text{-}p}}{2\sqrt{2}}$$

$$\text{Since, } P_{eff} = E_{eff} \times I_{eff} f$$

Then, using peak values:

$$P_{eff} = \frac{E_{peak}}{\sqrt{2}} \times \frac{I_{peak}}{\sqrt{2}}$$

$$= \frac{E_{peak} \times I_{peak}}{2}$$

And, using peak-to-peak values:

$$P_{eff} = \frac{E_{p\text{-}p}}{2\sqrt{2}} \times \frac{I_{p\text{-}p}}{2\sqrt{2}}$$

$$= \frac{E_{p\text{-}p} \times I_{p\text{-}p}}{8}$$

ANGULAR VELOCITY

Angular velocity is another way of expressing the speed of rotation. It is the rate of change of angular displacement expressed in RADIANS PER SECOND. A RADIAN is an angle of approximately 57.3 deg. In one complete revolution of a vector, the angular displacement will be 2π Radians or 360 deg. Assuming a frequency of 100 Hz, the ac vector would rotate 100 revolutions per second. If there is an angular displacement of 2π for each revolution, then the vector velocity would be:

$$2\pi \times 100 \text{ Radians/sec.}$$

$$2\pi f = \text{angular velocity}$$

Angular velocity is a common term used in electronic computations and is represented by the Greek letter omega (ω). When you see "ω", it means "$2\pi f$".

IMPROVING THE SIMPLE AC GENERATOR

From earlier discussion, output of a generator can be increased by increasing the number of conductors moving through the field. This is accomplished easily by winding the armature with many turns of relatively small wire and connecting the coils to appropriate slip rings.

Also, the output may be increased by increasing the magnetic field of the generator. In the earlier example, only permanent magnets of limited strength were used. These field magnets can be made electromagnets and either excited by a part of the generator output or by a separate power source.

SINGLE-PHASE GENERATORS

Although not commonly used, the single-phase generator produces an output of a single sine wave voltage exactly like the output of the generator described in Fig. 6-4. It is easier and more efficient to produce the single-phase current by more complex generators.

TWO-PHASE GENERATORS

By placing two separate windings on the armature of the generator and orienting the coils at an angle of 90 deg., two separate outputs can be obtained. See Fig. 6-10. Coil A

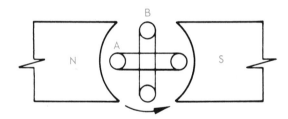

Fig. 6-10. Separate out-of-phase voltages are produced by each of the coils 90 deg. apart.

produces a sine wave output and coil B produces a sine wave output, but the two waves are 90 deg. out-of phase. Study Fig. 6-11.

A single-phase current may be obtained from either of the two windings or the two windings

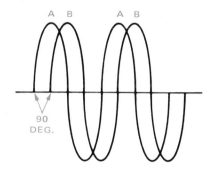

Fig. 6-11. Two-phase output sine waves.

may be connected in series to produce a peak output voltage of about 1.414 times the peak voltage of a single coil.

THREE-PHASE GENERATORS

The most practical and widely used generator is the three-phase unit. Three coils are the armature and spaced at angles of 120 deg. apart, Fig. 6-12. Each coil produces its own sine

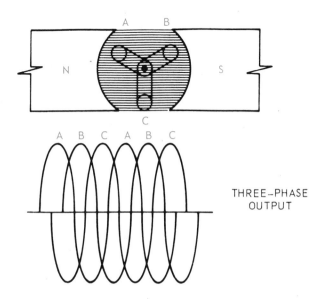

Fig. 6-12. A sketch of the armature coils and output wave forms of three-phase generator.

wave voltage. Three separate, out-of-phase voltages are generated and can be used singularly or together.

ALTERNATORS

Alternators or ac generators have a distinct disadvantage. The high voltages and currents produced in the rotating armature must be connected to the external circuit by some form of sliding contacts like slip rings and brushes. Sparking and burning can result. It is more satisfactory to induce the voltages in fixed coils called STATOR WINDINGS and revolve the magnetic field by ROTOR WINDINGS. In this form, the generator becomes the ALTERNATOR or DYNAMO. It is then only necessary to

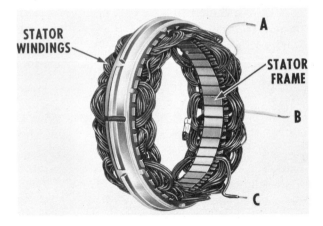

Fig. 6-13. The stator coils of a typical automotive alternator. (Delco – Remy)

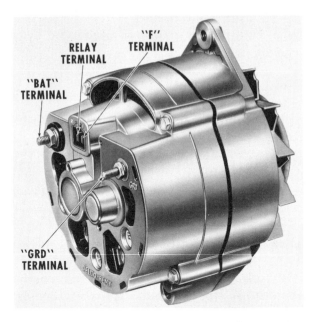

Fig. 6-15. The complete alternator found on many modern cars.

supply the rotor with sufficient current to produce a strong magnetic field. Figs. 6-13 and 6-14 illustrate the STATOR and ROTOR of a typical ALTERNATOR, Fig. 6-15, used in the modern automobile electrical system.

The six output leads may be connected so that they will not interfere with each other. In Fig. 6-17, two methods are illustrated. In the STAR method, all coils are connected to a common point which serves as the ground return for each phase. An advantage of this hookup is that the voltage between any two of

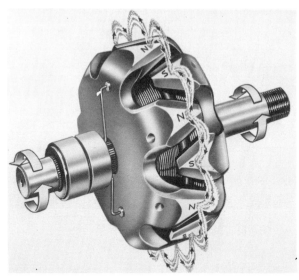

Fig. 6-14. The rotor of a typical alternator. (Delco – Remy)

The rotor coils are excited by an external dc source to produce a magnetic field. See Fig. 6-16. As the rotor passes each of the three stator poles, a voltage is induced. It is apparent that the three voltages of stator coils A, B and C are 120 deg. out of phase with each other.

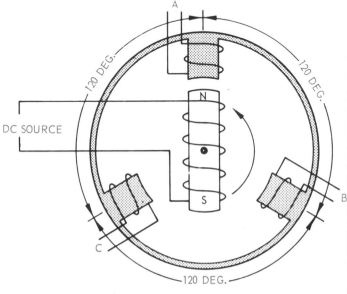

Fig. 6-16. A sketch showing the theory of operation of a three-phase alternator.

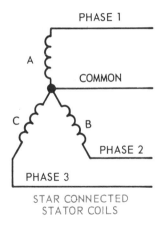

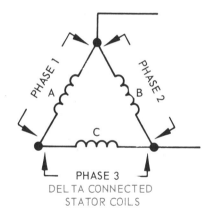

Fig. 6-17. Star and delta connections for stator coils.

the phase outputs if equal to 1.73 times the voltage from any one phase to neutral. That is:

$$1.73 = \sqrt{3}$$

In the DELTA connection, a single-phase voltage may be taken from any two points of the circuit. With this setup, the line current increases to 1.73 times the current produced by a single phase. In summary, the star connection gives increased voltage and the delta arrangement results in increased current.

ALTERNATOR REGULATION

A change in load current from an alternator also causes a change in the voltage drop across the stator windings and a change in output voltage. Generally, external regulation circuits are used to hold the voltage output at a constant level. A transistorized regulator for the automotive alternator is shown in Fig. 6-18. A transistorized regulator circuit is shown in Fig. 6-19.

Fundamentally, the regulator senses a change of voltage at the output and produces a corresponding and corrective change in the current through the rotor, which varies the magnetic field strength. After you have studied later chapters in this text on transistor action, return to this circuit for further understanding. Also note that diode rectifiers are mounted in the alternator. Automobile electrical systems

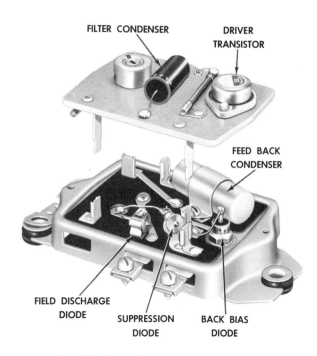

Fig. 6-18. A transistorized alternator regulator. (Delco — Remy)

require direct current, so the ac is converted to dc by these rectifiers. You will study this in Chapter 12.

TRANSISTORIZED ALTERNATOR REGULATION

When ignition switch (SW) in Fig. 6-19 is closed, relay K_1 is closed and the alternator rectified voltage output is applied to the collector and to the base of transistor Q_1. The

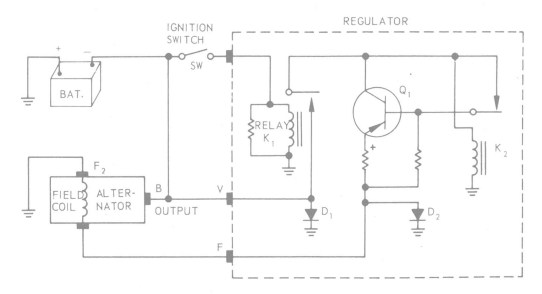

Fig. 6-19. A transistorized regulator circuit for an automotive alternator.

transistor is forward biased and conducts. It is in series with the alternator field coils, and collector current is the field current.

Consider what happens if alternator voltage output should rise above a preset value. Then the voltage coil in relay K_2 is sufficiently energized to open the points of K_2 which disconnects voltage to the base of the transistor. This cuts off the transistor and the collector current, and alternator field current is reduced. The output of the alternator also is reduced and the points of K_2 will reclose. These points vibrate many times per second and maintain the alternator voltage output at the fixed value. The diodes D_1 and D_2 prevent damage to the circuit that might be caused by transient spikes of voltages.

THE DC GENERATOR

Basically, the principle of a dc generator is the same as the ac generator. Each develops alternating voltage in the armature. The major difference is a mechanical switching device on the dc generator called a COMMUTATOR. The resulting action produces a pulsating direct current in the output rather than the ac. This is mechanical rectification. Diodes in the alternator perform the same function.

Now follow dc generator action in Fig. 6-20. Brush A is in contact with commutator section

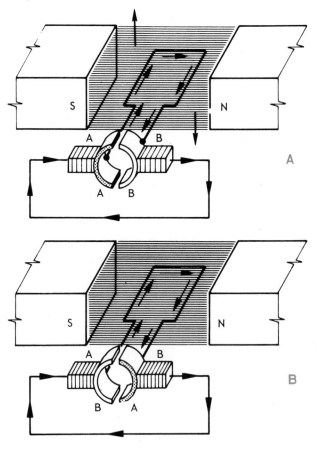

Fig. 6-20. The commutator changes alternating current in the armature to a pulsating direct current in the external circuit.

A and brush B is in contact with section B. Current flow in the armature and external circuit is indicated by arrows. Current flows IN from section A and OUT from section B. As the

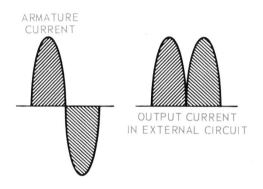

Fig. 6-21. Output of the dc generator is pulsating dc current.

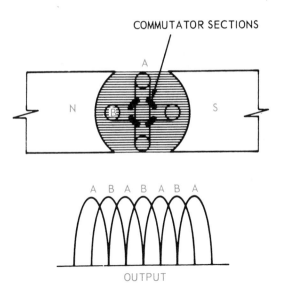

Fig. 6-22. By increasing the number of coils in the armature, the output approaches direct current.

armature turns 180 deg., or one half revolution, the current reverses in the armature as explained for the ac generator. This reverse current does not appear in the external circuit because now section A contacts brush B and section B contacts brush A. See view B in Fig. 6-20. By the switching action of the commutator, the output current appears as shown in Fig. 6-21.

Further improvements may be made on the dc generator by adding several coils to the armature. Each coil has its own commutator sections. The output becomes less pulsating and approaches a constant value dc. A generator with two coils and its output wave form is illustrated in Fig. 6-22. In practice, many coils are used in an automotive generator. See field coil and armature in Fig. 6-23.

The magnitude of generator output primarily depends upon the strength of the magnetic

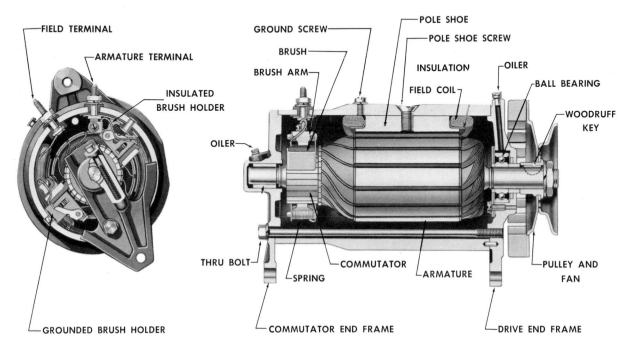

Fig. 6-23. Cutaway view of an automotive dc generator.
(Delco — Remy)

field. In order to increase this field strength and control it, the permanent magnets are replaced with electromagnets called FIELD COILS. The field coils are mounted on magnetic iron cores called POLE SHOES.

Fig. 6-24 illustrates a SHUNT GENERATOR, which is SELF-EXCITED. A small amount of residual magnetism in the pole shoes creates a weak magnetic field. As the generator is rotated, a small voltage is induced. This causes a

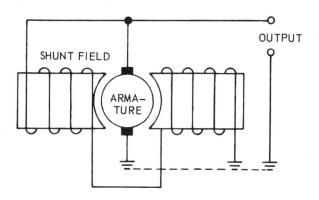

Fig. 6-24. Drawing of a shunt generator. It is self-excited.

small current to flow to the output and through the field coils (also connected to the output). This current increases the field strength and, consequently, the output is increased. This again increases the field strength. The action is cumulative and the generator quickly builds up to maximum output.

A generator also may have field coils in SERIES with its output. Compound generators will have both series and shunt field windings, Fig. 6-25, but their particular performances will not be discussed in this text.

LESSON IN SAFETY: Work with one hand only when testing or measuring voltages. A current between two hands can be more dangerous than a current from one hand to your foot. The wise TV service technician will learn to work on live sets with one hand in a pocket.

DC GENERATOR REGULATION

The practical method of maintaining both a constant output VOLTAGE and CURRENT for the dc generator is made possible by controlling the current through its field coils and thus controlling its magnetic field strength. When we wish to control current, we immediately think of resistance. Any automatic device that will vary the resistance of the field circuit when needed will act as a practical means of regulation. If either voltage or current output exceeds a predetermined value, the regulator will switch a resistance in the field circuit and lower the generator output. With no resistance in the field, the generator will produce its maximum output.

A typical regulator circuit is shown in Fig. 6-26. Follow the action. An excessive current flowing through the SERIES current relay will cause the points to open. The field current must then go through resistance R to ground. This

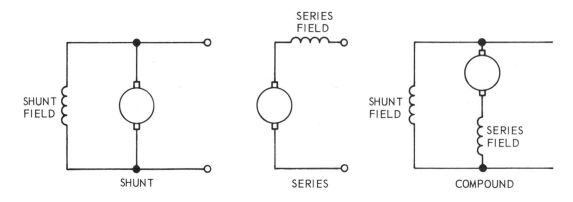

Fig. 6-25. Schematics for series, shunt and compound generators.

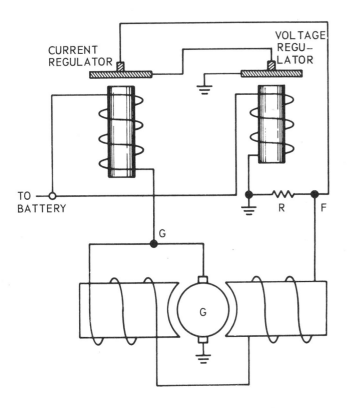

Fig. 6-26. A typical dc generator regulator circuit.

reduced current produces a reduced field and lowers the current output. If the voltage output exceeds a fixed value, the SHUNT voltage relay will be activated and again the field current must flow through R and the voltage output will be reduced. The generator is protected against excessive loads and burnout by the regulator.

GENERATOR LOSSES

Generator losses occur in several ways:

1. When current flows through the resistance of a wire, power is consumed. This power has served no useful purpose and is considered as lost to the generator output. It causes undesirable heat in the generator coils and may eventually cause destruction of insulation. This loss associated with the resistance of the wire in the coils is sometimes called the COPPER LOSSES or the $I^2 R$ power loss of a generator.

2. Not only are currents induced in the windings of the rotating generator armature, but also currents are induced with the iron core on which the armature coils are wound. These currents, called EDDY CURRENTS, also contribute nothing to the generator output and only cause the core to heat. To reduce this loss, the core is made up of thin sections or layers of iron, insulated from each other by varnish or lacquer. These are LAMINATIONS. The laminations form a high resistance path for any eddy currents and therefore reduce this loss to a minimal value.

3. A third loss is termed HYSTERESIS LOSS or loss due to molecular friction. The armature in a magnetic field becomes magnetized and the molecules tend to be lined up in a magnetized state. As the armature rotates, there is considerable friction between molecules as they attempt to follow the direction of the magnetizing force. This friction produces heat and a power loss. Special core irons of silicon steel with special heat treatment are now used to reduce this loss.

TEST YOUR KNOWLEDGE

1. What was Faraday's major discovery?
   a. Magnetic induction.
   b. Radio waves.
   c. Electron flow.
2. Lenz's Law states that the induced field around the moving conductor is opposed by the existing _____ .
3. Demonstrate to your instructor that you can use the left hand generator rule.
4. In a generator, what four factors determine the magnitude of its output?
5. In the study of electricity, the rise and fall and polarity of an induced voltage is represented by means of a _____.
6. Alternating current differs from direct current in that ac is continually changing _____ and _____ .

7. Frequency of an ac sine wave is the number of cycles occurring per _____ .

8. What is the average value of an 18 volt peak-to-peak wave?

9. What is the rms value of an 18 volt peak-to-peak wave?

10. What is the wavelength in metres of a 7 megahertz radio wave?
   a. 23 metres.
   b. 33 metres.
   c. 43 metres.

11. The period of a 60 Hz wave is _____ sec.

12. Using graph paper, construct a sine wave for a 100 volt peak ac wave. Plot the wave at intervals of 30 deg.

13. What is the angular velocity of a 1 KHz wave?

14. What three types of losses are associated with generators?

15. In an automotive alternator, what advantages are realized by rotating the magnetic field and taking the output from the stator?

16. Why is alternating current from an automotive alternator rectified?

Chapter 7

METERS AND INSTRUMENTS

The technical and scientific progress of any nation can be based upon its ability to measure. The success of a technician or engineer is judged by his ability to measure precisely and to interpret the results in circuit performance. Years ago, the scientist was required to think in fractions of inches. Today, measurements are made with sophisticated instruments to the millionth of a part or unit.

The techniques of understanding and using instruments are two of the most important lessons in your study of electronics. Without instruments, the technician becomes blind be-

cause instruments are the windows through which the operation and performance of electronic circuitry may be observed.

In this chapter, we will discuss the following concepts relating to meters and instruments:

1. Basic principles of how a meter movement works.
2. How one meter movement can be used to measure voltage, current and resistance.
3. Use of the oscilloscope as an important instrument in electricity and electronics.

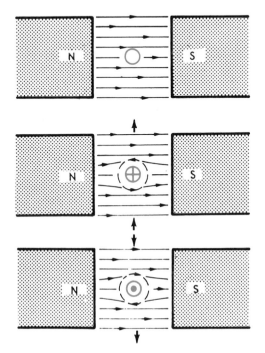

Conductor in magnetic field. No current in conductor.

Current in conductor flows away from you. Field of conductor opposes fixed field above conductor, but adds to field below. Conductor moves toward weakened field.

Current in conductor flows toward you. Field above conductor adds to fixed field. Conductor field below conductor opposes fixed field. Conductor moves toward weakened field.

Fig. 7-1. The action of a current carrying conductor in a magnetic field.

THE BASIC METER MOVEMENT

The voltmeter is one of the common instruments found on every technician's bench. It is wise to investigate its operation and its limitations. Most meters in use today employ the D'ARSONVAL meter movement, which depends upon the interaction of a moving magnetic coil in a fixed magnetic field. To understand this action, observe the diagrams in Fig. 7-1 and review Fig. 5-12 in the chapter on Magnetism.

Current flowing in a conductor produces a magnetic field around the conductor. When this current carrying conductor is placed in a fixed magnetic field, the two fields add together on one side of the conductor and oppose each other on the opposite side. As a result, the conductor will move toward the weakened field. This is called MOTOR ACTION. The principle will be used again in your study of MOTORS.

An improvement of the motor action experiment is shown in Fig. 7-2. Here, the conductors are wound on a rotating core or bobbin. The theory is exactly the same. When current flows in the coil as indicated in the drawing, the coil will attempt to rotate in a clockwise direction.

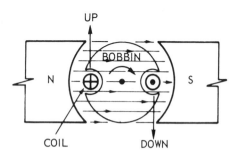

Fig. 7-2. Current flowing in the moving coil in the indicated direction will cause the coil to rotate in clockwise direction.

Now, examine the phantom view of a real meter movement in Fig. 7-3. The rotating coil is prevented from complete rotation by springs, which also provide the electrical connections to the coil. A current flowing through the coil will

Fig. 7-3. The D'Arsonval meter movement. (Weston Instruments, Inc.)

cause it to move clockwise against the tension of the springs. The stronger the current: the greater the rotating pressure against the springs. When current ceases to flow, the springs will return the moving coil to its original position.

Now, note that an indicating needle is attached to the moving coil and it will deflect from left to right as the coil turns. A scale placed under the indicating needle will measure the amount of rotation. Since current through the coil is a function of applied voltage, the scale can be made to read in volts.

> LESSON IN SAFETY: Wait for solder joints to cool. The practice of cooling soldered joints with a wet finger is ill advised.

THE VOLTMETER

We need to know some things about a meter movement before it can be used in a practical circuit.

1. What is the dc ohmic resistance of the moving coil? (All wire has a certain resistance per foot.)
2. What current is required in the moving coil to cause full scale deflection of the indi-

cating meter needle?

3. What voltage applied across the moving coil will cause a full scale deflection current to flow in the coil?

Answers to the first two questions will be supplied by the manufacturer. The third answer can be computed by Ohm's Law.

PROBLEM: Assume that our meter movement requires one milliampere for full scale deflection and its internal resistance is found from the manufacturer's specifications to be 50 ohms. We wish to use this meter as a voltmeter to measure from 0 to 1 volt.

1. Find the full scale deflection voltage.

$$E = I \times R_{int} \qquad E = .001A \times 50\,\Omega = .05 \text{ volt}$$

AT NO TIME CAN OVER .05 VOLT BE APPLIED WITHOUT DAMAGE TO THE METER.

2. In order to measure up to one volt, a resistor must be placed in series with the meter movement to drop a voltage of 1 volt by .05 volt to .95 volt. At a current of one milliampere, compute the value of the resistor by using Ohm's Law and check it against the circuit in Fig. 7-4.

$$R_M = \frac{.95V}{.001A} = 950 \text{ ohms}$$

The circuit will appear as in Fig. 7-4. This resistor is called a MULTIPLIER RESISTOR,

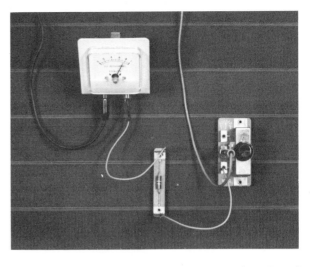

Fig. 7-5. A breadboard setup for determining the value of multiplier resistors. (Lab-Volt)

R_M, Fig. 7-5. It is also important to note that total resistance of meter and multiplier for a range of one volt is equal to 950 ohms + 50 ohms or 1000 ohms.

PROBLEM: Using the same meter as used before, compute multiplier resistor for the 0-10 volt range.

1. The voltage across the meter cannot exceed .05 volts and the meter current cannot exceed .001 amps. Therefore, the multiplier R_M must drop a voltage of 10 volts by .05 to 9.95 volts at a current of one milliampere.

2. Compute R_M by using Ohm's Law and check it against the circuit in Fig. 7-6.

$$R_M = \frac{9.95V}{.001A} = 9950 \text{ ohms}$$

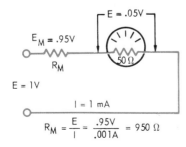

Fig. 7-4. Computation of a multiplier resistor to permit the use of the 0-1 volt range of the voltmeter.

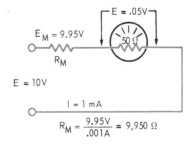

Fig. 7-6. Computation of a multiplier resistor for use of the 0-10 volt range of the voltmeter.

Again, note that total resistance of meter and multiplier in this range is 9950 ohms + 50 ohms or 10,000 ohms.

PROBLEM: Compute multiplier resistor to use the same meter movement to measure 0-50 volts.

1. The resistor must drop 50 volts − .05 volt or 49.95 volts in a current of .001 amp.
2. Compute R_M by using Ohm's Law and check it against the circuit in Fig. 7-7.

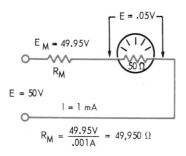

Fig. 7-7. Computation of R_M for 0-50 volt range of voltmeter.

$$R_M = \frac{49.95V}{.001A} = 49,950 \text{ ohms}$$

Again, note that total resistance of meter and multiplier in this range is 49,950 ohms + 50 ohms or 50,000 ohms.

The same procedure can be used to compute the multiplier for any range. NOTE: If the multiplier resistance is greater than 100 times the meter resistance, it generally is not necessary to subtract the meter resistance from the total resistance of the circuit. In the third PROBLEM, the multiplier resistance may be considered as 50,000 ohms.

SENSITIVITY

The sensitivity of a voltmeter can be used to determine how much the meter will load a circuit when it is used to measure voltage. It is expressed in OHMS PER VOLT and is equal to the reciprocal of the current required for full scale deflection. The meter used on the previous

problems was a one milliampere movement. Its sensitivity then is:

$$\text{Sensitivity} = \frac{1}{.001} = 1000 \text{ ohms/volt}$$

Referring back to the problems:

In 0-1 volt range, $R_M + R_{int} = 1000\,\Omega$

In 0-10 volt range, $R_M + R_{int} = 10,000\,\Omega$

In 0-50 volt range, $R_M + R_{int} = 50,000\,\Omega$

Therefore, if the sensitivity of a meter and the meter resistance are known, the multiplier resistor for any range can readily be found.

In Fig. 7-8, the three ranges computed for the voltmeter are drawn schematically. A switch is used to change the meter range. When the range is changed, the appropriate scale on the face of the meter must be used. For example: For 0-1 volt and 0-10 volts, use 0-10 scale. For 0-50 volts, use 0-50 scale.

Respect the quality of your instruments. Good meters and generators are expensive. Your school and your parents have provided your equipment so that you may learn about electronics. Take advantage of this opportunity. With care and intelligent use, the instruments will give years of service.

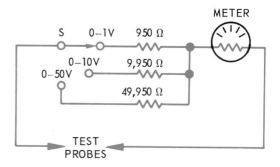

Fig. 7-8. This combined circuit permits the voltmeter to measure in three ranges.

TO MEASURE METER RESISTANCE

If the internal resistance of a meter movement is unknown, the simple circuit in Fig. 7-9

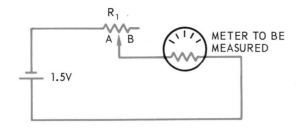

Fig. 7-9. Circuit used to measure meter resistance.

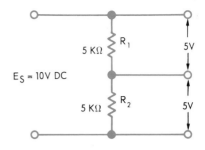

Fig. 7-10. The voltage divides equally across equal resistors.

may be used. Connect a fresh flashlight cell in series with a $5 K\Omega$ potentiometer and connect this setup to the meter. Adjust R for full scale deflection of meter. Remove R from circuit and measure its resistance and record it as $R_1 = \Omega$. Replace in circuit and adjust for exactly one-half scale deflection. Remove R and measure resistance. Record it as $R_2 = \Omega$.

$$R_{meter} = R_2 - 2R_1$$

PROOF: Assume that the meter movement has a resistance of 100 ohms. When R is set for full scale deflection and then measured, it is found to have a resistance of 900 ohms. The total circuit resistance is $R_1 + R_M$ or 1000 ohms. In order for the scale to read one-half deflection, the total resistance must be double or 2000 ohms. Since R_M is 100 ohms of this total resistance, then the value of R_2 must be 1900 ohms. Apply the following formula:

$$R_M = R_2 - 2R_1$$

$$R_M = 1900\,\Omega - 2\,(900\,\Omega) = 100 \text{ ohms}$$

CIRCUIT LOADING

Circuit loading by a meter causes incorrect measurements and should be avoided. This is part of knowing how to use a meter. The loading effect can best be demonstrated by a simple circuit in Fig. 7-10.

By Ohm's Law, it is a simple matter to compute the voltages across R_1 and R_2 as 5 volts respectively. The meter constructed as an example in the previous problems had a sensitivity of 1000 ohms/volt. To measure the

voltage across R_1 and R_2, the 10 volt scale would be used. In the 10 volt range, the resistance of the meter equals 10,000 ohms (10 x 1000 ohms = 10,000 ohms).

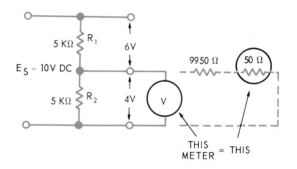

Fig. 7-11. The meter movement of 10 kilohms is in parallel with R_2.

In Fig. 7-11, the meter is connected to measure the voltage across R_2. It is in parallel with R_2. Therefore, the resistance of R_2 and R_M in parallel equals:

$$\frac{R_2\,R_M}{R_2 + R_M} = \frac{5\,K\Omega \times 10\,K\Omega}{5\,K\Omega + 10\,K\Omega}$$

$$= \frac{50\,K\Omega}{15\,K\Omega} \cong 3.3 \text{ kilohms}$$

The voltage across R_2 would become:

$$E_{R_2} = E_s \frac{3.3\,K\Omega}{3.3\,K\Omega + 5\,K\Omega}$$

$$= \frac{3.3\,K\Omega}{8.3\,K\Omega} \times 10 \cong 4 \text{ volts}$$

THE LOADING OF THE CIRCUIT BY THE VOLTMETER CAUSED A ONE VOLT ERROR IN MEASUREMENT.

Repeating this experiment, using a high quality meter with a sensitivity of 20,000 ohms/volt, the meter resistance in the 10 volt range would be 10 x 20 kilohms = 200,000 ohms. Placing a 200 kilohm meter in parallel with a 5 kilohm resistor would make an insignificant change in parallel resistance. The measured voltage would be very close to the expected 5 volts. It pays to use good instruments!

AMMETERS

In order to measure direct current in a circuit, the circuit must be opened and the meter inserted in series with the other components. Therefore, it is good common sense to keep the resistance of a current meter at a low value so that it will not upset the circuit performance.

Working with the same type of meter movement used for the voltmeter, let us review what we know about it.

1. Current required for full scale deflection is one milliampere.
2. Voltage required for full scale deflection is .05 volts.
3. Internal resistance of meter is 50 ohms.

This is a current meter and can be used, as is, to measure currents in the 0-1 mA range. The circuit is shown in Fig. 7-12.

IMPORTANT: The voltage drop across the meter is .05 volts. At no time can one milliampere of current be exceeded or the voltage across the meter be exceeded. The meter will be damaged.

In order to extend the range of the meter to 0-10 mA, it will be necessary to provide an alternate path or SHUNT for 9 mA of current. This is added to the circuit shown in Fig. 7-13.

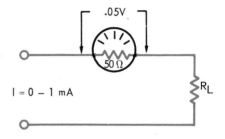

Fig. 7-12. The meter will measure from 0 to 1 mA without extra circuitry.

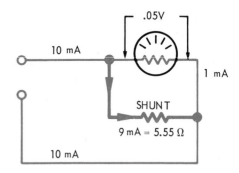

Fig. 7-13. A shunt increases the meter range to 10 mA.

Since we know the voltage across this shunt, it can be easily computed.

$$R_s = \frac{E_M}{I_s} = \frac{.05V}{.009A} \cong 5.55 \text{ ohms}$$

In order to measure currents up to 50 mA, another shunt must be used. See Fig. 7-14. The shunt must carry 49 mA of the current. The same voltage across the meter must be used.

$$R_s = \frac{.05V}{.049A} \cong 1.02 \text{ ohms}$$

In the 0-100 mA range, the shunt must carry 99 mA.

$$R_s = \frac{.05V}{.099A} \cong .505 \text{ ohms}$$

THE OHMMETER

An OHMMETER is a meter designed to measure the resistance of a circuit or compo-

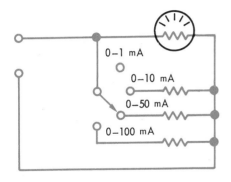

Fig. 7-14. Each current range of the meter will require switching to its shunt.

nent. Once again, the same meter movement is used as in the previous study of the voltmeter and ammeter. Before examining the circuit in Fig. 7-15, consider that:

An open circuit equals INFINITE OHMS.
A closed circuit equals ZERO OHMS.

Consequently, in Fig. 7-15, the meter is carrying zero current when test terminals A and B are open. The indicating needle would be far to the left at a point marked on the scale as ∞ ohms, which indicates infinity (an unlimited quantity).

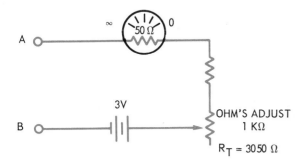

Fig. 7-15. The basic circuit of a series ohmmeter.

When test points A and B are shorted together, the indicating needle should deflect far to the right. This point is marked ZERO OHMS on the scale. Under what circuit conditions could this be possible? Only when one milliampere of current is flowing in the circuit. The current is limited to 1 mA by series resistor R and the Ohm's Adjust potentiometer.

The total resistance of the circuit for 1 mA of current must be:

$$R_{circuit} = \frac{3V}{.001A} = 3000 \text{ ohms}$$

The total resistance is R_M of the meter plus R plus Ohm's Adjust. The meter has an internal resistance of 50 ohms. Therefore the remaining resistance is 2950 ohms. Part of this resistance is made variable, so that any decrease in the battery voltage due to aging may be compensated.

IMPORTANT: Before using an ohmmeter on any range, the test prods should be shorted and the meter set at zero by the Ohm's Adjust knob.

If a resistor of unknown value is connected between test probes A and B, the needle will deflect to some position between zero and infinity. The dial is calibrated and marked with a group of precision resistors of known value, so that unknown values may be read directly.

VOLT-OHM-AMMETER

A VOLT-OHM-AMMETER (or volt-ohm-milliameter) is three meters combined in one case. It is called a VOM, Fig. 7-16. This is a basic

Fig. 7-16. A commercial volt-ohm-ammeter (VOM). (Triplett Corp.)

piece of test equipment used by many technicians. It is a very versatile instrument which deserves your respect and careful handling.

IMPORTANT: When using a meter, be certain to use the correct test leads and connect them to the circuit or equipment with the right polarity. Usually the color BLACK is negative or ground. The color RED is positive.

MEASUREMENT OF AC

In the voltmeter and ammeter circuits previously discussed, we have only considered direct current. It also is possible to use these meters for the measurement of alternating current and voltage. In order to do this, the alternating current must be rectified or changed to a pulsating direct current.

The theory of rectification will be studied in some detail in Chapter 12. For purposes of this instruction, we only need to know that a diode is an unidirectional (one direction) conductor. The circuit is arranged for current to flow in one direction but is prevented from flowing in the opposite direction. Therefore, an ac wave is converted to a pulsating dc wave.

Meter circuits may employ a full-wave or half-wave rectifier circuit. The output of each is illustrated in Fig. 7-17.

After the rectification, direct current or voltage is measured in the same manner as with

dc meters. The scales on the meter face, in most cases, will be calibrated to read in the rms or effective value of the ac voltage. However, some meters will also have scales which read in PEAK values and also PEAK-TO-PEAK values. Study the meter you are about to use and determine the correct scales for your purpose.

The circuit of the half-wave rectifier is schematically drawn in Fig. 7-18. When the ac input is negative, diode 1 conducts and diode 2 is cut off. On the positive ac input, D_2 conducts and D_1 is cut off. Only half-wave pulses of dc current flow through the meter.

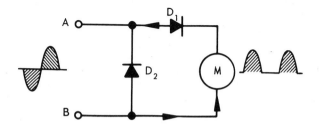

Fig. 7-18. Circuit of the half-wave rectifier.

A full-wave bridge rectifier for a meter is shown in the circuit in Fig. 7-19. This circuit is found in the more expensive meters. It has a greater dc output for any ac input.

When the ac input is positive, D_1 and D_2 will conduct. When the ac input is negative, D_3 and D_4 will conduct. As a result, there is a current pulse as indicated for each half cycle of ac input.

After rectification, the shunts and multipliers are used to adapt a particular meter movement to a selected range.

You should be aware of the fact that meters of this type which read in EFFECTIVE VALUES on their scales assume a sine wave input. Waves other than the ac sine wave will give incorrect readings.

LESSON IN SAFETY: Do not work with electricity when alone. It is good practice to have someone around to shut off the power in case trouble is experienced.

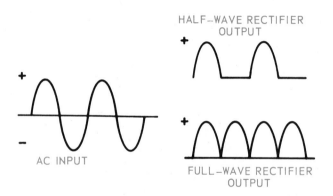

Fig. 7-17. A comparison of full-wave and half-wave rectifier outputs. Both are pulsating dc.

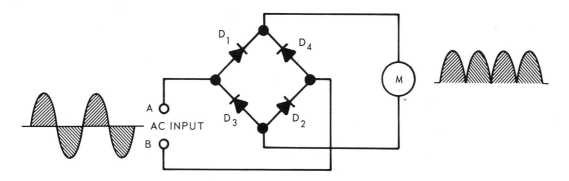

Fig. 7-19. A full-wave bridge rectifier for an ac metering circuit.

DIGITAL MULTIMETERS

Digital multimeters, Fig. 7-20, are being used more and more today in electronics. Advantages include: greater accuracy; easy readability; automatic polarity and zeroing; measurement of both ac and dc amperes.

Most digital multimeters use light emitting diode (LED) displays. However, some may use liquid crystal displays. These are similar to the kinds of displays used in digital watches.

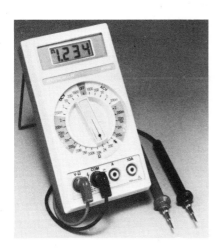

Fig. 7-20. A portable digital multimeter. (Beckman Instruments, Inc.)

THE OSCILLOSCOPE

The basic instrument of measurement of ac signal voltages and waves is the OSCILLO-SCOPE. It is at the right hand of the research scientist and electronic technician and engineer.

The oscilloscope will be used in many of your experiments and service procedures. Your familiarity with this instrument will increase to a great extent your understanding and utilization of electronic circuitry and equipment.

Oscilloscopes may be classified according to two basic types:

1. Recurrent sweep (free running).
2. Triggered sweep.

In the "recurrent sweep" oscilloscope, the sweep signal voltage is generated by a sawtooth oscillator. This oscillator usually is synchronized to the input signal by feeding it a sample of the input signal from the vertical circuits.

The oscilloscope has a visual readout that appears as a line or wave on the face of the CRT (cathode ray tube). We will investigate how this effect is produced. In many respects, the theory of the oscilloscope is similar to your TV set. A sketch of a cathode ray tube is shown in Fig. 7-21.

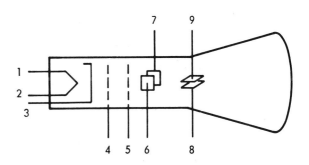

Fig. 7-21. The construction of the cathode ray tube.

In Fig. 7-21, a heater voltage is applied to pins 1 and 2. This is used to heat the cathode attached to pin 3. When heated, the cathode, which emits a cloud of electrons that are directed toward the face of the tube. The inside of the tube is coated with a phosphor that glows when bombarded with the stream of electrons. Grid number 1 (pin 4) controls the flow of electrons in the stream which, in turn, controls the intensity or brightness of the glowing spot on the tube face.

Grid number 2 (pin 5) is a voltage control on the electron beam that causes it to focus or converge and produce a sharp pattern on the tube face. The two deflection plates attached to pins 6 and 7 cause the beam to be deflected either to the right or left. Voltages applied to these plates are adjusted by the HORIZONTAL POSITION CONTROL, and the beam is made to sweep from left to right by voltage waves from the horizontal sweep oscillator. The wave from the horizontal oscillator usually is a sawtooth wave as shown in Fig. 7-22.

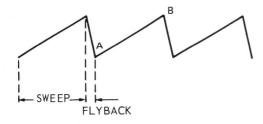

Fig. 7-22. Wave from horizontal sweep oscillator.

The vertical deflection plates attached to pins 8 and 9 will cause the trace to move up or down according to voltages applied to the plates. The VERTICAL POSITION CONTROL permits the operator to move the beam up or down on the screen or tube face. By using both the vertical and horizontal positioning controls, the spot can be located any place on the face of the CRT.

Newer laboratory scopes employ "triggered sweep" instead of recurrent sweep design. See Fig. 7-23. The sweep generator is inactive until a trigger signal, derived from the input signal, starts it operating. When the input signal reaches the selected polarity and amplitude, the trigger circuit delivers a pulse to the sweep generator, which then produces one cycle of sweep. The sweep generator then "rests" until the input is again at the selected polarity and amplitude.

If the input signal is a continuous sine wave, a continuous sawtooth is generated, as in a recurrent sweep scope. However, it is in sync with the input signal so that the display "stands still." If there is no input signal, no sweep occurs.

Triggered sweep scopes can be set so that they do not produce a trace in the absence of a vertical signal. Also, if the input consists of random pulses, the sweeps occur only when there are pulses.

A TIME/CM control on the scope can be used to set the width of the displayed image. The calibration markings on this control permit

Fig. 7-23. A typical oscilloscope used in electronics. (Hickok Teaching Systems)

the elapsed time between any two points on the display to be determined. This is done by multiplying the horizontal spacing in divisions by the numberical value of the TIME/DIV control setting.

The VOLTS/CM control on the scope is used to set the height of the display. The vertical distance tells the magitude of the voltage of the displayed waveform. This is done in much the same manner as the horizontal distance tells the time between any two points.

The oscilloscope shown in Fig. 7-23 features many other controls. These controls and their functions follow.

INTENSITY CONTROL: Varies the brightness of the electron beam as it strikes the screen.

FOCUS CONTROL: Sets the voltage on the focus ring in the CRT. By changing this voltage, the spot on the screen can be adjusted for minimum size.

ASTIGMATISM CONTROL: Sets the voltage on the second anode of the CRT. This control should be adjusted for the roundest spot on the CRT screen.

HORIZONTAL GAIN SWITCH: Multiplies the horizontal gain by five (x5) or leaves it normal (x1).

HORIZONTAL LEVEL (external): Controls the gain of the horizontal input from an external source.

TRIGGER LEVEL: Controls the gain of the trigger circuits. Consult your specific oscilloscope manual to determine the function of all of the other trigger controls.

INPUTS: Usually are fed into the vertical input. However, oscilloscopes may have external horizontal inputs, Z axis inputs or external trigger inputs.

PROBE CALIBRATE: Provides a calibration control voltage to the probe.

The great advantage of the oscilloscope is its ability to STOP a rapidly oscillating wave of a given frequency. This allows the operator not only to measure the amplitude of the wave, but also to examine the waveform or shape.

There are many other special features of oscilloscopes of different manufacturers. It is always wise to consult the INSTRUMENT MANUAL for correct adjustment and application of the oscilloscope you are using.

One significant advantage of measuring by the oscilloscope is its high INPUT IMPEDANCE, usually in the range of 10 megohms or more. This means that the instrument will not seriously affect the measurements and performance of the circuit. It draws insignificant power from the circuit under test.

VOLTAGE AND CONTINUITY TESTER PROJECT

You can build a handy voltage and continuity tester by assembling and wiring it around a light emitting diode (LED). This tester will show polarity (+ or −, 6V or 12V). It will check for dc at any terminal or connector, and it will test the continuity of a bulb, fuse, wire or cable.

Examine the tester circuit in Figs. 7-24 and 7-25. When the toggle switch is in the "voltage" position, the LED is in a simple series circuit along with the diode (D_1) and limiting resistor (R_1). You can then check for voltage polarity and presence of voltage. The LED will light only if the positive (red) test lead is connected to the positive voltage source, and the negative

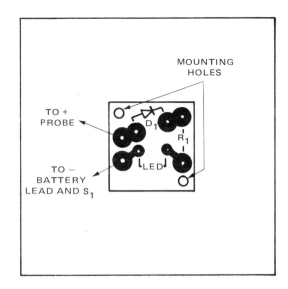

Fig. 7-24. Printed circuit layout for LED tester.

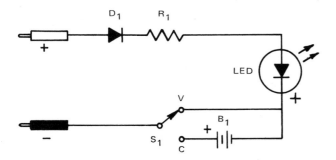

PARTS LIST FOR A VOLTAGE
AND CONTINUITY TESTER

R_1 — 470 ohm, 1/4 watt resistor
LED — Sprague ED150 (or ED 155)
     (cont. fwd. current = 40 mA)
     (cont. fwd. voltage = 1.65V)
     or Motorola MLED50
     or Monsanto type MV50
D_1 — Diode, Sprague RT218 or RCA SK 3100
B_1 — 9V transistor battery
S_1 — SPDT switch
     Miniature box, grommets, battery connector

Fig. 7-25. Schematic of LED voltage and continuity tester.
(Sprague)

(black) lead is connected to the negative terminal. If the leads are reversed, the LED will not light. The circuit will react to any voltage from 3-15 volts. Do not use the tester to check out circuits carrying more than 15 volts or you will ruin the LED.

When the toggle switch is in the "continuity" position, a 9 volt battery is added to the circuit. The LED will light when the test leads are connected to a conductor, provided the bulb, fuse, wire or cable is electrically sound.

The printed circuit layout, Fig. 7-24, and the simple schematic, Fig. 7-25, illustrate how easy it is to assemble and wire this electrical testing tool. A picture of the completed project is shown in Fig. 7-26.

IMPORTANT: When soldering LED devices, use standard transistor and IC techniques. That is, place a heat sink on the lead being soldered, between the device and the point at which the lead is being soldered.

Fig. 7-26. LED voltage and continuity tester.

TEST YOUR KNOWLEDGE

1. Most meters in use today employ the _____ meter movement which depends upon the interaction of a moving magnetic coil in a fixed magnetic field.
2. The _____ of a voltmeter can be used to determine how much the meter will load the circuit when it is used to measure voltage.
3. In order to measure dc in a circuit, the circuit must be opened and an ammeter inserted in series with other components. True or False?
4. The ohmmeter is designed to measure the _____ of a circuit or component.
5. The _____ is the basic instrument of measurement of ac voltage signals and waves.
6. Refer back to Fig. 7-9 in this chapter. The measured resistance of R_1 when the meter is at full scale deflection is 500 ohms. At half scale, the resistance R_1 is measured at 1200 ohms. What is the internal meter resistance?
7. A given meter requires 1 mA for full scale deflection and has an internal resistance of 100 ohms. What is its sensitivity?
8. What value multiplier is required for the 50 volt range?
9. What value multiplier is required to measure up to 300 volts?

10. Compute the shunt to measure in the 100 mA range.

11. A voltage divider in the accompanying illustration has the resistance values indicated. Using Ohm's Law, compute the voltage expected across resistor R_2.

12. Again referring to the accompanying illustration, compute the voltage expected across R_2 if measured with a 5000 ohms/volt meter.

13. Compute voltage expected across R_2 if measured with a 20,000 ohms/volt meter.

14. What is the purpose of the focus control on an oscilloscope?

15. What is the purpose of the V positioning control on an oscilloscope?

16. What is the purpose of the focus and astigmatism control on an oscilloscope?

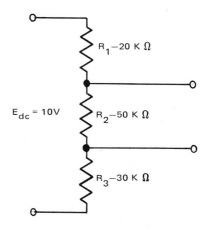

Chapter 8

INDUCTANCE

Inductance is a most interesting and important topic in electricity and electronics. In this chapter, you will:

1. Study the principles of electromagnetic induction.
2. Become familiar with various kinds of inductors and their uses in electronic circuits.
3. Analyze the performance of inductors in dc and ac circuits.
4. Study the theory and applications of transformers.

In your study of electricity, you probably will find the property of inductance in a circuit more fascinating than many other electrical components. You will want to know a great deal about inductors, and that will be to your advantage. A thorough understanding of inductance will lead to competence as a technician or an engineer.

DEFINITION OF INDUCTANCE

Inductance is that property of a circuit which opposes a CHANGE in current. It may also be described as that property of a circuit in which energy is stored in a magnetic field.

For review and further emphasis, a current through a coil of wire produces a magnetic field around the coil. Its strength and density will depend upon the number of turns of wire in the coil and the value of current flowing through the coil. Its polarity is determined by the LEFT HAND RULE.

Usually, an INDUCTOR is a coil of wire, with or without a core. However, a single loop of wire will exhibit properties of inductance under certain conditions.

Chapter 6, you will recall, covered Michael Faraday's experiments in magnetic induction. Remember that as a conductor moves through a magnetic field, a voltage and current will be induced in the conductor. The induced current will produce a magnetic field around the conductor in a direction which will oppose the movement of the conductor. An external force is required to move the conductor.

SELF-INDUCTANCE

In Fig. 8-1, an inductor or coil is connected to a variable power source. As the circuit is "turned on," the magnetic field, starting at zero, expands to its full strength and density. As the field moves outward, it cuts across the windings of the coil.

NOTE: As Faraday's experiments proved, relative motion between a magnetic field and a conductor will induce a voltage in the conductor. Either the field may be fixed and the conductor moved, or the conductor may be fixed and the field made to move. Either would result in relative motion.

In our circuit coil, the expanding or moving field will induce a voltage in the coil which will OPPOSE THE SOURCE VOLTAGE and OPPOSE THE RISE IN CURRENT through the

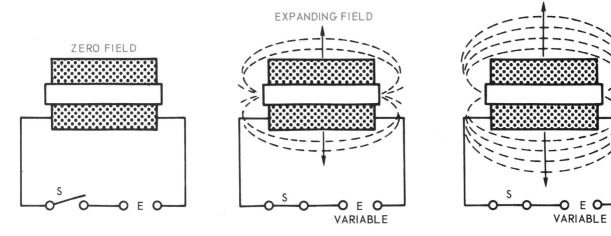

Fig. 8-1. As current is increased, the expanding magnetic field cuts across the coil winding.

coil. This induced voltage is called "counterelectromotive force" or CEMF.

After the current has risen to its full value, the magnetic field remains constant. There is no longer a change in current and the CEMF is zero.

By changing the source voltage to a lower value, the circuit current will tend to decrease. However, the decreasing magnetic field will again induce a voltage which will tend to hold the current at its original value. That is, the induced CEMF will oppose the change in circuit current.

The opposite is also true. An increase in source voltage will tend to increase the circuit current and further expand the magnetic field. This induces a CEMF of opposite polarity which tends to oppose the rise in circuit current. Therefore, an inductor in a circuit opposes any change in current in the circuit.

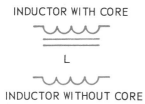

Fig. 8-2. Symbols used on schematic drawings for inductors and coils.

The electronic symbols for an inductor and an inductor with a core are illustrated in Fig. 8-2. The letter L is the symbol for inductance.

LENZ'S LAW

The opposition to a change in current by inductance can be summarized by Lenz's Law:

"The polarity of an induced EMF is such that it tends to set up a current, the magnetic field of which always opposes the change in the existing field caused by the original current."

THE MAGNITUDE OF CEMF

How great will be the opposition to a change in current of a specific inductor? In the chapter on magnetism, we found that the magnitude and density of a magnetic flux will depend upon the number of turns of wire in the coil. Consequently, the induced EMF will also depend upon the density and magnitude of the flux. However, the relative motion also must be considered. A rapid movement or a rapid change in flux will induce a larger CEMF than a slower change. Then RATE OF CHANGE (roc) must be taken into account.

The average value of CEMF of an inductor may be found by the formula:

$$CEMF_{average} = \frac{N\Phi}{10^8 t}$$

where:

N = number of turns of the coil
Φ = number of magnetic force lines
t = time in seconds to cut all lines
10^8 = a constant

The constant is derived from the fact that if a single conductor cuts across 10^8 or 100,000,000 lines of magnetic flux in one second, a voltage of one volt will be induced.

It is important to note that in the equation CEMF is inversely proportional to time. That is, the faster rate of change (less time) will produce a greater CEMF.

THE HENRY

The unit of measurement of inductance is the HENRY. In electronic circuitry, inductors with values in henrys, millihenrys (1/1000 henry) and microhenrys (1/1,000,000 of a henry) will be used. If a coil will produce ONE VOLT OF CEMF WHEN THERE IS A CHANGE OF CURRENT OF ONE AMPERE PER SECOND, it has a value of ONE HENRY of inductance.

Mathematically:

$$L \text{ (in henrys)} = \frac{e}{\Delta i / \Delta t}$$

where:

e = volts of CEMF
Δi = change of current in amperes
Δt = change of time in seconds

PROBLEM: What is the inductance of a coil that produces 100 volts of CEMF when the current changes one ampere at a frequency of 100 Hz?

$$L = \frac{100 \text{ volts}}{1/.01 \text{ sec}} = 1 \text{ H} \quad \text{roc} = 100$$

The inductance of a coil is determined by:

1. Number of turns of wire.
2. Permeability of the core material (if a core is used).
3. Cross-sectional area of the core enclosed by one turn of wire.
4. Length of the core.

Fig. 8-3 shows examples of several inductors, which are named according to their purpose.

MUTUAL INDUCTANCE

If two coils are placed close together so that the expanding and collapsing magnetic field of one coil will cut across the windings of the second coil, a voltage will be induced in the second coil. See Fig. 8-4. These two coils are described as having MUTUAL INDUCTION. The connection between the coils is by a magnetic field only.

The assembly of the two coils, shown in Fig. 8-4, is called a TRANSFORMER. The input coil is named the PRIMARY winding and the output coil is the SECONDARY winding. By definition, a transformer is a component used to transfer electrical energy from one circuit to another by means of a varying magnetic field.

Mutual induction, like self-induction, is measured in units of HENRYS. If a primary coil induces one volt in the secondary coil when the current in the primary coil is changing at the rate of one ampere per second, the mutual inductance is ONE HENRY. The letter M designates mutual inductance.

To calculate mutual inductance, use this equation:

$$M \text{ (in henrys)} = \frac{e}{\Delta i / \Delta t} \frac{\text{(secondary voltage)}}{\text{(roc)}}$$

PROBLEM: The current in a primary coil changes 2 amperes per second and induces 20 volts in a secondary coil. What is the mutual inductance?

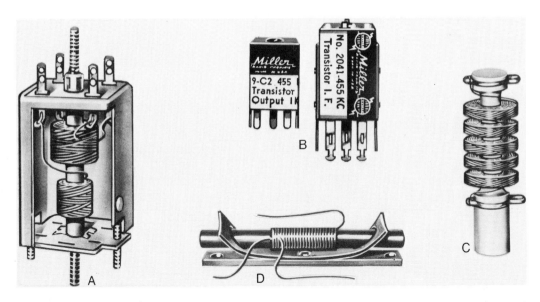

Fig. 8-3. Examples of inductors. A—Horizontal oscillator coil used in television. B—Intermediate frequency transformers. C—Radio frequency choke. D—Antenna coil for transistor radio.

$$M = \frac{20 \text{ volts}}{2/1 \text{ sec}} = \frac{20}{2} = 10 \text{ henrys}$$

COUPLING

It is reasonable to assume that if the primary and secondary coils are very close together, many lines of magnetic force will cut across the secondary coil and a greater secondary voltage will be induced. Mutual induction will be greater. Conversely, if the coils are separated, mutual inductance will decrease.

The effect of the position of the coils to each other is called COUPLING. If all the lines of force of the primary cut across the secondary, the coupling is called UNITY COUPLING, with a value of 1. If the coils are separated so that only half the flux lines cut across the secondary, the coupling is .5. The decimal number, which indicates the number of flux lines available for mutual induction, is called the COEFFICIENT OF COUPLING. It is assigned the letter k.

You should be aware of the angular position of the coils and how it affects mutual induction. To produce a maximum EMF in the secondary, the flux lines must cut across the secondary windings at right angles. This can

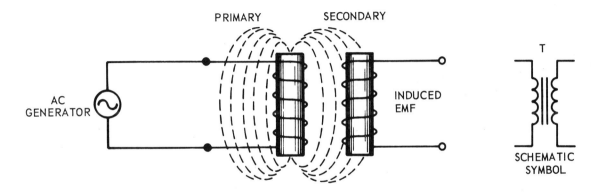

Fig. 8-4. In a transformer, energy is transferred by a moving magnetic field.

only happen when the axes of the coils are parallel to each other. If one coil is turned so that there is an angular difference between the axes, the mutual induction will be reduced. See Fig. 8-5.

With consideration of the coefficient of coupling and the inductance of each coil in mutual relationship, the mutual inductance may be found by the equation:

$$M = k \sqrt{L_1 L_2} = \text{henrys of mutual induction}$$

PROBLEM: Two 4H coils have a coupling coefficient of .5. What is their mutual inductance?

$$M = .5 \sqrt{4 \times 4} = .5 \times 4 = 2 \text{ henrys}$$

INDUCTORS IN SERIES

When two or more coils of any inductance are connected in series, as in Fig. 8-6, and there is no coupling between the coils, the total inductance of the circuit will be the SUM of the individual inductances:

$$L_T = L_1 + L_2 + L_3 \ldots$$

If coupling should exist between the coils, then mutual inductance must be considered. In Fig. 8-7, both of the coils produce magnetic fields with the SAME polarity and combine to form a total inductance of:

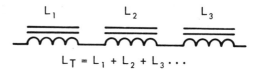

$$L_T = L_1 + L_2 + L_3 \ldots$$

Fig. 8-6. Inductors in series, zero coupling.

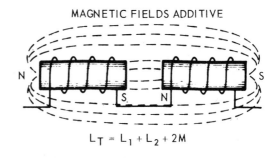

MAGNETIC FIELDS ADDITIVE

$$L_T = L_1 + L_2 + 2M$$

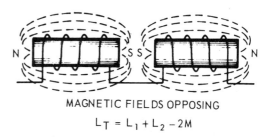

MAGNETIC FIELDS OPPOSING

$$L_T = L_1 + L_2 - 2M$$

Fig. 8-7. Positions of coils to produce additive or subtractive mutual inductance.

$$L_T = L_1 + L_2 + 2M$$

In the second part of Fig. 8-7, the two coils are producing opposing magnetic fields. Therefore, mutual inductance is subtracted from the total series inductance:

$$L_T = L_1 + L_2 - 2M$$

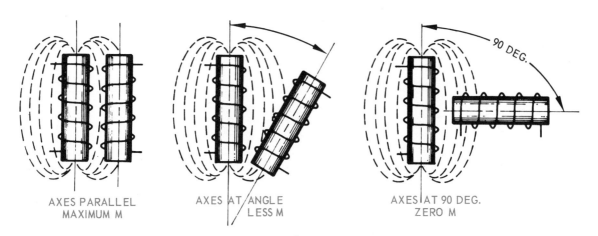

AXES PARALLEL
MAXIMUM M

AXES AT ANGLE
LESS M

AXES AT 90 DEG.
ZERO M

Fig. 8-5. Mutual inductance (M) depends on angular relationship between coils.

PROBLEM: Two coils of 4 henrys and 9 henrys are connected in series-opposing have a coupling coefficient of .5. What is the total circuit inductance?

1. Find mutual inductance:

$$M = k \sqrt{L_1 L_2} \qquad M = .5 \sqrt{4 \times 9}$$

$$.5 \times 6 = 3 \text{ H}$$

2. Find L_T:

$$L_T = L_1 + L_2 - 2M$$

$$= 4 + 9 - (2 \times 3)$$

$$= 13 - 6 = 7 \text{ henrys}$$

INDUCTORS IN PARALLEL

We will consider inductors in parallel only when no coupling exists between coils. When two or more inductors are connected in parallel, Fig. 8-8:

$$L_T = \cfrac{1}{\cfrac{1}{L_1} + \cfrac{1}{L_2} + \cfrac{1}{L_3} \cdots}$$

If only two inductors are in parallel, this simplified equation may be used:

$$L_T = \frac{L_1 L_2}{L_1 + L_2}$$

Note that the formula for series and parallel inductors are the same as series and parallel resistors.

SERIES RL CIRCUITS

The voltage and current relationships in a RL series circuit are of particular interest in our studies.

In Fig. 8-9, when switch S_1 is closed to position A, these circuit conditions exist:

1. Since current at that initial instant is zero, the voltage drop across R is zero.
2. Starting from zero, the greatest rate of change of current will occur and a CEMF

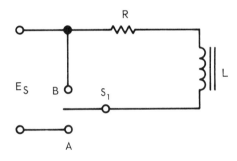

Fig. 8-9. Circuit used to demonstrate voltage and current relationships in a RL circuit.

will develop across L almost equal to E_s and opposing E_s.
3. Current then rises in the circuit. The voltage across R increases. The CEMF across L decreases.
4. When current reaches maximum, there is no

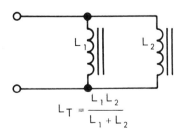

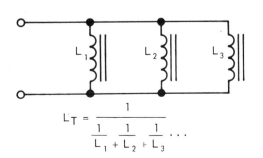

Fig. 8-8. Inductors in parallel and total inductance.

longer a CHANGE IN CURRENT so the CEMF across L is zero. The voltage drop across R is equal to $E_R = I \times R$ and is equal to the source voltage.

NOTE: Coil L is assumed to have no dc resistance.

5. Under equilibrium conditions with no change in current value, the circuit appears as only a resistor. The effect of the inductor has disappeared.

When switch S_1 is turned to position B in Fig. 8-9, the circuit will discharge.

1. Maximum voltage of opposite polarity will appear across L due to rapid change from maximum current toward zero. This voltage will try to keep the current flowing in the circuit.
2. Current will decay toward zero, and the voltage across R will decay toward zero.
3. Voltage across L will decay toward zero as the rate of change becomes less.

The RISE and DECAY of circuit current and voltages are illustrated graphically in Fig. 8-10.

RL TIME CONSTANT

The rise and fall of the current in the series circuit does take an interval of time, depending upon the relationship of the inductance and the resistance of the circuit. The time required for the current to rise to 63.2 percent of its maximum value, or decrease to 36.8 percent from maximum value, is the TIME CONSTANT of the circuit.

The time constant may be found by the equation:

$$t \text{ (in seconds)} = \frac{L \text{ (in henrys)}}{R \text{ (in ohms)}}$$

It is generally accepted that after FIVE TIME CONSTANT PERIODS, the circuit will

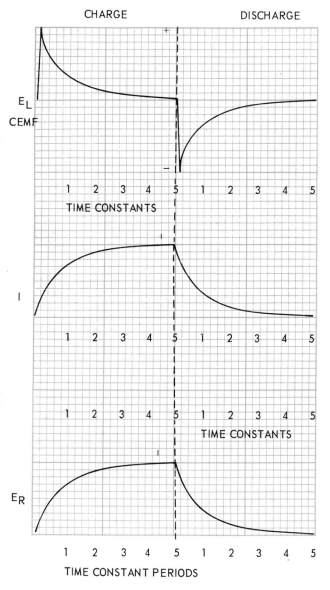

Fig. 8-10. Current and voltage rise and decay in a RL circuit.

be at equilibrium state with no further change. In the table in Fig. 8-11, the percentages of current or voltage at the end of each time

TIME CONSTANT PERIOD	PERCENT OF MAXIMUM RISE	PERCENT OF MAXIMUM DECAY
1	63.2	36.8
2	86.5	13.5
3	95.0	5
4	98.0	2
5	99.0	1

Fig. 8-11. Table shows percentages of current or voltage at end of each time constant period. Tabulated for both rise and decay.

constant period are tabulated for both rise and decay.

In Fig. 8-12, these percentage values are plotted on a graph for each interval of time. This chart is the UNIVERSAL TIME CONSTANT CHART.

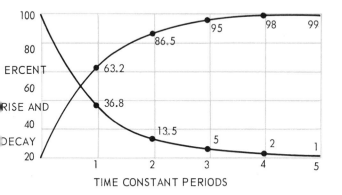

Fig. 8-12. The Universal Time Constant Chart for RL and RC circuits.

INDUCTIVE KICK

In Figs. 8-13 and 8-14, the NE2 is a neon glow lamp which presents a high resistance until a firing voltage of about 65 volts is applied to its terminals. Then it will glow.

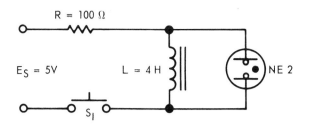

Fig. 8-13. Circuit for "inductive kick" demonstration.

The source voltage of 5 volts cannot light the lamp. When switch S is closed, the current and voltages rise and decay according to the circuit time constant.

$$t = \frac{L}{R} = \frac{4}{100} = .04 \text{ sec}$$

At the end of five periods (5 x .04 = .2 seconds), the circuit is fully charged.

$$E_R = 5 \text{ volts}, E_L = 0$$

and

$$I = \frac{E}{R} = \frac{5}{100} = .05$$

Recall the formula for inductance of a coil:

$$L = \frac{e}{\Delta i / \Delta t} \text{ or } e = L \times \frac{\Delta i}{\Delta t}$$

Induced voltage depends directly upon the rate of change of current per second.

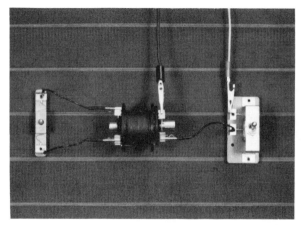

Fig. 8-14. A breadboard experiment to demonstrate "inductive kick." (Lab-Volt)

When switch S_1 is opened in Fig. 8-13, the power source is disconnected and the self-induced voltage attempts to maintain the circuit current of .05 amperes. But the current must drop instantaneously to zero because the NE2 has a very high resistance. The rate of change will be exceedingly high.

Assuming L = 4 henrys and the resistance of the NE2 is 100K ohms, then:

$$t = \frac{4}{1 \times 10^5} = 4 \times 10^{-5} \text{ secs}$$

During 4×10^{-5} seconds, the current will drop 63.2 percent of its maximum value:

.632 x .05 amp = .0316 amp

The rate of change will be:

$$\frac{\Delta i}{\Delta t} = \frac{.0316 \text{ amps}}{4 \times 10^{-5} \text{ secs}} = .0079 \times 10^5$$

The induced EMF is calculated as:

$$e = 4 \times 790 = 3160 \text{ volts}$$

This voltage is sufficient to make the NE2 flash. This high induced voltage, when an inductive circuit is switched open, causes:

1. Arcing and burning of switch contacts.
2. Dissipation of heat which will break down coil insulation.
3. Possible danger to personnel.

Care must be taken when interrupting inductive circuits because of the INDUCTIVE KICK. The energy stored in the magnetic field of the coil must be dissipated. Application of inductive kick is found in the automotive ignition system where the coil must supply 15,000 to 20,000 volts to fire the spark plugs.

REACTANCE

Fig. 8-15 demonstrates an alternating current characteristic known as INDUCTIVE REACTANCE. In this circuit, a small 6 volt lamp is connected in series with a resistor and an inductor.

When a dc voltage is applied, the lamp burns brightly. When the same circuit is connected to an ac voltage of the same magnitude, the lamp glows dimly. Something must be limiting the current in the ac circuit. And this is true. The continually varying alternating current has created a CEMF in the coil which constantly opposes the source voltage and holds the current at a lower value.

This opposition to the flow of an alternating current caused by an inductor or coil is called "inductive reactance." Its symbol is X_L and, like resistance, it is measured in OHMS.

In alternating current theory (review Chapter 6); we discovered that the rate of change is determined by the angular velocity of a vector representing the magnitude of the applied voltage. Angular velocity is measured in RADIANS PER SECOND. One complete revolution of a vector (one cycle of ac) represents 2π radians. If frequency (f) is stated in hertz (cycles per second), then the angular velocity is equal to $2\pi f$. This is the rate of change for an ac current. It is assigned the Greek letter ω (omega).

The CEMF of an inductor is:

$$e = L \times \frac{\Delta i}{\Delta t} \text{ or } L \times (\text{rate of change})$$

The reactance of an inductor is:

$$X_L = 2\pi fL \text{ or } \omega L$$

where:

2π = radians for 360 deg. cycle of ac wave.
f = Hertz (Hz) (cycles per second)
L = inductance of coil in henrys.
ω = angular velocity or radians per second.

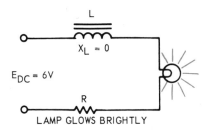

LAMP GLOWS BRIGHTLY

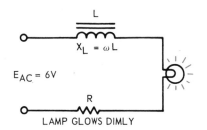

LAMP GLOWS DIMLY

Fig. 8-15. A demonstration of INDUCTIVE REACTANCE in a series RL circuit.

PHASE RELATIONSHIPS IN INDUCTIVE CIRCUITS

Examination of the curves in Fig. 8-16 will reveal some interesting information about a circuit containing only inductance. This is theoretically impossible since the wires contain some resistance. Notice the CEMF curve is 180 deg. out of phase with the applied EMF. Maximum CEMF can only be induced when the CURRENT RATE OF CHANGE IS MAXIMUM. This point must be where current crosses the ZERO line. At the peak of the current wave, there is an instant where the roc is minimum or zero.

CONCLUSION: In a pure inductive circuit, the current lags behind the applied voltage by an angle of 90 deg. It will always be slightly less due to resistance.

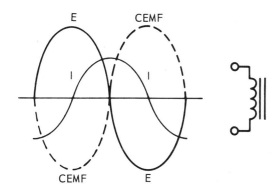

Fig. 8-16. Current, voltage and CEMF in an inductive circuit.

POWER IN INDUCTIVE CIRCUITS

The voltage and current wave forms are again shown in Fig. 8-17. A third curve which represents consumed power is plotted over these curves.

Power above the zero line is power consumed by the circuit. Power below the line is power returned by the circuit. The power curve represents the products of many instantaneous voltages and currents. Positive values of voltage and current equal positive power. When either voltage or current is negative, the power prod-

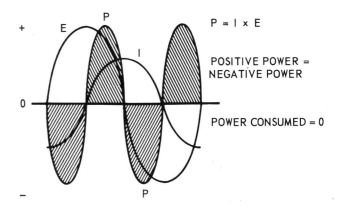

Fig. 8-17. A pure inductive circuit uses zero power.

uct is negative. Negative values of both voltage and current equal positive power. The positive power will equal the negative power. Power consumed by inductance is ZERO. Energy required to build the magnetic field is returned to the circuit. Negative power is sometimes called WATTLESS POWER.

PHASE RELATIONSHIPS IN THE SERIES RL CIRCUITS

In a pure RESISTIVE circuit, the current and voltages are always IN PHASE. In a pure inductive circuit, the current lags the voltage by an angle of 90 deg. You can assume then, in a circuit containing both resistance and inductance, that the current will lag the voltage by some angle between zero and 90 deg. The angle of phase displacement is called "theta" θ .

Restated for emphasis: In the RL series circuit, there are TWO resistive forces opposing the current flow. These forces are 90 deg. apart. One is the dc resistance in ohms; the other is inductive reactance X_L in ohms. To find the total opposing force, we must find the resultant force by combining these out-of-phase forces.

Using the circuit with assigned values in Fig. 8-18, we will work out the entire problem.

1. Compute the value of X_L .

$$X_L = 6.28 \times 100 \times 4 = 2500 \text{ ohms}$$

2. Draw vectors R = 1500 Ω on the horizontal line, Fig. 8-18. Use any convenient scale. R is drawn horizontally because the circuit current and voltage across R are in phase. It therefore becomes the reference line from which other angles are measured. Draw X_L

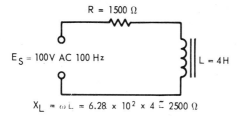

$$X_L = \omega L = 6.28 \times 10^2 \times 4 \doteq 2500 \,\Omega$$

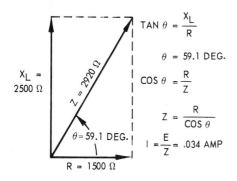

Fig. 8-18. Circuit and vector diagram for RL problem described in text.

upward at 90 deg. from R since voltage across X_L leads circuit current through R. Use the same scale.

3. The circuit impedance Z, which is also measured in ohms, is equal to the vector sum of X_L and R. This can be found by using the Pythagorean theorem:

$$Z^2 = R^2 + X_L{}^2 \text{ or } Z = \sqrt{R^2 + X_L{}^2}$$

It is more convenient to use trigonometry:

$$\tan \theta = \frac{X_L}{R} = \frac{2500}{1500} = \frac{5}{3} = 1.67$$

$$\theta = \text{arc tan } 1.67 = 59.1 \text{ deg.}$$
(Trig. Tables)

4. Find the cosine of θ = 59.1 deg.

$$\cos \angle \, 59.1 \text{ deg.} = .5135$$

Find Z, using the cosine function:

$$\cos \theta = \frac{R}{Z} \quad Z = \frac{R}{\cos \theta} = \frac{1500}{.5135} = 2920 \text{ ohms}$$

5. The circuit current is now computed by Ohm's Law for ac circuits.

$$I = \frac{E}{Z} = \frac{100 \text{ volts}}{2920 \,\Omega} = .034 \text{ amps}$$

6. Find voltage drops around circuit.

$$E_L = IX_L = .034 \times 2500 = 85 \text{ volts}$$

$$E_R = IR = .034 \times 1500 = 51 \text{ volts}$$

and,

$V_L + V_R \neq V_S$ because they are 90 deg. out of phase and must be added vertorially.

$$\tan \theta = \frac{E_L}{E_R} = \frac{85}{51} = 1.67 \; \theta = 59.1 \text{ deg.}$$

$$\cos \theta = \frac{E_R}{E_Z} \qquad E_Z = \frac{E_R}{\cos \theta}$$

$$= \frac{51}{.5135} = 100 \text{ volts}$$

(Slight errors are due to rounding off figures, but it is accurate enough for practical needs.)

In this inductive circuit, the current lags the applied voltage by angle θ = 59.1 deg.

It is very important to understand how this entire problem would be affected by a change in frequency. At a frequency of 1000 Hz:

$$X_L = 25,000 \text{ ohms}$$

Inductance

f	X_L	θ	Z	I	IX_L	IR
Frequency increases	increases ↑	increases ↑	increases ↑	decreases ↓	increases ↑	decreases ↓
Frequency decreases	decreases ↓	decreases ↓	decreases ↓	increases ↑	decreases ↓	increases ↑

Fig. 8-19. Chart shows the effect on various values in an RL circuit as frequency is changed and the inductance value is held constant.

θ = 86.6 deg.

Z = 25,300 ohms

I = .0039 amp

E_R = 5.8 volts

E_L = 97.5 volts

These effects are summarized in Fig. 8-19, which shows the change of values in an RL circuit as the frequency is changed and L is held at a constant value.

PARALLEL RL CIRCUITS

The voltage across all branches of a parallel circuit is the same. In schematic Fig. 8-20, the voltage across R is 100 volts. However, the branch currents are not in phase, and the current through L will lag behind the current through R.

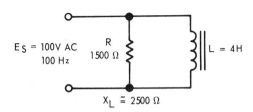

Fig. 8-20. Circuit used for parallel RL computations.

Follow the problem:

1. Compute X_L. X_L = 2500 ohms

2. Compute currents by Ohm's Law.

$$I_R = \frac{E}{R} = \frac{100}{1500} = .067 \text{ amp}$$

$$I_L = \frac{E}{X_L} = \frac{100}{2500} = .04 \text{ amp}$$

3. Draw according to any scale as in Fig. 8-21. I_R is in phase with circuit voltage. It is drawn horizontally as a reference. I_L is drawn downward since it is lagging by 90 deg.

4. Find θ by trigonometry.

$$\tan \theta = \frac{I_L}{I_R} = \frac{.04}{.067} \cong .6$$

$$\theta = \text{arc tan } .6 = 31 \text{ deg.}$$

$$\cos \theta = .8572$$

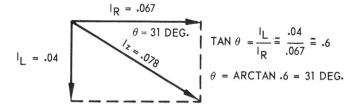

Fig. 8-21. The vector diagram for RL parallel problem described in text.

5. Find total circuit current I_z.

$$\cos \theta = \frac{I_R}{I_z} \qquad I_z = \frac{I_R}{\cos \theta}$$

$$= \frac{.067}{.8572} = .078 \text{ amp}$$

6. The impedance (Z) of the circuit may now be easily computed.

$$Z = \frac{E}{I_z} = \frac{100 \text{ volts}}{.078 \text{ amps}} \cong 1280 \text{ ohms}$$

The circuit has a total impedance of 1280 Ω and the circuit current lags the applied voltage by an angle of 31 deg.

As frequency is increased:

X_L will increase

I_L will decrease

I_R remains constant

θ will decrease

Z will increase

TRANSFORMERS

The basic principles of mutual induction and the transfer of energy between circuits by means of a transformer have already been discussed. Transformers are manufactured in a large variety of shapes and sizes, with and without cores, to fill the many requirements of electronic circuitry.

CONSTRUCTION OF THE POWER TRANSFORMER

As the name suggests, the power transformer is designed to be used at power frequencies, usually from 30 to 400 Hz. Both the primary and secondary windings are placed on a magnetic iron CORE. The coefficient of coupling must be as close to UNITY as possible, so various methods of winding the coils in layers or by interwinding are employed. Two basic core construction types are found in the modern transformer:

1. The CORE TYPE is made up of a package of thin laminate iron sheets in a rectangular

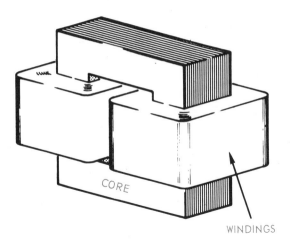

Fig. 8-22. Core type construction of a transformer.

shape. Each lamination is coated with an insulating varnish and the total core pressed together. The primary and secondary windings are placed on each side of the core. See Fig. 8-22.

2. A more popular method of construction, and also more efficient, is the SHELL TYPE CORE shown in Fig. 8-23. It, too, is made up of laminations covered with varnish and pressed together. The windings are wound in layers and fit over the center section of the core.

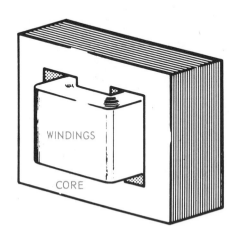

Fig. 8-23. Shell type core construction of a transformer.

Transformers may have one or more PRIMARY windings and one or several SECONDARY windings, all assembled on a single core. Some secondary windings may have center tap connections as well as taps at other points. Typical symbols used in electronic schematics are shown in Fig. 8-24 with their appropriate names. A variety of transformers used in electronic devices is illustrated in Fig. 8-25. Particular attention should be paid to miniature sizes used with transistors and on printed circuit boards.

TRANSFORMER THEORY

When an alternating current voltage is applied to the primary of a transformer and the secondary is OPEN, a small magnetizing current

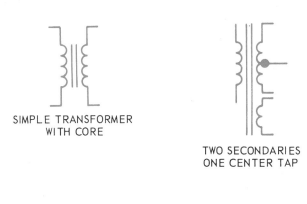

SIMPLE TRANSFORMER
WITH CORE

TWO SECONDARIES
ONE CENTER TAP

WITH ADJUSTABLE
CORE

AIR CORE
TRANSFORMER

Fig. 8-24. Some common transformer symbols.

will flow in the primary to establish the magnetic fields and to overcome any losses due to resistances. The magnetizing current is small and is limited by the amount of applied voltage, its frequency and the reactance of the primary winding. The expanding and collapsing magnetic fields of the primary cut across the secondary windings and induce a voltage in the secondary. The polarity of the secondary voltage will depend upon the direction of the secondary winding and the external connections. Generally, a transformer is considered as

Fig. 8-25. Typical miniature transformers used in transistor circuits. (United Transformer Co.)

inverting the signal input, but this is not always true. Wave forms are shown in Fig. 8-26. The black dot at the end of each winding indicates the polarity.

TRANSFORMER VOLTAGES

The magnitude and density of the flux lines of the primary coil will depend upon the current and the number of turns (I x N). Since both primary and secondary are wound closely together and on a common core, the flux for the secondary must be almost the same as the primary. This is stated in an equation as:

$$I_p N_p = I_s N_s$$

By rearranging the equation:

$$\frac{I_s}{I_p} = \frac{N_p}{N_s}$$

Notice that the CURRENT RATIO is INVERSELY proportional to the TURNS RATIO.

With unity coupling between the primary and secondary, the voltage induced in the secondary will be in DIRECT proportion to the TURNS RATIO.

$$\frac{E_p}{E_s} = \frac{N_p}{N_s}$$

If the turns of the secondary are more than the primary, it will increase the secondary voltage but decrease the current. It is called a STEP-UP transformer. If the secondary turns are less than the primary, the secondary voltage will decrease and the current increase. This is a STEP-DOWN transformer. A special application requires the same number of turns on both primary and secondary. The turns ratio is one-to-one. It is used as an ISOLATION transformer.

PROBLEM: A transformer with an input of 100 volts ac has a turns ratio of 5. What is the output voltage?

$$\frac{E_p}{E_s} = \frac{5}{1} \qquad E_s = \frac{E_p}{5} = \frac{100V}{5} = 20 \text{ volts}$$

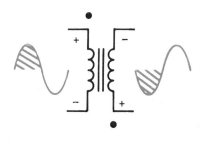

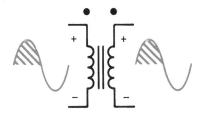

Fig. 8-26. Wave form relationships between input and output of a transformer.

It is a step-down transformer.

PROBLEM: A 500 volt ac output is required of a transformer which has a 100 volt input. What must the turns ratio be?

$$\frac{E_p}{E_s} = \frac{N_p}{N_s} \qquad \frac{100V}{500V} = \frac{N_p}{N_s} = \frac{1}{5}$$

A power transformer with several secondary windings, some step-up and some step-down, is illustrated in Fig. 8-27.

Fig. 8-27. A power transformer with high and low voltage secondaries. (Triad)

POWER RATIO

A transformer does NOT increase power. It has no way of creating electrical energy.

$$P_{in} = P_{out} \text{ and } P_{primary} = P_{secondary}$$

Power is the product of voltage and current, then:

$$I_p E_p = I_s E_s$$

In Fig. 8-28, the power relationship is described.

For the primary current, use formula:

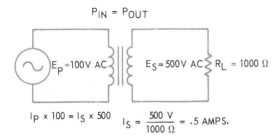

$$P_{IN} = P_{OUT}$$

$E_P = 100V\ AC$ $E_S = 500V\ AC$ $R_L = 1000\ \Omega$

$I_P \times 100 = I_S \times 500$ $I_S = \dfrac{500\ V}{1000\ \Omega} = .5\ AMPS.$

Fig. 8-28. This circuit describes the power relationships between the primary and secondary of transformer.

$$I_p E_p = I_s E_s$$

$$I_p \times 100 = .5 \times 500$$

$$I_p = \frac{250}{100} = 2.5\ amps$$

PROOF: $2.5 \times 100 = .5 \times 500 = 250$ watts.

TRANSFORMER LOSSES

Three types of losses are associated with transformer construction. All losses result in heat.

1. COPPER LOSSES. This loss is due to the resistance of the wire which is used to wind the transformer. It is in the form of heat and is computed by the formula, $P = I^2 R$. It is sometimes called the $I^2 R$ loss. It can be minimized by using wires of correct size to carry the expected currents. When a transformer is overloaded, this loss increases as the square of the current. The heat that results can seriously damage the transformer.
2. EDDY CURRENT LOSSES. Since the core is a conducting material, voltages can be induced in the core as well as in the secondary winding. Currents flowing within the core will heat up the core due to the resistance of the core. This eddy current loss is overcome by laminating the core. The thin insulated core stampings prevent the eddy currents from flowing excessively in the core. All power transformers have laminated cores.

3. HYSTERESIS LOSS. This loss is frequently called "molecular friction within the core material." As the core is magnetized, the domains are oriented in one direction. When the polarity of the magnetizing force reverses, then the domains must reverse. The constant aligning and realigning of the molecules cause heat and loss. This has been greatly reduced by using special alloys for transformer core stampings which have a minimal hysteresis loss.

AF AND RF TRANSFORMERS

One of the major uses of transformers, other than power transformation, is as a COUPLING DEVICE between stages of electronic circuits. The transformer will transfer a varying signal voltage and even increase its amplitude, yet it will block dc voltages between stages. Very often either the primary or secondary or both will be tuned to a specific frequency. This will be studied in a later chapter. Fig. 8-29 shows examples of the versions used in transistor circuits.

Fig. 8-29. Small transformers mounted on printed circuit boards. (United Transformer Co.)

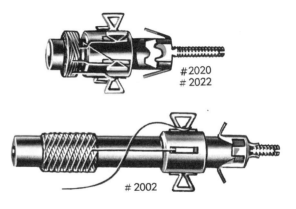

Fig. 8-30. Antenna Vari-Loopsticks used in AM radios.

At radio frequencies, the iron core is not used since core losses would be excessive. These transformers may have AIR CORES or special powdered metal cores designed for low loss. Antenna transformers with adjustable cores are shown in Fig. 8-30.

One very important application of the transformer is IMPEDANCE MATCHING or Z MATCHING.

POWER TRANSFER

A generator or power source has an internal resistance. In more advanced studies of electronics, the internal impedance of a source must always be considered. It has purposely been omitted in this text for simplification while

learning principles. But to understand power transfer, it must be considered. In Fig. 8-31, a generator is diagrammed.

Its internal resistance or impedance is represented by a SERIES resistor. Remember this resistance is, in fact, inside of the generator and a part of it. Voltages developed by the generator are the voltages across the resistance R_G. In Fig. 8-32 are tabulated the various currents, voltages and powers appearing in the circuit as the load R_L is changed. The voltage is held constant at 100 volts. The internal Z of the generator is 100 ohms, so R_G = 100 ohms.

CONCLUSIONS

1. For maximum transfer of power, the load resistance must equal the internal resistance of the generator. This is called MATCHING IMPEDANCES.
2. A low load resistance produces a high current; also poor efficiency and low power output.
3. A high load resistance produces a high output voltage; also high efficiency, but very low power output.
4. As load increases, current decreases and power loss due to internal generator resistance decreases. Efficiency will increase.

TRANSFORMER MATCHING

If the internal resistance of a generator, a signal source or a stage of an electronic circuit is to match the external load to secure a maxi-

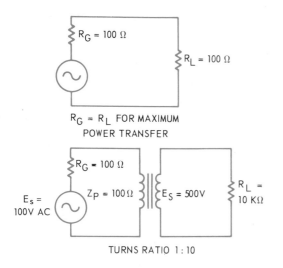

$R_G = R_L$ FOR MAXIMUM
POWER TRANSFER

TURNS RATIO 1 : 10

Fig. 8-31. These circuits describe the need for Z matching.

$R_L \, \Omega$	$I = \dfrac{E_G}{R_G + R_L}$	IR_G volts	IR_L volts	P_{load} watts	P_{R_G} watts	P_{total}	Eff. P_L / P_R
10 Ω	.909	90.9V	9.09V	8.26W	82.6W	90.86W	9%
50 Ω	.666	66.6V	33V	21.8W	43.5W	65.3W	33%
100 Ω	.5	50V	50V	25W	25W	50W	50%
500 Ω	.166	16.6V	83V	13.8W	2.75W	16.55W	84%
1,000 Ω	.09	9V	90V	8.1W	.8W	8.9W	91%
10,000 Ω	.01	1V	99V	1W	.01W	1.01W	99%

$E_G = 100V$ $R_G = 100 \, \Omega$

Fig. 8-32. Table showing transfer of power with selected load resistance.

mum power transfer, a transformer may be used to transform impedances. The impedance ratio of a transformer will vary directly as the square of its turn ratio. In equation form:

$$\frac{Z_p}{Z_s} = \left(\frac{N_p}{N_s}\right)^2 \quad \text{or} \quad \frac{N_p}{N_s} = \sqrt{\frac{Z_p}{Z_s}}$$

NOTE: An advanced study of electronics would include the derivation of this formula.

PROBLEM: A transformer is required to match a 300 ohm output impedance of a transistor power amplifier to a 3 ohm speaker coil. What will be the turns ratio?

$$\frac{N_p}{N_s} = \sqrt{\frac{300}{3}} = \sqrt{\frac{100}{1}} = \frac{10}{1}$$

PROBLEM: Using the figures in the table in Fig. 8-32, what turns ratio of a transformer would match R_G to an R_L of 10,000 ohms? Prove it.

$$\frac{N_p}{N_s} = \sqrt{\frac{100}{10,000}} = \sqrt{\frac{1}{100}} = \frac{1}{10}$$

This problem is diagrammed in Fig. 8-31. The impedance of the transformer primary must match the generator impedance. Then:

$$R_G = Z_p = 100\,\Omega \text{ and } R_G + Z_p = 200\,\Omega$$

At this match, the primary circuit current is:

$$I = \frac{E}{R_G + Z_p} = \frac{100}{200} = .5 \text{ amps}$$

$$E_p = IZ_p = .5 \times 100 = 50 \text{ volts}$$

The voltage across the secondary is:

$$\frac{E_p}{E_s} = \frac{N_p}{N_s} \quad \frac{50}{E_s} = \frac{1}{10} \quad \text{and } E_s = 500 \text{ volts}$$

The current in the load is:

$$I_{load} = \frac{E_s}{R_L} = \frac{500}{10,000} = .05 \text{ amps}$$

The power to the load is:

$$P_{load} = I_{load} \times E_s = .05 \times 500 = 25 \text{ watts}$$

NOTE: In the table in Fig. 8-31, the power in a load of 10 kilohms without matching is only one watt. In effect, as far as the generator is concerned, the transformer made 100 ohms look like 10 kilohms.

TRANSFORMER EFFICIENCY AND RATINGS

No perfect machine has yet been built. There are always some losses between input and output. For a transformer:

$$\text{Percent of efficiency} = \frac{P_{out}}{P_{in}} \times 100$$

Losses are dissipated in heat. These losses have been discussed earlier in the chapter.

In selecting a transformer for a circuit, the power handling capabilities of the transformer must be considered. Required secondary voltages and currents must not exceed the specifications.

The voltage handling capability of a transformer is determined by the type and thickness of insulation used on its windings. The current carrying capacity is limited by the size of the wire used in the windings. Small wires with high resistance will produce heat and destruction. The capacity to handle certain power requirements also depends upon the cross-sectional area of the core and the ability to safely dissipate heat.

HIGH INTENSITY LAMP PROJECT

A popular project that involves a working knowledge of transformer principles is the

Fig. 8-33. High intensity desk lamp.

1. Voltage to Primary   — 117 volts ac
2. Voltage of Secondary   — 13 volts
3. Current required for lamp — 1.04 amps
4. Output power required   — 13 watts

The size of the core required for a desired power output may be found in an electrical engineer's handbook. An abbreviated form is given in the table in Fig. 8-34.

POWER OUTPUT IN WATTS	CORE CROSS–SECTIONAL AREA IN SQUARE INCHES
10	.25
15	.37
20	.5
30	.7
40	.85
50	1.0
100	2.6

Fig. 8-34. An abbreviated table showing size of core required for a desired power output.

To be on the safe side, we will select the core size for a 20 watt output which is .5 square inches.

To determine the number of turns in the primary coil, see that the coil has sufficient inductive reactance at 60 Hz so that it will not create a short circuit across the 120 volt line. This demands at least 4 turns per volt and, as a margin of safety, 6.25 turns per volt is usually used. The core area must also be considered in the desired amount of reactance. Use this formula:

$$N_p = \frac{6.25 \times 120 \text{ Volts}}{\text{cross-sectional area of core}}$$

$$N_p = \frac{6.25 \times 120}{.5} = 1500 \text{ turns}$$

Use the factor 1.1 to determine the number of turns in the secondary. This will offset losses due to imperfect coupling between primary and secondary.

HIGH INTENSITY DESK LAMP, Fig. 8-33. These lamps of a commercial variety are found in most department and appliance stores. You will enjoy making one of your own. It is an excellent study lamp.

The mechanical construction of the lamp can be designed to fit your individual needs and the availability of materials. You will need to purchase a socket and a No. 93, 12.8 volt, 1.04 amperes lamp from an automotive parts store. The actual mechanical design of the lamp will depend upon your personal preference. The lamp in the photo is only a suggestion.

The construction of the transformer will be the only problem for us to solve before proceeding. First, look at the requirements:

$$\frac{N_p}{N_s} = \frac{E_p}{E_s} \text{ or } N_s = \frac{N_p \times E_s \times 1.1}{E_p}$$

and

$$N_s = \frac{1500 \times 13 \text{ volts} \times 1.1}{120 \text{ volts}} = 183 \text{ turns}$$

To determine wire sizes, we must find the primary and secondary currents. Again the factor of 1.1 is used to take care of losses.

$$P = I \times E \text{ and } I = \frac{P}{E}$$

$$I_p = \frac{13W \times 1.1}{120} = .120 \text{ amp}$$

$$I_s = \frac{13W}{12.8V} \cong 1 \text{ amp}$$

Referring to the B & S wire table, Fig. 8-35, we find that the minimum size wires for these currents should be:

WIRE SIZE	CURRENT CARRYING CAPACITY IN AMPERES
12	8.7
14	5.4
16	3.4
17	2.7
18	2.1
19	1.7
20	1.3
21	1.1
22	.85
23	.75
24	.54
30	.13

Fig. 8-35. Wire table.   (Brown and Sharpe)

For secondary (1 amp) — Size 21
For primary (.12 amp) — Size 30

One final consideration in constructing a transformer is the length of the coil to be used. In Fig. 8-36, coil lengths for various power

POWER OUTPUT IN WATTS	COIL LENGTH IN INCHES
10	1.25
20	1.625
25	1.875
30	2.00
50	2.25
100	2.75

Fig. 8-36. Coil lengths for various power outputs.

outputs are given. Our coils will be 1.625 in. long.

Now that the design has been completed, the coils must be wound on wooden forms Since the core will have .5 sq. in. cross-sectional area, a core material about .7 in. wide will be used. The wooden coil forms, Fig. 8-37, should be made about 1/8 in. oversize so that the finished coils will fit over the core.

Remember, the primary coil will have 1500 turns of No. 30 wire and the secondary will have 183 turns of No. 21 wire. (In constructing this project, No. 24 and No. 18 wire were used for added safety.) Wind each layer of the coils carefully and neatly. Between each layer, place a single layer of cloth or paper as insulation.

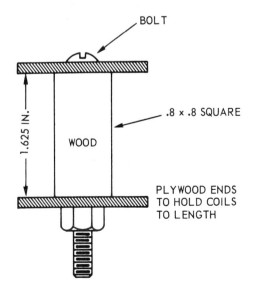

Fig. 8-37. A wooden coil form or mandrel is used to wind both secondary and primary coils.

Remove the finished coil from the form, dip it in insulating varnish and hang it up to dry. After drying wrap the coil with a layer of insulating tape.

The best source of laminated core iron is from old transformers. However, black iron or even galvanized sheet iron from 22 to 26 gauge can be used. Cut and stack it to form the core, Fig. 8-38. Each lamination should be dipped in varnish before stacking together.

After the core is stacked, slip out the laminations at one end and put your coils in place. Then, return the laminations to their original places. Drill holes at each corner and bolt the core together securely with stove bolts. Connecting leads on the finished transformer are shown in Fig. 8-39.

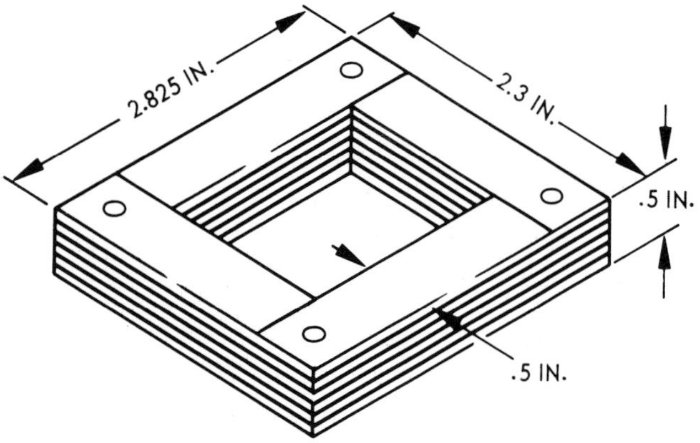

LESSON IN SAFETY: When desoldering parts and components, do not shake or brush off the hot solder in a way that it might hit you or your working partner in the eyes or on your clothes. If you have ever had a drop of hot solder fall inside your shoe, you will know that this rule of safety cannot be overemphasized.

LESSON IN SAFETY: Remember that the metal edges and sharp corners on chassis and panels can cut and scratch. File them smooth.

TEST YOUR KNOWLEDGE

1. Inductance is that property in a circuit which opposes a change in _____.
2. Explain the reason for increased CEMF when frequency of current is increased.
3. What is the inductance of a coil which produces a CEMF of 50 volts when the current changes 100 millamps at a frequency of 100 hertz?

Fig. 8-38. The laminations are coated with varnish and stacked with overlapping ends.

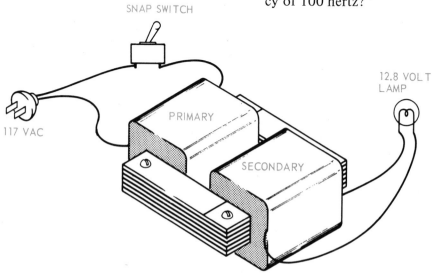

Fig. 8-39. The assembled transformer, ready for a lamp of your own design.

4. What is the mutual inductance between two coils of 2 and 4.5 henrys, when connected in series additive with coupling coefficient of .6?

5. The angular relationship between coils affects mutual inductance. The fields of one coil must cut across the conductors of the second coil at _____ for maximum coupling and mutual inductance.

6. What is the total inductance of 2, 4 and 8 henry coils in parallel without coupling?
   a. 115.0 Hz.
   b. 11.5 Hz.
   c. 1.15 Hz.

7. A series RL circuit has an R = 1000 ohms and L = 8 hertz. The applied voltage is 100 volts at a frequency of 1 kilohertz. Draw the circuit and vector diagram and compute Z. By what angle does the current lag the voltage?

8. A parallel circuit of 1000 ohms resistance and 8 hertz of inductance is connected to a 100 volt source at 1 kilohertz. Find Z. By what angle are the voltage and current out of phase?

9. A transformer has a turns ratio of .1 and the primary has 600 turns. How many turns in the secondary? What is secondary voltage when 10 volts ac are applied to primary?

10. A power amplifier with an output Z of 2500 ohms must match the input Z of 100 ohms of another circuit. What turns ratio will the matching transformer have?

a. 250/1
b. 25/1
c. 5/1

11. Can a transformer be connected to a dc source? Explain.

12. If a circuit has an inductance of 4 H at 10 KHz, then X_L = _____.

13. If a circuit has an inductance of 2.5 mH at 100 KHz, then X_L = _____ .

14. If a circuit has an inductance of 8 H at 60 Hz, then X_L = _____ .

15. If a circuit has an inductance of 8 H at 120 Hz, then X_L = _____ .

16. Draw a chart similar to the one shown in Fig. 8-19, but consider frequency as a fixed value. Enter the effect (decrease or increase) in each square as inductance is decreased or increased.

17. In a parallel RL circuit, will the circuit become _____ (more or less) inductive if the frequency is increased? Why?

18. A transformer used on the 117 volt ac power line has a turns ratio of 20 to 1. What is the output voltage?

19. Assume that a transformer has a turns ratio of 1/5 and is connected to a 100 volt ac source. A load across the secondary draws 100 milliamps of current. What is the primary current?

20. Referring to the transformer and circuit arrangement described in question 19, what is the secondary voltage?

21. What is the secondary power?

22. What is the primary power?

Chapter 9

CAPACITANCE

Now, the phenomenon of capacitance in an electronic circuit will take its place beside resistance and inductance in our studies. Only these three properties exist in any circuit. The more simple and the most complex combinations of components, vacuum tubes and transistors may be reduced to an equivalent circuit composed of resistance, inductance and capacitance.

In this chapter, we will discuss:

1. The theory of capacitance in a circuit.
2. The construction of the various types of capacitors.
3. The application of capacitors in series and parallel circuits.

In our previous study of inductance, we found that inductance stored energy in an electromagnetic field and accepted or returned the energy whenever a change in current appeared. An inductor, therefore, opposed a change in current in a circuit.

A capacitor (C) stores energy in an ELECTROSTATIC FIELD and accepts a charge of electricity or returns electrons to the circuit in an attempt to maintain a CONSTANT voltage. Capacitance is that property of a circuit which opposes a CHANGE IN VOLTAGE in the circuit.

THE SIMPLE CAPACITOR

The most elementary capacitor is illustrated in Fig. 9-1. It consists of two metal plates separated by an insulator or air, called the DIELECTRIC. A potential voltage connected across these plates produces the following action:

Electrons from the negative battery terminal flow instantaneously to one plate (A) of the capacitor. Plate B is very close but not in contact with plate A, and plate B is within the established electrostatic field. The electrons on plate B are repelled by the electrons on A and are attracted to the positive terminal of the battery. This creates a potential difference between the two plates and the capacitor is CHARGED. Note in Fig. 9-1 that the charge on the capacitor is opposite in polarity compared to the source voltage. A current does flow in the circuit but only during the charging of the capacitor. When the capacitor is fully charged, the current is zero.

If you removed the capacitor from the circuit, it would remain charged and would be a source of voltage in its own right. In high voltage circuits, capacitors can retain charges which are dangerous to personnel working with the circuits. For safety, capacitors should be shorted and discharged before handling.

In Chapter 1, you studied the properties of an electrostatic field. Demonstrations were made to prove the existence of invisible lines of force surrounding a charged body. This applies to the capacitor. A strong electrostatic field exists in the dielectric between the plates of the charged capacitor. Electrons existing in this field would have a tendency to move toward

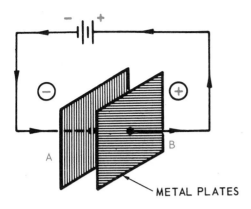

Fig. 9-1. A simple capacitor.

the positive plate. Fig. 9-2 shows one theory which explains how energy is stored in a dielectric.

The electrons associated with each atom of dielectric material are distorted out of their regular orbits and assume new locations in the direction of the positive plate. They remain in these positions under the influence of the static field.

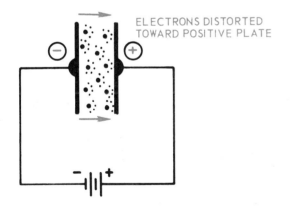

Fig. 9-2. Electrons in the dielectric are distorted toward the positive plate when capacitor is charged.

When the capacitor is discharged, the electrons return to their original orbits and, by doing so, return the energy to the circuit.

If a very high electrostatic field exists between the plates, excessive voltage applied to the capacitor will not only distort the electrons in the dielectric but may tear them from their orbits and permit conduction through the di-electric. This results in a spark or arcing between the plates.

All capacitors have a WORKING VOLTAGE DIRECT CURRENT (WVDC) rating, which is the specified voltage the capacitor can withstand without destruction of the dielectric. This specification must always be considered by the technician when using or replacing capacitors.

LESSON IN SAFETY: Be on guard for electrolytic capacitors which may remain highly charged long after the equipment has been "turned off." Use an insulated screwdriver to short out all capacitors to ground. Short them not only once but twice. Be certain they are discharged before touching the circuit with your bare hands.

UNITS OF MEASUREMENT OF CAPACITANCE

In early experimentation, it was found that the charge deposited on a plate of a capacitor was proportional to the voltage required to move this charge, for any given value of capacitance. Capacitance (C) is measured in FARADS in honor of Michael Faraday. In electronics, much smaller values of capacitance will be found. Most values will be in microfarads (1/1,000,000 farad) or picofarads (1/1,000,000 of 1/1,000,000 of a farad). The older term for picofarad is micromicrofarad.

A FARAD (F) is the capacitance which will cause one ampere of charging current to flow when the applied voltage is changing at a rate of one volt per second.

Expressed mathematically:

$$C = \frac{i}{\Delta v / \Delta t}$$

where,

C is in farads
  i = charging current
Δv = the change in volts
Δt = the change in time in seconds

EXAMPLE: What is the capacitance of a capacitor if a charging current of 100 milliamperes flows when the voltage changes ten volts at a frequency of 100 hertz.

1. Change 100 millamps to .1 amp. The time period of 100 hertz is .01 seconds.

2. $C = \dfrac{.1}{10/.01} = .0001$ farads or 100 microfarads

Another method of defining capacitance is by the charge or quantity of electrons or coulombs which a capacitor will accept per volt of potential applied to it. This is expressed as:

$$C = \frac{Q}{V}$$

where,

C is in farads
Q = coulombs
V = volts

EXAMPLE: A certain capacitor will accept a charge of .1 coulomb with an applied voltage of 100 volts. What is its capacitance?

$$C = \frac{.1}{100} = .001 \text{ farad or } 1000 \text{ microfarads}$$

Fig. 9-3. A variable capacitor. (J.W. Miller Co.)

TYPES OF CAPACITORS

Various kinds of capacitors are shown in Figs. 9-3 through 9-9. They are described in the following paragraphs.

VARIABLE CAPACITOR

The variable capacitor, Fig. 9-3, consists of one or more metal plates in a fixed position called the STATOR. A second set of plates called the ROTOR can be rotated so that they interleave between each stator plate. The amount of capacitance is controlled by a knob which turns the rotor. The dielectric is AIR.

Variable capacitors are widely used in tuning circuits. The tuning knob on your radio turns a capacitor of this type. Two or more of these capacitors may be connected to a single shaft so that all may be rotated together. This assembly is called a GANGED CAPACITOR.

A typical variable capacitor will have a screw on its side, which is another small capacitor in parallel with the larger variable capacitor. This little TRIMMER CAPACITOR is used to make fine adjustments on the total capacitance of the device. The trimmer capacitor will have small flexible metal plates separated by mica or some other dielectric. By turning the screw inward, the plates are compressed and its capacitance is increased.

PAPER CAPACITOR

A paper capacitor, one of the more common types of fixed capacitors, is shown in Fig. 9-4. Two layers of thin metal foil are separated by waxed paper or other paper dielectric. The sandwich of foil and paper is then rolled into a

Fig. 9-4. A paper capacitor.

cylindrical shape and enclosed in a paper tube or encased in a plastic or wax capsule. Leads extending from each end of the capacitor are attached to the metal foil plates. These are made in hundreds of capacitance values and working voltages. These specifications usually are printed on the capacitor case.

MICA CAPACITOR

The mica capacitor, Fig. 9-5, is a sandwich of thin metal plates separated by thin sheets of mica. Alternate plates are connected together and leads attached. The total assembly is encased in a plastic capsule. These capacitors have small capacitance values, yet high voltage ratings.

Fig. 9-5. Mica capacitors.

CERAMIC DISC CAPACITORS

Ceramic disc capacitors, Fig. 9-6, are about the size of a dime and smaller. They have a metal film deposited on each side of the ceramic to act as the capacitor plates. The ceramic is the dielectric. Leads are attached to each side of the disc and the component is encapsulated in a moisture proof coating. You will find dozens of ceramic disc capacitors used in electronic circuits. They have small capacity values but can be made to operate at very high voltages.

ELECTROLYTIC CAPACITORS

The tubular type of electrolytic capacitor is shown in Fig. 9-7. An entirely different process

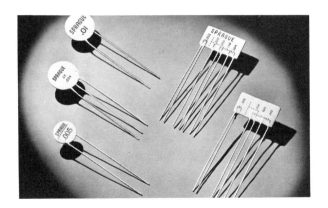

Fig. 9-6. Ceramic disc capacitors.

is used to make these capacitors. A thin strip of aluminum is coated with an oxide by an electrochemical process. This is covered with a second layer of paper or gauze impregnated with a paste-like electrolyte. On top of this is another metal plate in contact with the electrolyte. In an electrolytic capacitor, the first plate is the positive plate, the oxide film is the dielectric and the electrolyte is the other capacitor plate. The second metal plate serves only as a contact to the electrolyte. It is the negative terminal.

This entire sandwich is rolled up, similar to a paper capacitor, and placed in a metal cylinder or can. This metal container contacts the outside metal foil and serves as the negative terminal. These are made as single capacitors and sometimes with two or three capacitors in a single can with a common negative terminal. The outside metal container is usually enclosed in a paper tube to insulate it from other components and to protect personnel from shocks.

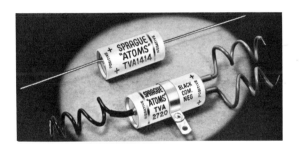

Fig. 9-7. Tubular type electrolytic capacitors.

CAN TYPE ELECTROLYTICS

Another version of the electrolytic capacitor is called the "can type," Fig. 9-8. This capacitor is assembled in a metal can, which is the negative terminal. One or more capacitors, which have different capacitance and/or working voltage, may be placed in one can. The capacitor is mounted on a metal plate, which is bolted to a chassis. The capacitor stands up-

Fig. 9-8. Can type electrolytic capacitor.

right. If the negative terminal is not to be connected to chassis ground, the capacitor is mounted on a fiber insulating plate. The value of the capacitors and their voltages will be printed or stamped on the outside of the can. The one or more positive terminals of the can type capacitor are brought out at the bottom end by prong terminals. They are identified by printed or cut out geometric figures beside each terminal, such as: ▲ ● ■ ▬

Electrolytic capacitors are used when a lot of capacitance is required in a small space. Disadvantages include:

1. They are polarity sensitive and must be installed with correct polarity in a circuit. If connected in reverse, the oxide film will dissolve and the capacitor will short circuit. A distinctive cloud of peculiar smelling white smoke will inform you of your error.

2. The electrolytic capacitor has a low leakage resistance. This resistance can be measured with an ohmmeter. Excessively leaky capacitors are a source of trouble to the service technician. They must be replaced.

There is a special type of electrolytic capacitor which can be used in ac circuits. These are NONPOLARIZED and specified as NP. They will be found in motor starters and in some electronic circuits.

MINIATURE ELECTROLYTICS

With the development of transistorized circuits and compact equipment, a whole new line of very small electrolytic capacitors was brought on the market. They will be used in many projects in this text. These miniatures are sold under various trade names such as, TINYLYTTICS, LITTLELYTTICS, etc., Fig. 9-9.

The construction of miniature electrolytics involves many trade secrets. Companies compete with each other to get the most capacitance in the smallest package, with a high

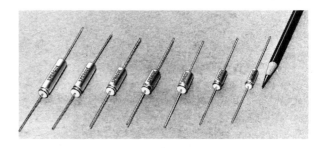

Fig. 9-9. Tinylyttics. (Sprague Products Co.)

degree of accuracy and reliability. Since these capacitors are used in transistor circuits, the working voltages can be low. They are made in a range between one volt and 50 volts.

CHARACTERISTICS OF CAPACITORS

What determines the capacitance of a capacitor? There are several factors to be considered:

1. PLATE AREA. The capacitance is increased with an increase in plate area. The variable capacitor is a working example of this principle. As the rotor is turned so that its plates mesh with the stator plates, a larger

plate area is being used. As a result, a variable capacitor, in a fully meshed position, is at its maximum capacity. Larger plates can store more electrons than smaller plates, and a larger dielectric area can store a larger amount of energy.

2. SPACING BETWEEN PLATES. If the plates of a capacitor are very close, the field of one plate has a very pronounced effect on the field of the other plate. As the plates are separated, the ability of the fields to produce molecular distortion and store a charge within the dielectric will decrease. Therefore, as the plates move closer together, the capacitance is INCREASED. An increase in the spacing between plates will DECREASE the capacitance.

3. KIND OF DIELECTRIC. When plates are separated by air, a certain value of capacitance is realized. If a piece of material such as glass or paper is placed between the plates, the capacitance will increase. Due to the molecular formation of different materials, they will have different abilities to store electrical energy when used as a dielectric in a capacitor.

These abilities are rated in respect to air (more accurately, a vacuum) which has been assigned the value of ONE. The number given

MATERIAL	DIELECTRIC CONSTANT (k)
Vacuum	1.0000
Air	1.0006
Waxed Paper	3 to 5
Glass	5 to 10
Mica	3 to 6
Rubber	2.5 to 30
Wood	3 to 8
Pure Water	81

Fig. 9-10. Constants of several materials used in capacitors.

each material is called its DIELECTRIC CONSTANT. The constants of several materials used in capacitors are given in the table in Fig. 9-10.

Example of the use of this table: A capacitor using waxed paper as a dielectric will have from 3 to 5 times more capacitance than a capacitor with air as a dielectric.

The previous factors are summarized in this equation:

$$C \text{ (in picofarads)} = .2249 \frac{kA (n-1)}{d}$$

where,

k = dielectric constant.
A = area of one plate in square inches.
n = number of plates.
d = distance between plates in inches.
.2249 = a factor used to convert from metric to British units of measurement.

This equation means that capacitance increases directly with the value of the dielectric constant and the plate area, but it is inversely proportional to the distance between plates.

CAPACITORS IN SERIES

Two capacitors are drawn in series in Fig. 9-11. Capacitor C_1 starts to charge by electrons from the negative source. Plate A becomes

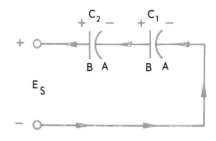

Fig. 9-11. The electric charge divides between the capacitors.

negative and plate B positive. Electrons repelled from plate B of C_1 travel to plate A of C_2 and make it negative. Electrons from plate B of C_2 continue on to the positive source. No electrons can pass through either C_1 or C_2 so the total charge of C_1 in series must be divided between C_1 and C_2.

Since,

$$C = \frac{Q}{V} \quad \text{or} \quad V = \frac{Q}{C}$$

the total charge Q_T divided by the total capacitance C_T must equal source voltage V, and

$$\frac{Q_T}{C_T} = \frac{Q_1}{C_1} + \frac{Q_2}{C_2} \quad \text{and} \quad E_s = E_{c_1} + E_{c_2}$$

By cancelling out the Q in the above equation, we have:

$$\frac{1}{C_T} = \frac{1}{C_1} + \frac{1}{C_2} \cdots \qquad C_T = \frac{1}{\frac{1}{C_1} + \frac{1}{C_2} \cdots}$$

The total capacitance of any number of capacitors in series may be found by this formula. When only two capacitors are in series, this simplied formula may be used:

$$C_T = \frac{C_1 C_2}{C_1 + C_2}$$

CAPACITORS IN PARALLEL

In Fig. 9-12, two capacitors are connected in parallel.

In this case, a charging current flows to C_1 and charges it to E_s. A charging current also charges C_2 to E_s. The total charge (Q_T of the circuit) must be equal to $Q_1 + Q_2$. The voltage across each capacitor is the same and equal to E_s. The total charge is more than the charge of any one capacitor, so the total capacitance must have increased to accomodate the total charge, Q_T. That is, the total quantity of stored electrons has increased, and the only way this can happen is by an increase in capacity. THE TOTAL CAPACITANCE OF A PARALLEL CIRCUIT OF CAPACITORS IS EQUAL TO THE SUM OF THE INDIVIDUAL CAPACITORS.

$$C_T = C_1 + C_2 + C_3 \cdots$$

You can look at this formula another way. For each capacitor added to the circuit, there is an increase in plate area. Two equal capacitors have twice the plate area of one. An increase in plate area results in an increase in capacitance.

CURRENT AND VOLTAGES IN A SERIES RC CIRCUIT

A typical series RC circuit is shown in Fig. 9-13. With switch S open, the voltages across C and R are zero. When S is turned to position A, a maximum current will instantaneously flow in the circuit because there is no opposing voltage from C. A maximum current will produce a maximum voltage drop across R. As C becomes charged, the current also decreases and the voltage across R will decrease. When C is fully charged to the value of the source voltage, the

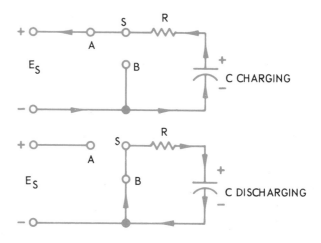

Fig. 9-13. This circuit shows current paths while C is charging and discharging.

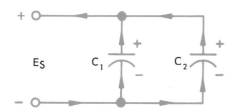

Fig. 9-12. Capacitors in parallel may be added together to find the total circuit capacitance.

current becomes zero and the voltage across R is zero. See the wave forms of the voltages and current in Fig. 9-14.

When S is switched to position B, maximum discharge current will flow, decaying to zero as C becomes discharged. The voltage across R starts at maximum and decreases to zero.

RC TIME CONSTANT

The time required for capacitor C to charge or discharge will depend upon the resistance in the circuit. One period of time, in which the

voltage across C will increase to 63.2 percent of E_s, will be found by:

t in seconds = R in ohms x C in farads

A total time of FIVE TIME CONSTANT PERIODS is required for C to be considered fully charged. On discharge, the opposite is true. During the first period the charge will decrease to 36.8 percent of its fully charged value. In five periods, the capacitor will be fully discharged. Currents and voltages rise and decay as plotted on the UNIVERSAL TIME CONSTANT CHART. See Fig. 9-15. Voltages and currents at time constant intervals may be computed from this chart.

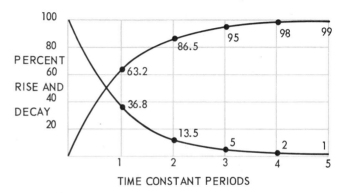

Fig. 9-15. The Universal Time Constant Chart for RL and RC circuits.

PROBLEM: What is the time constant of a circuit containing an 8 microfarad capacitor in series with a 100 kilohm resistor?

$$t = 8 \times 10^{-6} \times 1 \times 10^{5}$$

$$= 8 \times 10^{-1} = .8 \text{ second}$$

What is the total charging time?

5 x .8 = 4 seconds to completely charge C.

A practical demonstration of the time constant may be observed in the circuit shown in Fig. 9-16. The NE2 glow lamp has a high resistance when not glowing. If a voltage of about 65 volts is applied to the NE2, it will

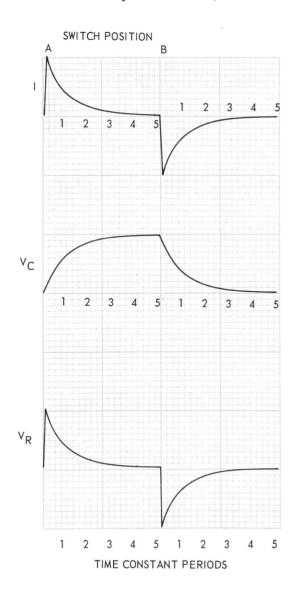

Fig. 9-14. Voltage and current wave forms in a RC circuit.

"fire" or glow and become a very low resistance while burning.

As the circuit is turned on, the 100 volt E_s is applied to the circuit and C starts to charge. When the voltage across C reaches the 65 volt

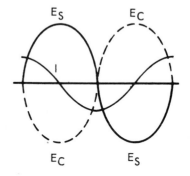

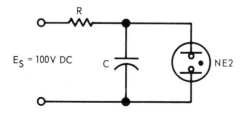

Fig. 9-16. The circuit of a sawtooth generator or relaxation oscillator.

point, the lamp glows and C rapidly discharges through the NE2. The lamp goes out and C starts to charge again.

The NE2 glow lamp will continue to flash each time the voltage reaches the "firing" point. You have constructed a flasher or blinker. Technically, it is called a RELAXATION OSCILLATOR. By changing the values of either C or R, the rate of flash may be controlled since the rate depends upon the circuit time constant.

CAPACITANCE IN AN AC CIRCUIT

In Fig. 9-17, an alternating current voltage symbolized as a generator is connected to a capacitor. As the ac voltage builds in a positive direction, the CEMF (opposing voltage) is small and the charging current is maximum. When the charge on C is equal and opposite to the applied voltage, the current is at zero. As the applied voltage decreases toward zero, capacitor C also discharges toward zero. At zero voltage, the discharge current is maximum due to no counter EMF.

In effect, the source voltage and capacitor voltage are alternately changing polarity and rising to maximum in opposite directions. When both are equal at the peak of the waves, there is no change of voltage. At the point where it

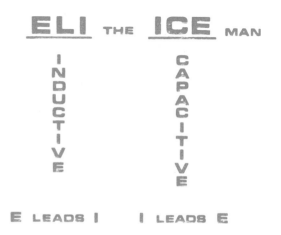

Fig. 9-17. The voltage and current in a pure capacitive circuit.

crosses the zero line, the change of voltage is most rapid and the current must be at maximum to satisfy the change.

CONCLUSION: The current and voltage are 90 deg. out of phase in a circuit containing only capacitance and the CURRENT LEADS THE VOLTAGE.

The following device, used to remember the effect of either inductance or capacitance in a circuit, was used in some of our Armed Forces Service Schools.

In an inductive circuit, voltage E leads current I.

In a capacitive circuit, current I leads voltage E.

It is better for you to understand these relationships rather than depend upon a memory device.

CAPACITANCE IN AC AND DC CIRCUITS

In Fig. 9-18, a lamp is connected to a dc voltage source in series with a capacitor. The lamp does not glow. A capacitor is an OPEN CIRCUIT as far as dc is concerned. Stating it another way: A CAPACITOR BLOCKS DC. Capacitors are used in this manner in many electronic circuits.

When the source voltage is changed to ac, the lamp will glow. This is not because current is flowing through the capacitor; it is the alternating displacement current as the voltage of the capacitor changes polarity.

The brightness of the lamp in the circuit will depend upon the size of the capacitor. Currents required to charge a small capacitor may be insufficient to light the lamp. In effect, the current in the ac circuit seems to be limited by

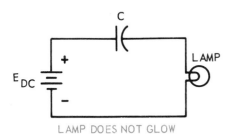

LAMP DOES NOT GLOW

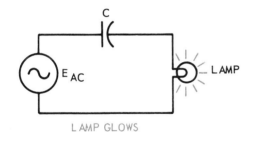

LAMP GLOWS

Fig. 9-18. A capacitor blocks dc and passes ac.

the capacitor. This opposing force is named CAPACITIVE REACTANCE. Its symbol is X_c. Like resistance, it is measured in ohms. THE MAGNITUDE OF CAPACITIVE REACTANCE IS INVERSLEY PROPORTIONAL TO THE SIZE OF THE CAPACITOR IN FARADS AND THE FREQUENCY OF THE AC CURRENT.

Expressed in the equation:

$$X_c = \frac{1}{2 \pi fC} \quad \text{or} \quad \frac{1}{\omega C}$$

where,

$$2 \pi f = \text{angular velocity} = \omega$$
$$C = \text{capacitance in farads}$$
$$X_c = \text{capacitive reactance in ohms}$$

It is easy to see that a larger capacitor requires more current to charge it. Therefore, it offers less opposition. Also, as frequency is increased or decreased, the capacitor will be charged more or less times per second. An increase of frequency would cause more current to flow. Therefore, the reactance is less.

In fact, a given capacitor may have a very high reactance path at low frequencies, but become a very low reactance path, or an effective short circuit, at high frequencies. The technician may say, "This circuit is at ground potential for signal voltages." This means it is short circuited to ground through a capacitor. For dc voltages, however, the capacitor appears as an open circuit.

In Fig. 9-19, the graph shows how X_c decreases with an increase in frequency at a constant value of C in farads.

In Fig. 9-20, the curve shows the effect on reactance X_c as the size of the capacitor is increased while holding the signal frequency at a constant value.

PROBLEM: Compare the reactance of a .01 μF capacitor at 100 Hz, 1 KHz and 10 KHz.

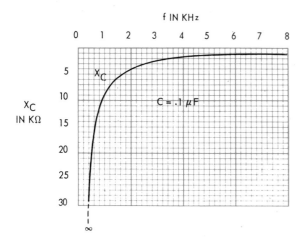

Fig. 9-19. At zero frequency or dc, X_C equals infinity. As frequency increases, X_C decreases.

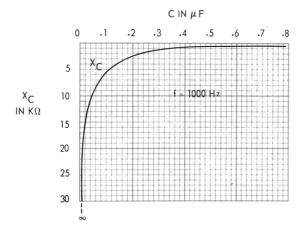

Fig. 9-20. At a fixed frequency, X_C decreases when the value of C is increased.

At 100 Hz, $X_c = \dfrac{1}{\omega C} = \dfrac{1}{6.28 \times 10^2 \times 10^{-8}}$

$$= .159 \times 10^6$$

$$= 159,00 \text{ ohms}$$

At 1 KHz, $X_c = \dfrac{1}{6.28 \times 10^3 \times 10^{-8}}$

$$= .159 \times 10^5 = 15,900 \text{ ohms}$$

At 100 KHz, $X_c = \dfrac{1}{6.28 \times 10^5 \times 10^{-8}}$

$$= .159 \times 10^3 = 159 \text{ ohms}$$

NOTE: In each of the above equations, $\dfrac{1}{6.28} = .159$ and it can become a constant.

$$X_c = \frac{.159}{fC} \quad \text{or} \quad \frac{.16}{fC} \text{ (approx.)}$$

The small error is insignificant.

THE SERIES RC CIRCUIT

In a circuit containing resistance only, the current and voltage are in phase. In a circuit of pure capacitance, the current leads the voltage by an angle of 90 deg. As you can guess, in a circuit which has both resistance and capacitance, the current will lead the voltage by some angle between zero and 90 deg.

In Fig. 9-21, a series RC circuit is drawn and values assigned to the components. We will work out this problem to determine the phase angle, impedance and voltages of the circuit.

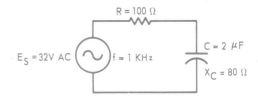

Fig. 9-21. The series RC circuit used in the text problem.

1. Find the reactance of C at 1 KHz.

$$X_c = \frac{.16}{10^3 \times 2 \times 10^{-6}} = .08 \times 10^3 = 80 \text{ ohms}$$

2. Draw the vectors for R and X_c. Use any convenient scale. In the series circuit, the current is constant and voltages are out of phase. R is used as a horizontal reference and X_C is plotted downward by 90 deg. since the voltage across C will lag the current. See Fig. 9-22.

3. The impedance Z can be found by the Pythagorean Theorem:

$$Z^2 = R^2 + X_c^2 \quad \text{or} \quad Z = \sqrt{R^2 + X_c^2}$$

Capacitance

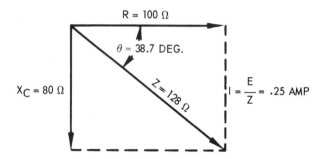

Fig. 9-22. Vector diagram used for text problem.

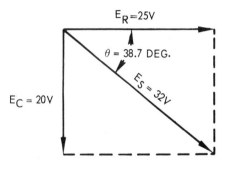

Fig. 9-23. The vector sum of the voltages will equal the supply voltage.

or use trigonometry:

$$\tan \theta = \frac{X_c}{R} = \frac{80}{100} = .8$$

$$\theta = \text{arc tan } .8 \qquad \theta = 38.7 \text{ deg.}$$

$$\cos \theta = .7804$$

4. Find Z, using cosine function:

$$\cos \theta = \frac{R}{Z} \quad Z = \frac{R}{\cos \theta} = \frac{100}{.7804} \cong 128 \text{ ohms}$$

5. Find the circuit current by Ohm's Law:

$$I = \frac{E}{Z} = \frac{32}{128} \cong .25 \text{ amp}$$

6. Find voltages across R and C:

$$E_R = IR = .25 \times 100 = 25 \text{ volts}$$

$$E_c = IX_c = .25 \times 80 = 20 \text{ volts}$$

and

$$E_R + E_c \neq E_s, \quad 25 + 20 \neq 32$$

They must be added vectorially because they are 90 deg. out of phase. Refer to Fig. 9-23.

7. Prove that vector addition of voltages does equal the source voltage. Refer to Fig. 9-23.

$$\cos \theta = \frac{E_R}{E_s} \qquad E_s = \frac{E_R}{\cos \theta}$$

$$E_s = \frac{25}{.7804} \cong 32 \text{ volts}$$

$$\sin \theta = \frac{E_c}{E_s} \qquad E_s = \frac{E_c}{.6252} \cong 32 \text{ volts}$$

This may also be proved by:

$$E_s = \sqrt{E_R{}^2 + E_c{}^2}$$
$$= \sqrt{25^2 + 20^2} = 32 \text{ volts}$$

NOTE: Computations in above equations have been taken to the whole number. The error is insignificant.

LESSON IN SAFETY: As you work in your laboratory or shop, you will learn how to use tools correctly and safely. These will help you develop good working habits. Study the job at hand and think through your procedures, your methods and the application of tools, instruments and machines.

POWER IN AN AC CIRCUIT

Capacitive reactance uses no power. The energy stored in a capacitor is returned to the circuit. Resistance is the only component of a circuit which will use power. The power curve for a pure capacitive circuit is shown in Fig. 9-24.

Note that power is the product of I and E. The power above the zero line is positive power

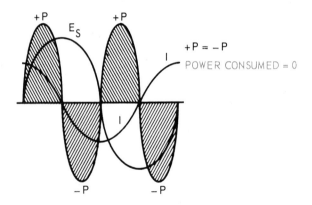

Fig. 9-24. A pure capacitive circuit uses no power.

and below the line it is negative or wattless power. They are equal and opposite. In these curves, the current leads the voltage by 90 deg.

The power curve in Fig. 9-25 is drawn for the problem just completed when R = 100 ohms and the current leads the voltage by θ = 38.7 deg. Apparently the power used by the circuit should be:

$$I^2 Z = (.25)^2 \times 128 = 8 \text{ watts}$$

But the TRUE POWER CONSUMED is:

$$I^2 R = (.25)^2 \times 100 = 6.25 \text{ watts}$$

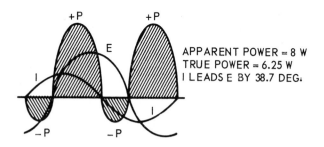

Fig. 9-25. The RC circuit uses power.

The relationship of:

$$\frac{\text{TRUE POWER}}{\text{APPARENT POWER}} = \text{POWER FACTOR}$$

In this problem, the PF = $\frac{6.25}{8}$ = .78

In step 3 in the previous problem, you discovered that cos θ = .78.

CONCLUSION: The power factor of a reactive circuit is equal to:

$$\cos \theta \qquad \text{or} \qquad \frac{R}{Z}$$

POWER FACTOR OF A CAPACITOR

The determination of the quality of a capacitor (Q) is rated by its leakage which permits a small current to flow. This causes wasted power and heat. The capacitor is better if the power factor is at a very low value. Since the leakage is through the dielectric, this is sometimes called the dielectric loss or the power factor of the dielectric. Capacitor checking instruments will measure this loss and make compensations for it when measuring capacitance values.

APPLICATIONS OF RC CIRCUITS

The circuit in Fig. 9-26 has a dc power source of 10 volts in series with an ac signal source of 10 volts.

Consider first the circuit without the ac voltage. C will become charged to 10 volts. If the voltages across R and C were then measured, E_R = 0 volts, E_C = 10 volts. The capacitor blocks the dc, and there is no dc output across R. Now the ac source is applied in series with the dc power source. If the reactance of C is 1000 ohms and R is 1000 ohms the voltage across C will be approximately 5 volts ac. It will also be 5 volts ac across R.

NOTE: Any out-of-phase condition is omitted to simplify the calculations.

If the frequency of the ac is decreased or the value of C is decreased so that the reactance becomes 10,000 ohms, then the ac voltage division may be approximated by:

$$E_c = E_s \times \frac{X_c}{X_c + R} = 10 \times \frac{10,000}{11,000} = 9 \text{ volts ac}$$

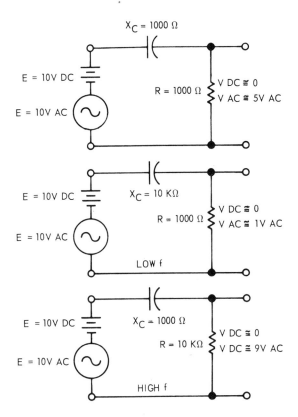

Fig. 9-26. These circuits demonstrate the effect of frequency on the output of a high-pass filter.

and only about one volt of signal voltage will appear across R for the circuit output.

Now, assume that X_c is returned to 100 ohms and R is increased to 10 kilohms. Then, the signal output across R will be:

$$E_R = E_s \times \frac{R}{X_c + R} = 10 \times \frac{10,000}{11,000} = 9 \text{ volts ac}$$

If frequency is increased, X_c will decrease and even a greater proportion of signal voltage will appear across R. This circuit passes high frequencies better than low frequencies. It is a HIGH-PASS FILTER.

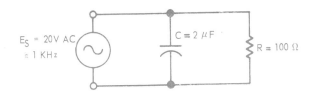

Fig. 9-27. A parallel RC circuit used for text problem.

This circuit will be used later in the RC coupling of amplifers. To assure maximum output, the value of R should be ten or more times greater than the value of X_c at operating frequency.

RC IN PARALLEL CIRCUITS

In a parallel circuit, the voltage across each component is the same. In Fig. 9-27, it is the source voltage E_s.

However, the currents are out of phase. The total effect is computed in the following problem, with assigned values of R and C.

1. Compute value of X_c:

$$X_c = \frac{.16}{fC} = \frac{.16}{10^3 \times 2 \times 10^{-6}}$$

$$= .08 \times 10^3 = 80 \text{ ohms}$$

2. Find branch currents:

$$I_c = \frac{E_s}{X_c} = \frac{20V}{80 \, \Omega} = .25 \text{ amp}$$

$$I_R = \frac{E_s}{R} = \frac{20V}{100 \, \Omega} = .2 \text{ amp}$$

and

$$I_c + I_R \neq I_{line}$$

because I_c is 90 deg. out of phase with I_R.

3. Find I_{line} or I_z. The current vectors are drawn to a convenient scale in Fig. 9-28.

Again, the current I_R is drawn horizontally as a reference. Because I_c leads the current through R, the I_c vector is drawn upward at an angle of 90 deg.

I_z may be found by $I_z^2 = I_R^2 + I_c^2$

$$\text{or } I_z = \sqrt{I_R^2 + I_c^2}$$

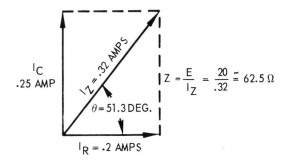

Fig. 9-28. The vector diagram for parallel RC problems in text.

But trigonometry will be used:

$$\tan \theta = \frac{I_C}{I_R} = \frac{.25}{.2} = 1.25$$

$$\text{arc} \tan 1.25 = 51.3 \text{ deg.}$$

$$\cos 51.3 \text{ deg.} = .6252$$

$$\cos \theta = \frac{I_R}{I_Z} \qquad I_Z = \frac{I_R}{\cos \theta}$$

$$= \frac{.2}{.6252} = .32 \text{ amp}$$

4. Find Z.

$$Z = \frac{E}{I_Z} = \frac{20}{.32} = 62.5 \text{ ohms}$$

CONCLUSION: The total impedance of the parallel RC circuit is 62.5 ohms and the circuit current leads the applied voltage by an angle of 51.3 deg.

RC BYPASSING

In Fig. 9-29, combination circuit of R and C is connected to a dc and an ac source voltage. First, consider the dc only. There is a 5 volt dc drop across R_1 and a 5 volt drop across R_2. No dc currents will flow through C. Now, consider the ac only. The ac voltage will divide according to the resistance of R_1 and the impedance of R_2 and C in parallel. When the ratio of R_2 to X_C is ten times or more, the total parallel impedance is very close to the value of X_C only

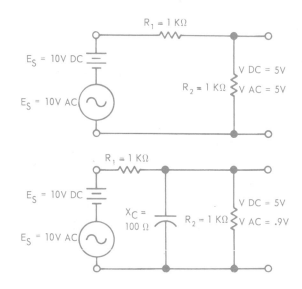

Fig. 9-29. This circuit shows the effect of bypassing an ac signal.

and may be considered as 100 ohms. Then, 10/11 of the ac voltage will appear across R_1 and only 1/11 of the ac voltage across R_2.

$$E_{R_1} = E_{ac} \times \frac{R_1}{X_C + R_1} = 10 \times \frac{1000}{1100} = 9.1 \text{ volts}$$

$$E_{X_C} = E_{ac} \times \frac{X_C}{X_C + R_1} = 10 \times \frac{100}{1100} = .9 \text{ volt}$$

When both ac and dc voltages are applied to the circuit, the voltage change across R_2 is only .9 volt and can be considered almost constant. The signal ac has been BYPASSED around R_2 and has a minimum effect on the voltage across R_2.

LESSON IN SAFETY: Use the right tool for the job at hand. The improper tool selection may cause damage to the equipment and personal injury.

LESSON IN SAFETY: Check and re-check replacement components to be sure they are the correct wattage or power rating. Be certain that components such as electrolytic capacitors are connected with the correct polarity. Errors in connection can produce smoke, fumes, short circuits and burns.

TEST YOUR KNOWLEDGE

1. An increase in plate area of a capacitor will _____ its capacitance.
2. An increase in distance between plates of a capacitor will _____ its capacitance.
3. What is the capacitance of .01 μF, .02 μF and .05 μF capacitors connected in series?
4. Using the same capacitor values in question 3, what is the total capacitance if connected in parallel?
5. Calculate the RC time constant of a series circuit of a 2.2 megohm resistor and a 10 μF capacitor.
   Compute the reactance of a .05 μF capacitor:
6. At 100 Hz, X_c is _____
7. At 1 KHz, X_c is _____
8. At 10 KHz, X_c is _____
9. At 100 KHz, X_c is _____
   A series circuit of a 10 K Ω resistor and a .01 μF capacitor are connected to a 50V, 1 KHz source. Draw the circuit and vector diagrams, then:
10. Find X_c _____
11. Find Z _____
12. Find I_T _____
13. Find θ _____
14. Find E_L _____
15. Find E_R _____
16. Find PF _____
17. A 500 ohm resistor is bypassed by a capacitor. The operating frequency is 100 Hz. What is the value of C in microfarads?
18. If 100 volts is applied to the circuit in Fig. 9-16, the NE2 lamp should flash at what approximate frequency if R is 1 megohm and C is 5 μF? Build the circuit and prove your calculations.

Chapter 10

RCL CIRCUITS

In the previous two chapters, the properties and effects of both RL and RC circuits in series and parallel circuits have been studied. In many typical electronic circuits, inductance, capacitance and resistance will be found at one time, either in series or parallel. At first, the computation of these circuits may seem difficult, but really it is only an extension of the work which you have already accomplished.

In this chapter, you will:

1. Study the voltage and current relationships in a Series RCL Circuit.
2. Acquire an understanding of Series Resonant Circuits and their applications.
3. Study the voltage and current relationships in Parallel RCL Circuit.
4. Become familiar with Parallel Resonant Circuits and their applications.

SERIES RCL CIRCUITS

A series circuit containing R, C and L is diagrammed in Fig. 10-1.

Values have been given to each component. We will find the opposing forces to the flow of current, such as R, X_L and X_C, and then add them together vectorially to discover the total circuit impedance.

1. Find X_L and X_C.

L = 8H and C = .01 μF

$X_L = 2\pi fL = 6.28 \times 10^3 \times 8 = 50,240$ ohms

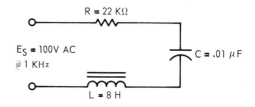

Fig. 10-1. The series RCL circuit used in the text problem.

$$X_C = \frac{1}{2\pi fC} = \frac{.159}{10^3 \times 1 \times 10^{-8}}$$

$$= .159 \times 10^5 = 15,900 \text{ ohms}$$

2. In Fig. 10-2, these vectors are drawn:

R = 22 kilohms horizontally as a reference for current and voltage.

X_L = 50,240 ohms upward at 90 deg. This voltage leads circuit current by 90 deg.

X_C = 15,900 ohms downward to 90 deg. This voltage lags circuit current by 90 deg.

(Any convenient scale may be used.)

Now that X_L and X_C are 180 deg. out of phase, they may be added together algebraically.

$$X = X_L - X_C$$

$$= 50,240 \ \Omega - 15,900 \ \Omega = 34,340 \text{ ohms}$$

The vector diagram is redrawn to show this change.

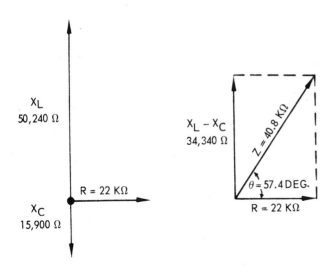

Fig. 10-2. The vector diagram used in the text problem.

3. The circuit impedance Z and θ are found exactly in the same manner demonstrated in Chapter 8.

$$\tan \theta = \frac{X}{R} = \frac{34,340}{22,000} = 1.56$$

arc tan 1.56 = 57.4 deg. = θ

cos θ = .5388

$$Z = \frac{R}{\cos \theta} = \frac{22,000}{.5388} \cong 40,800 \text{ ohms}$$

4. Find the circuit current and voltage drops.

$$I = \frac{E}{Z} = \frac{100}{40,800} = .0024 \text{ amp}$$

$$E_R = IR = .0024 \times 22,000 = 52.8V$$

$$E_L = IX_L = .0024 \times 50,240 = 120V$$

$$E_C = IX_C = .0024 \times 15,900 = 38V$$

5. Proof.

$$E_L - E_C = 120 - 38 = 82V = E_X$$

$$E_S = \sqrt{E_R{}^2 + (E_L - E_C)^2} \text{ or } \cos \theta = \frac{E_R}{E_Z}$$

and

$$E_Z = \frac{E_R}{\cos \theta} \qquad E_Z = \frac{52.8}{.5388} \cong 100V$$

NOTE: Slight errors are due to rounding off figures.

It is to your advantage to realize what a change of frequency of the input voltage will do to this circuit:

1. As frequency is increased, X_L will become larger and X_C smaller. As a result, the circuit becomes even more INDUCTIVE and angle theta increases and the voltage across L will increase.
2. As frequency decreases, X_L will become smaller and X_C larger and angle theta will decrease toward zero.
3. At one frequency X_L will equal X_C and theta will be zero. This will be discussed later in this chapter under RESONANCE.
4. A further decrease in frequency will make X_C larger than X_L and the circuit will become CAPACITIVE and theta will increase in a negative direction.

These changes are summarized and illustrated in Figs. 10-3 and 10-4.

SERIES RESONANCE

In the graph of Fig. 10-5, the reactance of L and C are plotted as a function of frequency. X_L increases as frequency is increased and X_C decreases as frequency is increased. The frequency at which the curves X_L and X_C cross is called the RESONANT FREQUENCY designated as f_o. At this point $X_L = X_C$.

Before discussing series resonance, another look at Fig. 10-5 will disclose other information frequently overlooked or not understood by the beginning student.

1. X_C at zero frequency (dc) approaches INFINITY. It is an open circuit for dc.

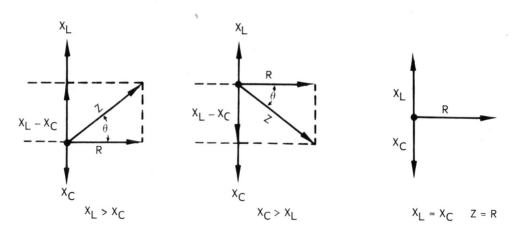

Fig. 10-3. Vector diagrams describing the change of frequency effect on a RCL circuit.

2. X_c at a very high frequency approaches zero, which is a short circuit.

3. X_L is zero, or equivalent to a short circuit at zero frequency, or dc.

4. At high frequencies, X_L approaches infinity, or an open circuit.

These fundamental relationships are basic to circuit design and understanding.

In order to determine the frequency at which $X_L = X_c$ (resonant frequency), follow these steps:

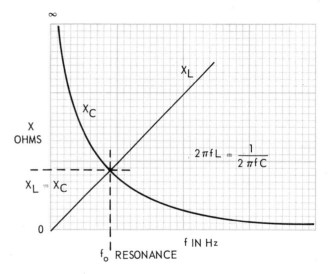

Fig. 10-5. At resonance, X_L equals X_C.

1. At resonance, $X_L = X_c$. However:

$$X_L = 2\pi fL \text{ and } X_c = \frac{1}{2\pi fC}$$

Therefore, $2\pi fL = \frac{1}{2\pi fC}$ at resonance.

2. Transpose f to the left side of the equation and $2\pi L$ to the right side.

$$f \times f = \frac{1}{2\pi L \times 2\pi C} \text{ and } f^2 = \frac{1}{(2\pi)^2 LC}$$

3. Extract the square root of both sides of the equation.

FOR FREQUENCIES ABOVE RESONANCE

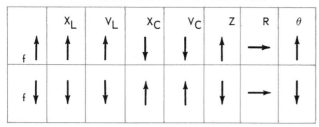

	X_L	V_L	X_C	V_C	Z	R	θ
f ↑	↑	↑	↓	↓	↑	→	↑
f ↓	↓	↓	↑	↑	↓	→	↓

FOR FREQUENCIES BELOW RESONANCE

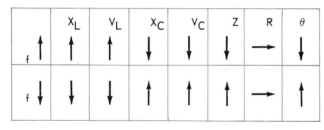

	X_L	V_L	X_C	V_C	Z	R	θ
f ↑	↑	↑	↓	↓	↓	→	↓
f ↓	↓	↓	↑	↑	↑	→	↑

Fig. 10-4. Table for frequencies above and below resonance.

$$f_o = \frac{1}{2\pi\sqrt{LC}}$$

This is the resonant frequency equation. For practical purposes, the equation may be simplified to:

$$f_o = \frac{.159}{\sqrt{LC}} \quad \text{or} \quad \frac{.16}{\sqrt{LC}}$$

By rearrangement of the resonant frequency equation, the value of either L in henrys or C in farads may be found, which will produce resonance at a given frequency.

$$C = \frac{1}{4(\pi)^2 f^2 L} \quad \text{or} \quad L = \frac{1}{4(\pi)^2 f^2 C}$$

PROBLEM: What is the resonant frequency of an inductor of 200 μH and a 200 pF capacitor?

$$f_o = \frac{.16}{\sqrt{2 \times 10^{-4} \times 2 \times 10^{-10}}} = \frac{.16}{2 \times 10^{-7}}$$

$$= .08 \times 10^7 = 800,000 \text{ hertz}$$

or 800 kilohertz or .8 megahertz

IMPEDANCE OF SERIES RESONANT CIRCUIT

Fig. 10-6 shows a RCL circuit and a vector diagram of the resistance and reactances at a frequency when $X_L = X_C$, or at resonance. Since X_L and X_C are equal and 180 deg. out of phase, the algebraic sum of X_L and X_C is ZERO. The only opposition of current flow in the circuit is R. At series resonance:

$$Z = R$$

At frequencies above resonance, X_L is greater than X_C. The circuit becomes inductive and Z increases. At frequencies below resonance, X_C is greater than X_L. The circuit becomes capacitive and Z increases. The point of lowest impedance of the circuit is at resonance.

Since the circuit permits the maximum current flow at resonant frequency, it is called an ACCEPTOR CIRCUIT. This is graphically illustrated in Fig. 10-7, which shows minimum Z and maximum current I at f_o. At frequencies

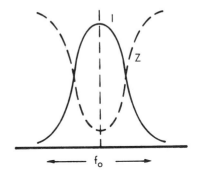

Fig. 10-7. Curves showing current and impedance of a series tuned circuit.

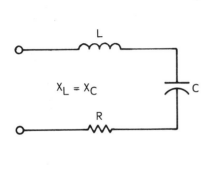 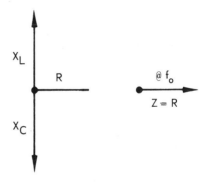

Fig. 10-6. At resonance, Z equals R.

either above or below resonance, the impedance increases and the current decreases.

Q OF SERIES RESONANT CIRCUIT

The letter Q is used to designate the quality of a tuned circuit. It is an indication of its maximum response as well as its ability to respond within a band of frequencies.

To secure maximum currents and response, the resistance must be kept at a low value. At resonance, R is the only resistance in the circuit. The Q of a circuit is the relationship of the reactance of the circuit to its resistance. X_L usually is used for reactance.

$$Q = \frac{X_L}{R} = \frac{X_C}{R}$$

This may be stated as the relationship between the stored energy in the circuit to the energy used by resistance.

$$Q = \frac{E_L}{E_R}$$

This equation also tells us that the voltage gain in the resonant circuit depends upon the Q of the circuit.

ANALYSIS OF SERIES RESONANT CIRCUIT

A circuit with assigned values is shown in Fig. 10-8. The current and voltage relationships will be computed:

1. Compute resonant frequency of circuit. This was done in an earlier problem and found to be 800 KHz (approx.).
2. Find reactances of L and C. They are equal at f_o, and the reactance X_L is easier to find than X_C.

$$X_L = 2\pi fL = 6.28 \times 8 \times 10^5 \times 2 \times 10^{-4}$$

$$\cong 1045\Omega \quad X_L = X_C \cong 1045 \text{ ohms}$$

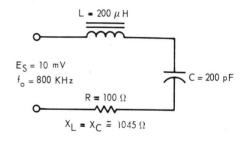

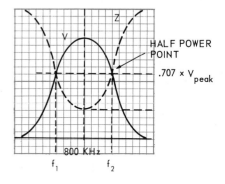

Fig. 10-8. The circuit and response curves for text problem.

3. Find circuit current.

$$I = \frac{E}{Z} = \frac{10mV}{100\ \Omega} = .0001 \text{ amp or } .1 \text{ milliamp}$$

4. Find voltage drops around circuit.

$$E_R = IR = .0001 \times 100 = .01 \text{ volt} = E_S$$

$$E_L = IX_L = .0001 \times 1045 = .1045 \text{ volt}$$

$$E_C = IX_C = .0001 \times 1045 = .1045 \text{ volt}$$

Where did we get these higher voltages? It is the result of the exchange of energy between L and C at resonance.

5. Find circuit Q.

$$Q = \frac{X_L}{R} = \frac{1045}{100} = 10.45$$

And Q has already been expressed as:

$$Q = \frac{E_L}{E_R} \text{ and } Q \times E_R = E_L$$

and $Q \times E_s = E_L$

so,

$$E_L = Q \times E_s = 10.45 \times .01V = .1045 \text{ volt}$$

$$E_c = Q \times E_s = 10.45 \times .01V = .1045 \text{ volt}$$

6. Find the bandwidth.

$$Bw = \frac{f_o}{Q} = \frac{8 \times 10^5}{10.45}$$

$$= .765 \times 10^5 = 76,500 \text{ hertz}$$

Lower frequency =

$$f_1 = f_o - \frac{Bw}{2} = 800,000 \text{ Hz} - 38,250 \text{ Hz}$$

$$= 761,750 \text{ hertz}$$

Upper frequency =

$$f_2 = f_o + \frac{Bw}{2} = 800,000 \text{ Hz} + 38,250 \text{ Hz}$$

$$= 838,250 \text{ hertz}$$

The BANDWIDTH of a tuned circuit is defined as those frequency limits above and below resonant frequency where the response of the circuit will drop to .707 of its peak response. If current or voltage drops to .707 of its peak value, the power drops to 50 percent.

Bandwidth, then, is the frequency above and below resonance where power drops to one-half of its peak value. These are called the HALF POWER POINTS. Expressed in decibels, the technician says "3 db down."

EXAMPLE: Given I = 1 amp, R = 10 ohms,

$$P = I^2 R = (1)^2 \times 10 = 10 \text{ watts}$$

If current drops to .707 x 1, then:

$$P = (.707)^2 \times 10 = .5 \times 10 = 5 \text{ watts}$$

A signal with a frequency either above or below the band limits is considered as RE-JECTED by the tuned circuit.

BANDWIDTH AND Q

Using the same circuit, the response and bandwidth for two values of R in the circuit are plotted and computed in Figs. 10-9 and 10-10. When the resistance is reduced: the Q increases;

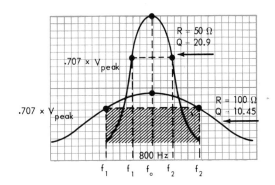

Fig. 10-9. These curves show the effect on response and bandwidth when the Q of the circuit is changed.

f	X_L Ohms	X_C Ohms	Z Ohms	I μA	E_{X_L} μV	E_{X_C} μV
200 KHz	250	4000	3750	.13	33	520
400 KHz	500	2000	1500	.33	165	660
600 KHz	750	1500	750	.66	495	990
800 KHz	1000	1000	10	5.0	50,000	50,000
1000 KHz	1250	800	450	1.1	1,375	880
1200 KHz	1500	666	833	.6	900	400
1600 KHz	2000	500	1500	.33	660	165

Fig. 10-10. Table shows changes that occur in the series tuned circuit as frequency is varied above and below resonance.

the response of the circuit increases; the band of frequencies falling within the half power points is greatly reduced.

A high Q circuit is sharply tuned for maximum response. A low Q circuit is BROAD BAND and has a low circuit gain and will pass a larger band of frequencies.

Consider the voltages developed across L or C in both Q and high Q circuits. In our first example:

$$E_c = E_L = QE_s = 10.45 \times .01$$

$$= .1045 \text{ volts}$$

When the Q is raised to 20.9 by reducing the value of R to 50 ohms, then:

$$E_c = E_L = QE_s = 20.9 \times .01$$

$$= .209 \text{ volts}$$

The bandwidth also becomes less.

$$Bw = \frac{f_o}{Q} = \frac{800 \text{ KHz}}{20.9}$$

$$= 38.3 \text{ KHz or } 38,300 \text{ hertz}$$

NOTE: The "skirts of the band response curve" become steeper, and circuit is much more selective.

TUNING THE SERIES RCL CIRCUIT

In Fig. 10-11, another RCL circuit is connected to a variable frequency generator. By measuring voltages, the resonant frequency may be determined. Apply the previous lessons.

1. Circuit current is maximum at resonance.
2. E_R = generator voltage E_G at resonance.
3. $E_L = E_C$ at resonance.

NOTE: These conclusions are only approximately true, since the resistance of the wire of

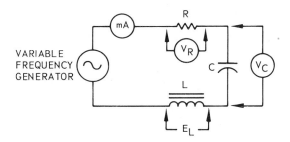

Fig. 10-11. The voltage and current values at series resonance.

the inductance is not considered. In Fig. 10-10, the voltages and currents in a series RCL circuit are tabulated as frequency is varied above and below resonance.

PARALLEL RCL CIRCUITS

A parallel RCL circuit is drawn schematically in Fig. 10-12. This is not a practical circuit because inductor L will have some resistance and C may have some leakage resistance. The

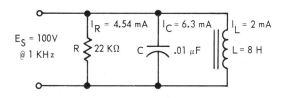

Fig. 10-12. A parallel RCL circuit used in text problem.

circuit is simplified, so that you will understand the principles involved in the solution of such a problem.

First, what do we already know about this circuit?

1. The voltage across each branch of the parallel circuit is the same and equal to E_s.
2. The current through R is in phase with circuit voltage.
3. The current through C is leading the circuit voltage.
4. The current through L is lagging the circuit voltage.
5. In order to find the total circuit current, we must make a vector addition of I_R, I_C and I_L.

Knowing thses facts, proceed with the problem. Values have been assigned to the components.

1. Find the reactance of L at circuit frequency.

$$X_L = 2\pi fL = 6.28 \times 10^3 \times 8 = 50,240 \text{ ohms}$$

2. Find the reactance of C at circuit frequency.

$$X_C = \frac{1}{2\pi fC} = \frac{.159}{10^3 \times 1 \times 10^{-8}}$$

$$= .159 \times 10^5 = 15,900 \text{ ohms}$$

3. Find branch currents.

$$I_R = \frac{E}{R} = \frac{100V}{22 \times 10^3} = .00454 \cong 4.54 \text{ milliamps}$$

$$I_L = \frac{E}{X_L} = \frac{100V}{50,240} = .002 \text{ amp} = 2 \text{ milliamps}$$

$$I_C = \frac{E}{X_C} = \frac{100V}{15,900} = .0063 \text{ amp} = 6.3 \text{ milliamps}$$

4. Plot the current vectors as in Fig. 10-13.

The current and voltage are in phase in R, so the R vector is drawn horizontally as a reference. Any convenient scale may be used for the vectors.

   a. R is drawn 4.54 units long horizontally.

   b. I_C is leading by 90 deg. and is drawn 6.3 units long upward.

   c. I_L is lagging by 90 deg. and is drawn 2 units long downward.

5. I_C and I_L are added algebraically since they are 180 deg. out of phase. The resultant and current is plotted and is 4.3 units long. $(6.3 - 2 = 4.3)$

The circuit current I_z may be found by the Pythagorean Theorem:

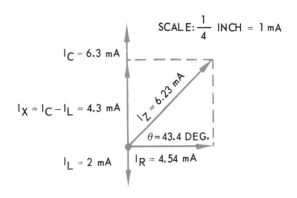

Fig. 10-13. The vector diagram for RCL circuit problem.

$$I_z^2 = I_R^2 + I_X^2 \text{ or } I_z = \sqrt{I_R^2 + I_X^2}$$

It may also be measured graphically, if you have accurately drawn your vectors to scale.

6. Find angle θ by trigonometry.

$$\tan\theta = \frac{I_X}{I_R} = \frac{4.3}{4.54} = .9471$$

$$\text{arc tan } .9471 = 43.4 \text{ deg.} = \theta$$

$$\cos\theta = \cos 43.4 \text{ deg.} = .7266$$

7. Find circuit current I_z.

$$\cos\theta = \frac{I_R}{I_z} \qquad I_z = \frac{I_R}{\cos\theta}$$

$$= \frac{4.54}{.7266} = 6.23 \text{ milliamps}$$

8. Find the circuit impedance.

$$Z = \frac{E}{I_z} = \frac{100}{6.23 \times 10^{-3}} \cong 16,000 \text{ ohms}$$

CONCLUSIONS:

1. The parallel RCL circuit has an impedance of 16,000 ohms and the circuit current is 6.23 milliamps, which leads the applied voltage by an angle of 43.4 deg. The circuit is capacitive.

2. $I_R + I_C + I_L \neq I_{circuit}$. I_R, I_C and I_L are out of phase. They must be added by vectors.

In Fig. 10-14, the values of voltage and current as a function of frequency are tabulated, above and below resonance. Before proceeding further, study the table and be sure that you understand these values.

PARALLEL RESONANCE

The frequency at which X_C and X_L become equal in a parallel circuit is resonance, which may be computed by the same formula used for series circuits.

$$f_o = \frac{.159}{\sqrt{LC}} \quad \text{or} \quad \frac{.16}{\sqrt{LC}}$$

Fig. 10-15 illustrates a simple RL parallel circuit and a voltage source. When switch S is closed, C charges instantaneously to the source voltage. Then S is opened. C starts to discharge through L and the discharge current creates a magnetic field around L. As the charge on C approaches zero, the current through L starts to

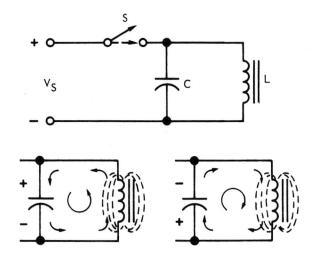

Fig. 10-15. Oscillation is produced by the discharge of C through L in the tank circuit.

decrease and the magnetic field collapses, causing an induced voltage which charges C in the opposite polarity. Now C attempts to discharge in the opposite direction. The discharge current again builds the magetic field, which collapses as the current approaches zero. The induced voltage charges C to its original polarity.

This action is called OSCILLATION, and one cycle of oscillation has been described. If it were not for friction (in this case, resistance),

FOR FREQUENCIES ABOVE RESONANCE

f	X_L	I_L	X_C	I_C	Z	R	θ	
↑	↑	↓	↓	↑	↑	→	↑	CIRCUIT IS CAPACITIVE
↓	↓	↑	↑	↓	↓	→	↓	

FOR FREQUENCIES BELOW RESONANCE

f	X_L	I_L	X_C	I_C	Z	R	θ	
↑	↑	↓	↓	↑	↑	→	↓	CIRCUIT IS INDUCTIVE
↓	↓	↑	↑	↓	↓	→	↑	

Fig. 10-14. Table shows values of voltage and current as a function of frequency tabulated above and below resonance.

the oscillations would continue indefinitely and perpetual motion would have been invented. But during each cycle of oscillation, some of the initial energy stored in C will be dissipated by the circuit resistance. Also, each succeeding oscillation will become less and less until they die out. This is called a DAMPED OSCILLATION.

A damped oscillation wave form of either voltage or current is displayed in Fig. 10-16. The circulation of current in the LC circuit is likened to the sloshing of water back and forth

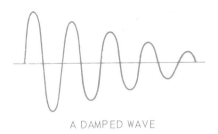

A DAMPED WAVE

Fig. 10-16. The wave decreases in amplitude due to energy used by circuit resistance.

in a tank. It is sometimes referred to as "FLYWHEEL ACTION" and the circuit is a "TANK CIRCUIT." The frequency of the tank circuit is found by the resonant frequency formula.

Remember that the energy required to maintain circuit oscillation depends upon the circuit resistance that uses the energy. Therefore, a parallel tuned circuit that has a very low series resistance will require very little current to sustain its oscillations at resonance. The impedance of the parallel circuit is very high at resonance. It is called a REJECT circuit. It offers a high impedance to signals at or near its resonant frequency.

DAMPED WAVE DEMONSTRATOR

Fig. 10-17 illustrates a circuit by which you can observe the tank action and damped wave oscillations of a tuned circuit. The tank circuit is energized by pulses of voltage from across C, which is part of a relaxation oscillator (Chapter 9, Fig. 9-16). The ouput from the tuned circuit may be observed on the oscilloscope. Since the resonant frequency of the tank is much higher than the operating frequency of the relaxation oscillator, the oscillations in the tank will die out between energy pulses. A damped wave will be seen.

Q OF A PARALLEL RESONANT CIRCUIT

As you may suspect, the quality or Q of this tuned circuit depends upon its resistance. Most

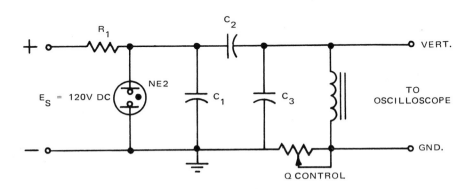

PARTS LIST FOR DAMPED WAVE OSCILLATOR

R_1 — 2.2 megohm resistor
R_2 — 1 K Ω potentiometer
C_1 — .05 μF paper capacitor
C_2 — .01 μF paper capacitor

C_3 — .1 μF paper capacitor
L — 1.5 H choke
NE2 — glow lamp
wires and terminals

Fig. 10-17. A damped wave demonstration circuit. (Charles Schuler, California State College, California, Pennsylvania)

of the resistance is found in the wire of the coil and, in Fig. 10-18, this is represented by a series resistor. Again the Q of the circuit is the relationship between the reactance of either X_c or X_L to the resistance and is expressed as:

$$Q = \frac{X_L}{R} \quad \text{or} \quad \frac{X_c}{R}$$

ANALYSIS OF RCL CIRCUIT AT RESONANCE

Fig. 10-18 is provided for use in analyzing a parallel RCL circuit at resonance. Assigned values that appear on the circuit were used in previous examples of series resonant circuits.

1. Find resonant frequency.

$$f_o = \frac{.159}{\sqrt{LC}} = 800 \text{ KHz}$$

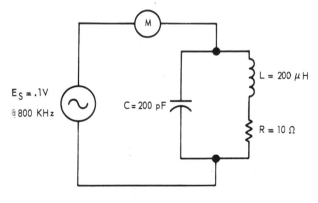

Fig. 10-18. Circuit used in text problem for analysis of tuned RCL circuit.

2. Find reactances of C and L.

$$X_L = 2\pi fL = 1045 \, \Omega = X_c$$

3. Since the value of R is very low in comparison to X_L, it will be disregarded in the computation of the branch currents. Then:

$$I_c = \frac{E_G}{X_c} \quad \text{and} \quad I_L = \frac{E_G}{X_L} \qquad I_L = I_c$$

and they are equal and cancel each other.

4. The Q of the circuit is:

$$Q = \frac{X_L}{R} = \frac{1045}{10} = 104.5$$

5. In the series resonant circuit, the voltage gain or magnification factor of the circuit was:

$$E_{X_L} \quad \text{or} \quad E_{X_c} = QE_s$$

In the parallel tank circuit, the magnification factor will apply to the impedance of the tank circuit and will determine the line current.

$$Z_T = Q \times X_L \quad \text{at resonance}$$

In the problem, the impedance of the tuned circuit will equal:

$$Z_T = 104.5 \times 1045 \, \Omega = 109,203 \text{ ohms}$$

The actual line current can be computed as:

$$I_{line} = \frac{E_s}{Z_T} = \frac{.1V}{109,203 \, \Omega} \cong .9 \text{ microamp}$$

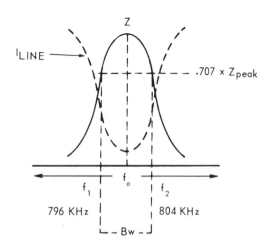

Fig. 10-19. A response curve of a tuned parallel circuit showing impedance, line current and bandwidth.

This is the only current required from the source at resonance. At frequencies above or below resonance, X_L and X_C are not equal and do not cancel. The difference in current must be supplied from the source.

A meter inserted in the line can be used to indicate the approximate resonant frequency of the circuit because when tuned to resonance, the meter will "dip" or read a minimum value. Current will rise sharply above or below resonance. This principle is widely used in tuning tank circuits, and you should be most familiar with it.

BANDWIDTH OF PARALLEL TUNED CIRCUIT

In Fig. 10-19, the response curve of the parallel tuned circuit is plotted. Note that at resonance, its impedance is maximum and the circuit current is negligible. At frequencies other than resonance, the impedance drops off sharply on each side. The points above and below resonant frequency at which the maximum impedance drops to .707 of its maximum value is considered to be the bandwidth of the circuit. This may be computed as:

$$Bw = \frac{f_o}{Q}$$

In our problem, this is:

$$Bw = \frac{800 \text{ KHz}}{104.5} \cong 7650 \text{ Hz}$$

The bandwidth is divided equally above and below resonance, so the lower frequency is found by:

$$f_1 = f_o - \frac{Bw}{2} = 800 \text{ KHz} - \frac{7650}{2}$$

$$= 796,175 \text{ hertz}$$

$$f_2 = f_o + \frac{Bw}{2} = 800 \text{ KHz} + \frac{7650}{2}$$

$$= 803,825 \text{ hertz}$$

EFFECT OF Q ON BANDWIDTH

If the circuit series resistance were increased to 100 ohms, then the Q would become:

$$Q = \frac{X_L}{R} = \frac{1045}{100} = 10.45$$

and the circuit impedance at resonance would drop to:

$$Z = Q \times X_L = 10.45 \times 1045 \cong 10,900 \text{ ohms}$$

and the bandwidth would increase to:

$$Bw = \frac{800 \text{ KHz}}{10.45} \cong 76,500 \text{ hertz}$$

This is shown on the graph in Fig. 10-20. The circuit becomes LESS SELECTIVE and HAS A LOWER IMPEDANCE.

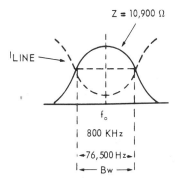

Fig. 10-20. The curves show the wider bandwidth produced by a lower Q circuit.

In Fig. 10-21, the figures are computed for voltages and currents at resonance and off-resonance.

SHUNT DAMPING

The response of a parallel tuned circuit may also be affected by a resistance in parallel with the circuit as in Fig. 10-22. It is designated as R_s. The current through R_s is independent of frequency and its effect cannot be cancelled by the resonant conditions of the tank circuit. The

f KHz	X_L Ω	X_C Ω	I_{X_L} μA	I_{X_C} μA	I_{line} μA
200	250	4000	2.0	.125	1.875
400	500	2000	1.0	.25	1.75
600	750	1500	.66	.33	.33
800	1000	1000	.5	.5	0
1000	1250	800	.4	.625	.225
1200	1500	666	.33	.75	.42
1600	2000	500	.25	1.00	.75

Fig. 10-21. Table showing figures computed for voltages and currents at resonance and off-resonance.

line current cannot "dip" below this value and the circuit appears to be more resistive. This does change the Q of the circuit.

$$Q = \frac{R_s}{X_L}$$

When R_s is 150 kilohms, then the circuit in Fig. 10-22 has a Q of:

$$Q = \frac{150,000}{1500} = 100$$

Recall that the formula for bandwidth of a tuned circuit is:

$$Bw = \frac{f_o}{Q}$$

and since Q for the shunt damped circuit is:

$$Q = \frac{R_s}{X_L} \quad \text{then} \quad Bw = \frac{f_o}{\dfrac{R_s}{X_L}}$$

or $\dfrac{f_o \times X_L}{R_s}$

The bandwidth for this circuit becomes:

$$Bw = \frac{800 \text{ KHz} \times 1500}{150 \text{ K}\,\Omega}$$

$$= \frac{8 \times 10^5 \times 1.5 \times 10^3}{1.5 \times 10^5}$$

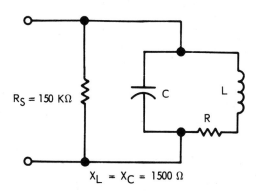

$R_S = 150$ KΩ

$X_L = X_C = 1500$ Ω

Fig. 10-22. A resistor R_S is connected for shunt damping the tuned circuit. The Q is lower and the bandwidth increased.

$$= 8 \times 10^3 = 8000 \text{ hertz}$$

The circuit response can be broadened by lowering the shunt resistance. A change of R_s to 100 kilohms produces:

$$Q = \frac{100,000}{1500} \cong 66$$

$$Bw = \frac{800 \text{ KHz} \times 1500}{100 \text{ K}\,\Omega}$$

$$= \frac{8 \times 10^5 \times 1.5 \times 10^3}{1 \times 10^5}$$

$$= 12 \times 10^3 = 12,000 \text{ hertz}$$

LC RATIO AT RESONANCE

An examination of the resonant frequency formula will disclose that this frequency is a function of the product of L and C. There would be an infinite number of combinations of

L and C which would produce resonance at a given frequency. It is possible to bring a circuit into resonance by varying either L or C or both. Although it is beyond the scope of this text to include all the mathematics necessary to prove the effects of varying C and L, it would only require the working out of the previous problems with other values of C and L.

TEST YOUR KNOWLEDGE

1. A series resonant circuit is called an _____ circuit.
2. A parallel resonant circuit is called a _____ circuit.
3. At resonance, $X_L = X_C$ and $Z =$ _____.
4. What is the Q of a tuned circuit?
5. What is the formula for finding the Q of a series resonant circuit?
6. What is the resonant frequency of a 100 mH coil and a .001 F capacitor?
7. In question 6, $X_L =$ _____ .
8. If the circuit in question 6 has 50 ohms of resistance, what is the circuit Q?
9. What is the effect of an increase in the frequency of input voltage in a series RCL circuit?
   a. X_L will become larger and X_C smaller.
   b. X_L will become smaller and X_C larger.
10. The point of lowest impedance of a series RCL circuit is at resonance. True or False?
11. The frequency at which X_C and X_L become equal in a parallel circuit is _____ .
12. _____ is produced by the discharge of C through L in a tank circuit.
13. What determines the bandwidth of a tuned circuit?
14. What is the formula for finding the bandwidth of a tuned circuit?
15. How can the response of a parallel tuned circuit be broadbanded?

Chapter 11

AC-DC MOTORS

How many electric motors do you have in your home? They are used to keep your food cool in the refrigerator, to drive the circulating warm air in your heating system and to operate your washing machine and clothes dryer. Motors are used to mix our foods, brush our teeth and ventilate the kitchen. But look at industry! The wheels of production would remain at rest if it were not for the motor. There probably is no more useful effect of electricity than the conversion of electrical energy into mechanical energy by means of THE ELECTRIC MOTOR.

In this chapter, we will:

1. Explore the reaction between magnetic fields which will produce mechanical motion.
2. Study the principles of operation of the direct current motor.
3. Discover how an alternating current can be used to operate a motor.
4. Learn what methods are used to control the speed and turning power of a motor.

MOTOR ACTION

Basic to the understanding of all electric motors is the reaction between a fixed magnetic field and a current carrying conductor in that field. Before examining Fig. 11-1, you may wish to study Figs. 5-12 and 5-13 in Chapter 5. You must be assured that there is a circular magnetic field surrounding a current carrying conductor. The direction of this magnetic field of force will depend upon the direction of current flow. Use the LEFT HAND RULE.

In Fig. 11-1, the current in the conductor is assumed to flow toward you. This causes the circular magnetic field around the conductor to oppose the direction of the fixed field below the conductor. It also reinforces the field above the conductor, causing the conductor to move DOWNWARD toward the weakened field.

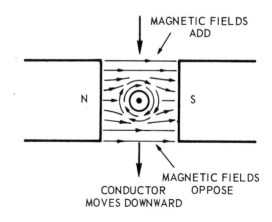

Fig. 11-1. The field of the conductor opposes the fixed field below the conductor. The conductor moves toward the weakened field.

In Fig. 11-2, the current in the conductor is assumed to flow away from you, and the circular field is in the opposite direction. In this case, the weakened field appears above the conductor, and the conductor will move UPWARD.

A demonstration of this action, called MOTOR ACTION, is simulated in the project shown in Fig. 11-3. This setup is only a suggestion. It can be constructed many other ways. The switch is a DPDT in order to reverse

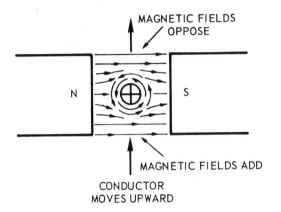

Fig. 11-2. The field of the conductor opposes the fixed field above the conductor. The conductor moves upward.

at a right angle to your fingers. Point your four fingers in the direction of the magnetic field, toward the SOUTH POLE. Point your thumb in the direction of current flow in the conductor. The conductor will move TOWARD the palm of your hand. Try this rule in the diagrams of Figs. 11-1, 11-2, 11-3 and 11-4, and prove to yourself that the right hand motor rule is true.

THE SIMPLE DC MOTOR

The construction of a simple dc motor is illustrated in Fig. 11-4. We will use this setup to

the current in the single conductor. However, do not let the circuit remain ON for more than a second or two. It is a direct short circuit across your battery or supply. The No. 8 copper wire is bent like a large letter U. Holes in the small brackets permit the copper U to swing back and forth. Any fixed magents may be used. The arrows indicate current flow in each position of the switch.

To determine the direction of movement of a conductor in a field, use this simple RIGHT HAND MOTOR RULE. Extend the fingers of your right hand in a flat plane with your thumb

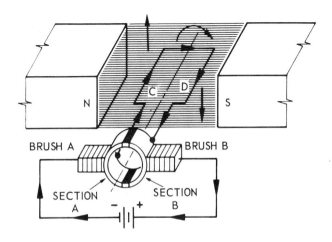

Fig. 11-4. The construction and theory of a simple dc motor.

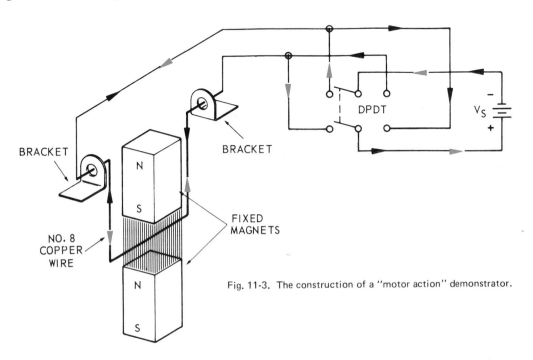

Fig. 11-3. The construction of a "motor action" demonstrator.

describe "motor action." Note that the single loop coil is arranged to rotate in the magnetic field. The rotating coil is the ARMATURE. The ends of the coil are connected to semicircular metal rings called the COMMUTATOR. The two half-sections of the commutator are insulated from each other, and the commutator revolves with the armature coil.

BRUSHES of carbon provide a sliding contact to the commutator sections to bring electric current from the battery into the armature coil. Note the current flow as indicated by the arrows in Fig. 11-4:

From battery negative to brush A.
Brush A to section A.
Section A into armature.
Out of armature to section B.
Section B to brush B
  to the battery positive.

The interacting fields will cause side C of the armature coil to move UPWARD and side D to move DOWNWARD. The coil will rotate 90 deg., and the coil sides will become parallel to the fixed field. At this moment, commutator section A comes in contact with brush B and section B comes in contact with brush A. With that, THE CURRENT IN THE ARMATURE IS REVERSED. Coil side C moves DOWNWARD and coil side D moves UPWARD. As the armature continues to rotate, the direction of armature current is changed each half revolution.

The position of the coil at each 90 deg. of rotation is diagrammed in Fig. 11-5. In this simple motor, the inertia of the rotating armature is used to carry the armature through the portion of rotation when the coil is parallel to the fixed field.

A practical motor armature will have many coils placed in slots around the armature core. Each set of coils will have its respective pair of commutator sections. This arrangement produces continuous torque and power, and inertia of the armature is no longer required to maintain revolution. Fig. 11-6 indicates typical construction of dc motor.

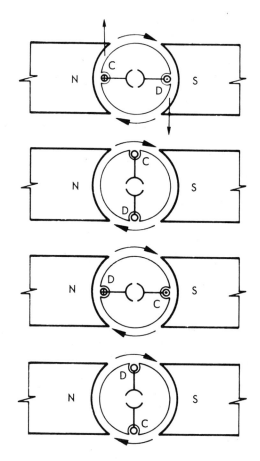

Fig. 11-5. The simple dc motor continues to rotate. The direction of current in the armature is changed each half revolution.

MOTOR TORQUE

In the case of the simple motor of Fig. 11-4, only the force of one wire in the fixed field was used to produce rotation. If this coil was

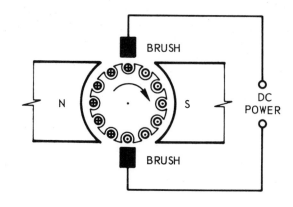

Fig. 11-6. Many armature coils are placed in slots around the armature core.

replaced by another with a large number of turns, the turning force or torque would greatly increase. In the case of Fig. 11-6, many coils are contributing continuously to the torque. The total torque of the motor would be the sum of all the contributing torques of individual conductors and coils.

FIELD STRENGTH

Since movement is produced by the interaction of a fixed field and the field around a current carrying conductor, it is only reasonable to assume that this force can be increased by INCREASING THE STRENGTH OF THE FIXED MAGNETIC FIELD. In order to do this, the permanent magnets previously used will be replaced with electromagnets. Mathematically speaking, the torque of a motor is in direct proportion to its field strength and the armature current. It is stated as:

$$Torque = K \Phi I_A$$

where,

K = a constant for a particular motor and represents the number of windings in the coils and the number of coils.

Φ = magnetic flux lines per field pole.

I_A = armature current.

In Fig. 11-17, the field windings of a shunt wound or parallel field motor are illustrated. The number of field coils is also increased to four, or two pair, and two sets of brushes are used. These improvements all contribute to the torque and power of the motor.

COUNTER EMF

A comparison of a motor and a generator reveals many similarities. In fact, a generator can be "motored" and a motor can "generate" if turned by mechanical power. However, certain design and construction features prohibit interchange for practical purposes. But to understand the motor theory, you must realize that as a motor is rotating, the armature and

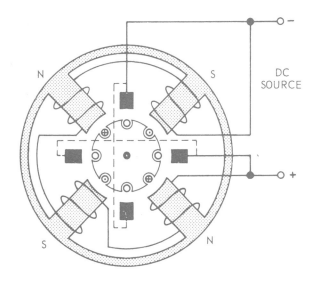

Fig. 11-7. The permanent field magnets are now replaced with electromagnets. Field coils are in parallel with armature.

conductors are cutting across the magnetic field and a VOLTAGE is induced. This is named COUNTER ELECTROMOTIVE FORCE or CEMF. This CEMF is proportional to the SPEED of the motor and its FIELD STRENGTH.

Consider why we are concerned with CEMF. Since it is in opposition to the source voltage used to drive the motor, the effective voltage applied to the armature is the algebraic sum of these two voltages.

$$E_{armature} = E_S - E_{CEMF}$$

Refer to Fig. 11-8. The current flow in any circuit is found by Ohm's Law. The dc resistance of the armature coils is the result of the length and size of the wire used for winding. The CURRENT through the armature is directly proportional to applied effective voltage.

$$I_{armature} = \frac{E_S - E_{CEMF}}{R_{armature}}$$

The voltage drop across the motor armature is:

$$E_{armature} = I_{armature} \times R_{armature}$$

$$= E_S - E_{CEMF}$$

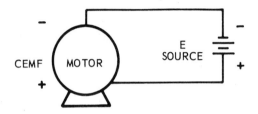

Fig. 11-8. The effective voltage applied to the motor is the source voltage minus the CEMF.

By rearranging this equation, we find that:

$$E_S = E_A + E_{CEMF}$$

Referring back to Chapter 6 on generators, you will discover that CEMF depends upon speed of rotation, the number of conductors and field strength. The formula is repeated as:

$$E_{CEMF} = \frac{(RPM)\,\Phi}{10^8}$$

Other variables have been omitted for simplification.

This theoretical discussion is extremely valuable from a practical point of view. Read these questions and try to "think through" to the correct answer. Then, read the explanation to check on your knowledge.

1. When does a motor draw the greatest amount of current from the line? Why?

Answer: A motor when stopped is developing no CEMF, and the armature current is at maximum, limited only by the resistance of the armature windings.

2. When does a motor draw the least current from the line?

Answer: When the motor is at full speed under no load. Then, the CEMF is almost equal to the line voltage and the current is minimum.

3. Why does a motor under load use more power?

Answer: When a motor slows down under load, the CEMF drops and the armature current increases.

4. Why does an overloaded motor become excessively hot?

Answer: The increased current due to lower CEMF produces heat by flowing through the resistance of the armature windings. P = I² R. This heat can eventually destroy the motor by burning off insulation.

5. Why is it necessary to start a motor with a series resistance between the motor and source voltage?

Answer: To limit the high currents when the motor is starting. The resistance is switched-out of the circuit when the motor reached its speed and CEMF is sufficient to limit the current.

MOTOR SPEED

It has been found that CEMF in a motor is directly proportional to speed and field strength.

$$E_{CEMF} = \Phi N$$

By rearranging this equation to:

$$N = \frac{E_{CEMF}}{\Phi}$$

we find the SPEED is inversely proportional to field strength. This will present some interesting theory as we continue our studies of methods of connecting motors.

MOTOR REGULATION

The term "motor regulation" is used to describe the performance of a motor under load and no load conditions. It is desirable, of course, for a motor to maintain a constant speed. The percent of regulation may be obtained from:

Percent of regulation =

$$\frac{\text{speed no load} - \text{speed full load}}{\text{speed full load}} \times 100$$

Assume a motor runs at 1700 rpm and drops to 1600 under load. Its percent of regulation is:

Percent of regulation =

$$\frac{1700 - 1600}{1600} \times 100 = 6.2 \text{ percent}$$

This term does not refer to SPEED CONTROL which is the varying of motor speed by external circuitry and devices.

THE SHUNT MOTOR

The SHUNT MOTOR, as you might suspect, has its field coils and armature in parallel. See Fig. 11-9. Assuming that a constant voltage is applied to this motor, the field strength tends to remain constant. If a load is applied and the speed of rotation is reduced, the CEMF becomes lower and the armature current increases.

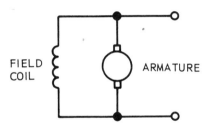

Fig. 11-9. The circuit of a shunt motor.

This consequently increases the torque of the motor to match the load. At this point, the motor remains at the reduced speed to match the load. The shunt motor is classified as a constant-speed machine and is desirable to drive machine tools which require a relatively constant speed.

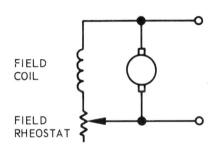

Fig. 11-10. Resistance in the field of the shunt motor will change its speed.

Although this shunt motor is constant speed, its speed may be changed by inserting resistance in the field circuit to change the field current and flux density. See Fig. 11-10. An increase in resistance will decrease the field current and increase the motor speed.

Another method of varying the speed of the shunt motor is by a variable series resistance in the armature circuit. Look at the equation:

$$N \text{ in rpm} = \frac{E_{CEMF}}{\Phi}$$

$$\text{and } E_{CEMF} = E_S - (IR_{armature} \text{ or } E_A)$$

You can see that a change in speed will also result from a change in armature voltage. This is easily accomplished by reducing the armature current by a series resistance as in Fig. 11-11.

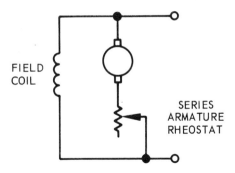

Fig. 11-11. Resistance in the armature circuit will change the speed of the shunt motor.

Although this is a widely used method of speed reduction, the high currents as a result of reduced CEMF produce appreciable power losses in the series resistance. You will construct light dimmer and motor speed control projects from plans in this text using SILICON CONTROLLED RECTIFIERS to limit currents.

SERIES MOTORS

The series motor is drawn schematically in Fig. 11-12. The field coils are in series with the armature circuit, and the field strength is proportional to the armature current. The series

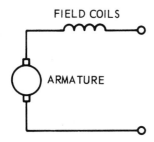

Fig. 11-12. The diagram of a series motor.

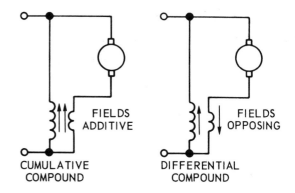

Fig. 11-13. The circuit diagrams of cumulative and differential motors.

motor must always have a load connected to it. Why? As the speed of the motor increases, the CEMF also increases and the armature current decreases.

A decrease in armature current also decreases the field current and the strength of the field. A decreasing field strength causes an increase in speed. The effect is cumulative, and the rpm of the motor will rise to a point of self-destruction. It literally may fly apart.

The series motor is most useful to turn heavy loads from a dead stop. At these speeds, tremendous torque is developed due to high armature and field currents. These motors are used on heavy equipment. They are always geared to the load. No belts are used. A broken belt would cause the motor to excessively speed up to self-destruction. The torque of this motor actually varies as the square of its armature current.

COMPOUND DC MOTORS

Motors may be wound and connected to have both series and shunt field coils. These are called COMPOUND MOTORS. If the series winding is connected so that its field reinforces or aids the shunt winding, the motor is said to be a CUMULATIVE COMPOUND MOTOR. On the other hand, if the series and shunt windings have opposing fields, it is called a DIFFERENTIAL COMPOUND MOTOR. The circuits of both of these types are illustrated in Fig. 11-13.

The reason for manufacturing compound motor circuits is to combine the desirable

features of both series and shunt motors. The detailed characteristics of compound motors will be found in more advanced texts on dc machinery.

DC MOTOR STARTERS

There are many different types of motor starters. The purpose of each starter is to insert a resistance in the motor circuit. This limits the armature current until the motor reaches its operating speed and develops sufficient counter electromotive force (CEMF) to hold the line current at a low value. Some starters are manually operated; others are automatic.

A simple step resistance motor starting circuit for a series motor appears in Fig. 11-14.

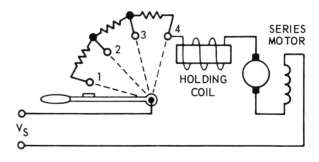

Fig. 11-14. A step motor starter circuit for a series motor.

With the lever in position 1, there is maximum resistance in the motor armature circuit. As the speed of the motor progressively builds up, the operator moves the lever to position 2, then 3 and 4. The lever is held at position 4 by the magnetic coil, which also is in series with the line. If the line voltage should fail, the lever would be released and return to the "off" position by spring action. This is a safety feature. It would be necessary to start the motor again, proceeding through each position of resistance.

MAGNETIC STARTING SWITCH

Many machines around the shop will have a push button starter consisting of a small box with a RED button and a BLACK button. See Fig. 11-15. This switch uses magnetic relays for very rapid on and off switching.

The LOCKED-OUT RELAY circuit is shown in Fig. 11-16. When the BLACK button is

Fig. 11-15. Push button magnetic safety switch using a locked-out relay circuit.

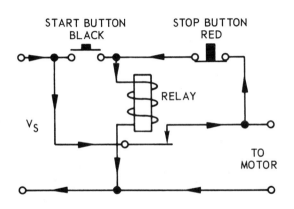

Fig. 11-16. The locked-out relay in ON position.

pushed, current flows through the relay coil and closes the contact points. Now current flows directly to the motor through relay contacts. It is not necessary to hold the BLACK button down, since current now flows through the closed RED button switch and through the coil. The relay remains closed. By pressing the RED button, the relay coil circuit is opened and the relay opens immediately and disconnects the power to the motor.

THE UNIVERSAL MOTOR

Can a dc motor be operated with ac current? Yes, with some minor design changes. You will recall the action of a conductor in a magnetic field. If the current in the conductor is reversed, the force and movement of the conductor is also reversed. But, if the field polarity is reversed at the same time that the armature current is reversed, then original conditions will exist and the motor will rotate. In a series motor, both field and armature currents change approximately together, and it is insensitive to the polarity of the applied voltage.

The series motor will have some serious defects when operated on alternating current. Chief disadvantages are the high currents induced in the armature and field windings by the rapidly changing flux. This causes excessive SPARKING at the brush contacts and commutator. These disadvantages are minimized by laminating the armature core to prevent eddy currents in the core, and by keeping the reactance of the field windings at a low value. Frequently, resistors are placed in the brush leads to limit high induced currents and prevent sparking.

The series UNIVERSAL MOTOR finds wide usage in small power tools such as drills, grinders and ventilating fans.

The SHUNT DC MOTOR does not operate well on ac. The reactance of the shunt field coils causes out-of-phase conditions which seriously impair the motor operation.

AC MOTOR THEORY

Since alternating current is readily available from the local power company, most motors used at home and in industry are ac motors. In order to understand the principles of induction and rotating magnetic fields, some additional background will be discussed.

Fig. 11-17 illustrates the magnetic field and a conductor loop. In this case, the conductor is FIXED and the field is passed across it. If the field moves from right to left, this is the same as the conductor moving from left to right. Apply the LEFT HAND RULE learned in Chapter 6.

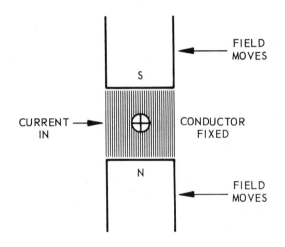

Fig. 11-17. A moving field cutting across a conductor induces a voltage and current in the conductor as indicated.

Point the four fingers of your left hand toward the south pole, turn your hand so that the conductor is moving toward your palm and your thumb will point inward. The conditions shown in Fig. 11-17 cause the induced current to flow inward as indicated.

Now look at Fig. 11-18 to discover the MOTOR ACTION resulting from the conditions previously described. This time use the RIGHT HAND RULE FOR MOTORS: Point the fingers of your right hand toward the south pole. Point your extended thumb in the direction of current flow in the conductor, and the movement of the conductor is TOWARD the palm of

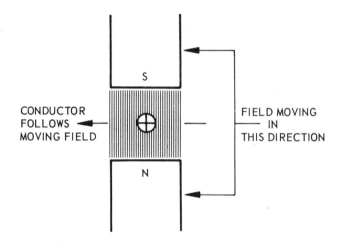

Fig. 11-18. The conductor tends to follow the moving field.

your hand. This movement is indicated in Fig. 11-18. The conductor follows the moving field.

An interesting project is developed around induced ac fields. A photograph of the ANIMATED RING is shown in Fig. 11-19. You should explain the difference between the action of the aluminum ring and the iron magnetic ring.

ROTATING MAGNETIC FIELDS

It would be easy to make a conductor or armature rotate, if it were possible to make a magnetic field rotate. This can be done with three-phase ac. See Fig. 11-20. Each phase is connected in sequence to field coils.

Follow the action: The magnetic field of coil A builds up to maximum. As the current through A drops off, the current in coil B is building up to maximum. As the current in coil B drops off, coil C is building up to maximum. Can you see that the magnetic field moves from A to B to C to A and so on? Is it not a rotating field?

THE INDUCTION MOTOR

To build the polyphase induction motor, it is now necessary to place a rotating armature in the rotating field. These armatures, called ROTORS, may be of two types. Either type will be built around a laminated iron core.

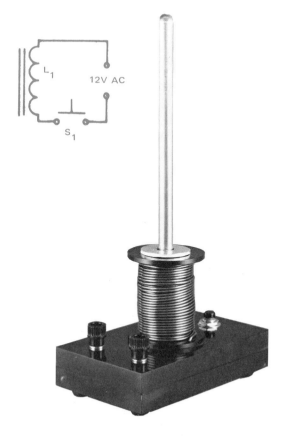

1. The SQUIRREL CAGE ROTOR will have heavy single copper wires imbedded in the iron core. See Fig. 11-21.
2. The FORM-WOUND ROTOR will have coils of wire fixed in slots around the rotor core similar to the dc motor armature.

Voltages and currents induced in these rotor coils will cause them to move with the rotating field. But consider this! In order to induce a voltage, there must be RELATIVE MOTION between field and conductor. If the conductor moves with the field, there is no relative motion. As a result, the conductors and rotor must rotate slightly slower than the field.

This type of movement is called SLIPPAGE. When no load is on the rotor, the rotor speed can approach the synchronous speed of the rotating field. As the motor is loaded, it slows down slightly, slippage increases and larger currents are induced in the rotor windings. This, in turn, increases the torque of the motor.

INDUCTION MOTOR SPEED

The speed of the induction motor, as you can readily see, is dependent upon the frequency of the applied voltage and the number

Fig. 11-19. The animated ring demonstrates motor action. Coil L_1 is wound with 325 turns of No. 18 enameled copper wire. The coil is connected to terminals through a push button switch S_1. The core is a low carbon steel rod about 8 in. long. Make copper, aluminum and iron washers to fit over the core. Connect 12 volts ac to coil. Press button and observe the action.

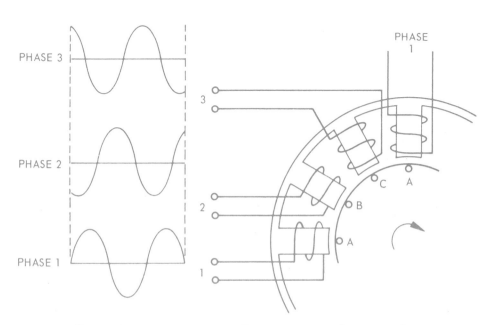

Fig. 11-20. A three-phase current will produce an ac rotating magnetic field.

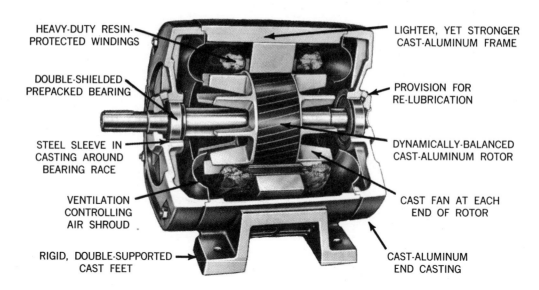

HEAVY-DUTY RESIN-PROTECTED WINDINGS

DOUBLE-SHIELDED PREPACKED BEARING

STEEL SLEEVE IN CASTING AROUND BEARING RACE

VENTILATION CONTROLLING AIR SHROUD

RIGID, DOUBLE-SUPPORTED CAST FEET

LIGHTER, YET STRONGER CAST-ALUMINUM FRAME

PROVISION FOR RE-LUBRICATION

DYNAMICALLY-BALANCED CAST-ALUMINUM ROTOR

CAST FAN AT EACH END OF ROTOR

CAST-ALUMINUM END CASTING

Fig. 11-21. Photo of a squirrel cage rotor in an induction motor. (W. W. Grainger, Inc.)

of poles around the field or STATOR windings. Assuming no slippage, the synchronous speed should be:

$$rpm = \frac{120\ f}{P}$$

where,

f = frequency of applied voltage.
P = number of poles produced in stator field.

EXAMPLE: A two-pole, three-phase motor connected to a 60 Hz source would rotate at:

$$rpm = \frac{120 \times 60}{2} = 3600\ rpm$$

SINGLE-PHASE MOTORS

The most popular type of fractional horse-power motor is the single-phase motor. The theory of operation is quite similar to the three-phase motor, except that the single-phase current cannot produce a rotating field. If we could get the single-phase motor up to synchronous speed, it would continue to run just like the three-phase motor. The problem is getting it started.

The starting problem is solved by placing two sets of windings in the STATOR. These windings have unequal reactances due to the number of turns in each coil and the size of wire used. Now, unequal reactances will result in unequal lags of current. In other words, the unequal reactance will produce two out-of-phase currents. These out-of-phase currents produce the rotating field which is so necessary to get the motor started.

When the motor reaches speed, a centrifugal switch disconnects the STARTING WINDINGS, and the motor operates only on its RUNNING or MAIN WINDINGS. This setup is shown schematically in Fig. 11-22. A split-phase motor is illustrated in Fig. 11-23.

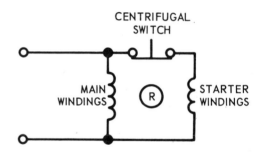

CENTRIFUGAL SWITCH

MAIN WINDINGS

R

STARTER WINDINGS

Fig. 11-22. Diagram of a split-phase motor showing starting and running windings.

Fig. 11-23. Fractional horsepower split-phase motor. (Delco Products Div., General Motors Corp.)

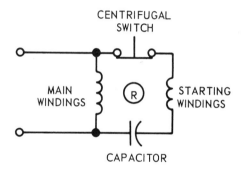

Fig. 11-25. Diagram of a capacitor-start, split-phase motor.

A vector diagram of the currents in the windings of this split-phase motor is drawn in Fig. 11-24. You must realize that magnetic strength is the result of current, so the magnetic fields of each stator coil will reach maximum at different times. This causes the moving or rotating field.

CAPACITOR-START MOTORS

If the phase difference between starting and running windings can be increased, the starting torque of the motor can be substantially improved. Basic theory tells us that capacitance in a circuit causes the current to lead the voltage, while inductance causes the current to lag. Both of these effects are incorporated in the capacitor-start motor in Figs. 11-25 and 11-26.

Some motors employ the centrifugal switch to disconnect the starting circuit when the motor approaches full speed. Capacitors used for this application range between 80 and 400 microfarads for motors under one horsepower.

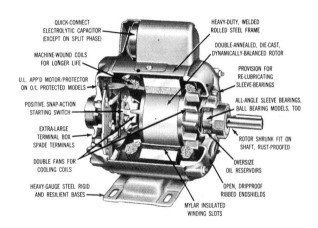

Fig. 11-26. A commerical capacitor-start, split-phase motor. Note snap-action, centrifugal starting switch and electrolytic capacitor for starting. (W. W. Grainger, Inc.)

SHADED POLE MOTORS

Another method of getting an ac induction motor started is by using a SALIENT-POLE STATOR. This is a method of SHADING a pole. It is accomplished by wrapping a single turn of heavy copper wire around a part of the pole face. See Figs. 11-27 and 11-28.

A rising current in the coil in position A, Fig. 11-28, causes an increasing magnetic field. In the shaded portion, a current is induced in the

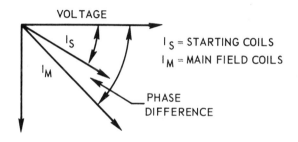

Fig. 11-24. Vector diagram of phase displacement between currents in starting and main windings of split-phase motor.

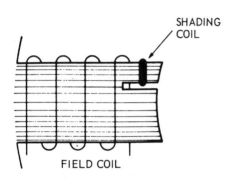

Fig. 11-27. Partial view of a field coil with a shaded-pole.

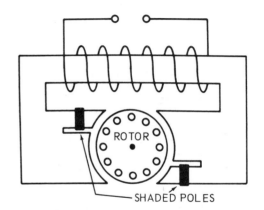

Fig. 11-29. A diagram of a two-pole, shaded-pole motor.

shading coil which produces an opposing magnetic field (Lenz's Law). Therefore, the field on that side is weaker. Position B is at the peak of the incoming ac wave. At this point, there is NO CHANGE in magnetic field and NO induced field in the shading coil. The magnetic field is EQUAL across the pole face.

Position C shows the effect of a decreasing current and magnetic field. The induced field in the shading coil opposes the reducing field (Lenz's Law). Therefore, the field becomes stronger than the left side of the pole where there is no opposition. This action is sufficient to start the rotor turning.

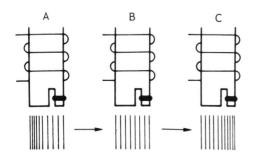

Fig. 11-28. The field on the shaded coil moves from left to right because of the shading coil.

After starting, the motor runs as an induction motor, approaching synchronous speed similar to the split-phase motor. This motor develops very little starting torque, but find application in electric clocks and small fans. It certainly is economical to manufacture. A diagram of a typical two-pole clock motor is given in Fig. 11-29.

REPULSION INDUCTION MOTOR

The repulsion start motor, in many ways, resembles the dc motor in construction. See Fig. 11-30. It has a form-wound rotor with coils connected to commutator sections and brushes.

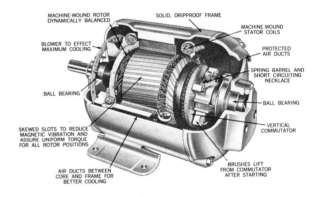

Fig. 11-30. A commercial repulsion induction motor. Note commutator sections shorted brushes. (W. W. Grainger, Inc.)

The stator coils are similar to the single-phase motor without starting windings. The two opposite brushes are connected together during starting. As the motor approaches speed, a centrifugal ring closes down on the commutator sections and shorts them all together. At this point, the motor runs as an induction motor.

The theory of the repulsion induction motor is described in Fig. 11-31. The alternating current in the stator windings induces a voltage in the rotor windings by transformer action. A current flows in the particular rotor coils shorted out by the commutator and shorted

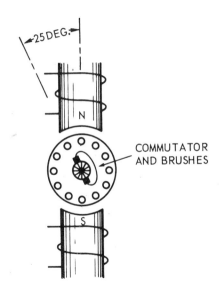

Fig. 11-31. This diagram shows how the repulsive fields are built up to cause the motor to start rotation.

the brushes are displaced from the field pole axis (usually about 25 deg.). In Fig. 11-31, the motor will turn in a counterclockwise direction.

The study of motors in this chapter is intended as an exploratory experience of these popular machines. If your interest is high, it is suggested that you study texts devoted to the subject of electric motors.

SCR MOTOR SPEED CONTROL

A useful project which applies many of the lessons in this text is the Silicon Controlled Rectifier (SCR) Motor Speed Control. You will need to refer to Chapter 12 to understand the operation of the SCR. Fig. 11-32 gives the schematic diagram and parts list for the SCR Motor Speed Control. The completed project is shown in Fig. 11-33.

brushes. A field is developed in the particular rotor coil which opposes the field coil. The rotor then starts to rotate in the direction that

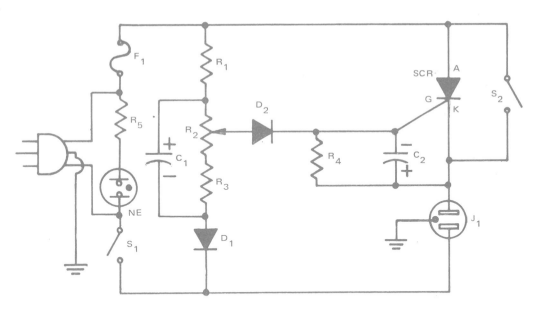

PARTS LIST FOR MOTOR SPEED CONTROL

R_1 — 2K Ω, 5W resistor
R_2 — 500 Ω, 2W potentiometer
R_3 — 10 Ω, 12W resistor
R_4 — 150 Ω, 1/2W resistor
R_5 — 100 Ω, 1/2W resistor
C_1, C_2 — 100 µF, 25V electrolytic capacitor
NE — NE2 neon lamp
heat sink, 1/8 x 2 4 1/2 in. aluminum

D_1, D_2 — rectifier, Motorola HEP156
or RCA SK 3016
SCR — Motorola HEP302 or RCA SK 3557
F_1 — 3A, slo-blo fuse
J_1 — receptacle, Amphenol 61-F1
S_1, S_2 — SPST switch
fuse holder
ac plug, 3 wire

Fig. 11-32. Schematic and parts list for SCR Motor Speed Control. (Graymark)

The construction is not critical and you should design your own circuit layout. Switch S_2 can be set for straight-through operation or controlled operation. The current through the SCR is controlled by the voltage applied to the gate which depends upon the time constant of the circuit and the setting of R_2. This circuit will control up to 3 amps maximum.

NOTE: Always read the instruction manual supplied by the manufacturer before using any meter, generator or test instrument.

Fig. 11-33. SCR Motor Speed Control or Light Dimmer.

TEST YOUR KNOWLEDGE

1. A motor converts _____ energy to _____ energy, and a generator converts _____ energy to _____ energy.
2. Describe the use of the left hand rule to determine direction of the magnetic field around a current carrying conductor.
3. Motor action is based on _____ toward the weakened field.
4. Draw diagrams of shunt and series dc motors.
5. CEMF is counter voltage induced by the rotating armature. True or False?
6. Why does a motor heat up if excessively loaded?
7. What is the desirable characteristic of a shunt motor?
8. Why does a series motor always have to be connected to its load?
9. A distinct advantage of the series motor is that it will attain _____ from a dead stop.
10. What is the purpose of a motor starter?
11. How does a split-phase motor start?
12. What is the speed of a four-pole induction motor connected to a 60 Hz power source?
    a. 1200 rpm.
    b. 1800 rpm.
    c. 2400 rpm.
13. Give two advantages of a capacitor start motor.
14. Compound motors may be wound and connected to have both _____ and _____ field coils.
15. The effective voltage applied to a motor is the source voltage minus the _____.
16. The term "motor regulation" is used to describe the performance of a motor under _____ and _____ conditions.

Chapter 12

SEMICONDUCTOR DIODES AND POWER SUPPLIES

The basic semiconductor device is the diode. It is the simplest component of the family of semiconductors. Diodes perform a very important role in electronics today.

In this chapter, you will:

1. Study the theory and characteristics of semiconductor diodes.
2. Explore methods of rectification using diodes.
3. Understand the principles and application of filtering circuits.
4. Study the requirements of a low ripple, well-regulated power supply.
5. Explore some of the more common special diode devices.

THE DIODE

A DIODE IS A TWO ELEMENT UNILATERAL CONDUCTOR. This means that a diode has two connections, "anode" and "cathode," and that current will flow through this component in ONLY ONE DIRECTION. It will have a high resistance to a current in its reverse direction.

The theory of semiconductor conduction should be reviewed at this time in Chapters 1 and 3. Remember that conduction through an N type material is by ELECTRONS; conduction through a P type material is by HOLES. Also, remember that both N and P type crystals will have a small quantity of minority carriers.

There are several methods of forming a diode, and they are used individually or sometimes in combination. The desired result is to produce a surface which separates the N type and P type crystals. Diodes are made by alloying an N type germanium crystal with a metal base such as aluminum or indium.

In the "rate growth method," a crystal is drawn from a molten semiconductor material containing both N and P type impurities. The concentration of impurities in the drawn crystal depends upon the rate at which the crystal is formed. PN junctions, for example, can be made by changing the growth rate periodically. See Fig. 12-1.

In the "diffusion method" of forming a diode, a solid crystal is heated in an atmosphere containing the desired impurity. The gas dif-

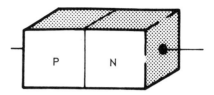

Fig. 12-1. A diagram of a PN junction crystal.

fuses gradually into the base crystal to form a layer of the opposite type. The diffusion method is used extensively in the manufacturing of integrated circuits.

POTENTIAL HILL OR BARRIER

Diffusion of a junction diode is shown in Fig. 12-2. At the junction, some of the electrons from the N crystal are attracted across the junction to fill holes in the P crystal. Also, some holes from the P crystal are attracted across the junction to the electrons in the N crystal. Because of this exchange of carriers, an area on each side of the junction becomes void of current carriers. This is called the DEPLETION AREA, TRANSITION AREA or SPACE CHARGE REGION.

Due to the diffusion of electrons and holes across the junction, an electrical potential develops. This is illustrated in Fig. 12-2 by a small battery in dotted lines across the junction. This voltage is called the POTENTIAL HILL or BARRIER. Its polarity is indicated in the illustrations. Since electrons leave the N side of the junction, that side becomes positive. Holes leaving the P side of the junction cause that area to become negative. In Fig. 12-2, a graph below the junction shows the height and polarity of the potential hill as well as the width of the depletion area.

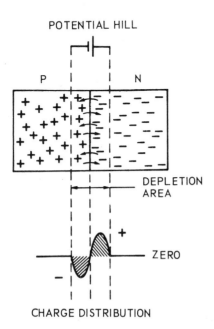

Fig. 12-2. Diagram shows charge distribution and formation of potential hill by diffusion across a PN junction.

NOTE: The overall net charge on the PN diode does not change due to the formation of the potential hill. The overall charge on the diode is zero. The total component has neither gained nor lost electrons because of diffusion across the junction.

REVERSE BIAS

The diode is connected in a circuit diagrammed in Fig. 12-3. In this case, the negative terminal of the battery is connected to the P crystal and the positive terminal to the N crystal. The holes, being positive, are attracted

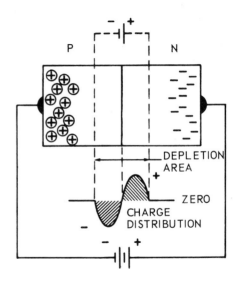

Fig. 12-3. The diode is reverse biased. The potential hill increases and the depletion area increases in width.

to the negative source; the electrons are attracted to the positive source. The depletion area increases in width. The potential hill and the source voltage join together to prevent current carriers from crossing the junction.

NOTE: The diode is REVERSE BIASED and no current flows through the crystal, except a small leakage current due to minority carriers.

FORWARD BIAS

In Fig. 12-4, the diode is connected in a forward bias circuit. Positive battery to P

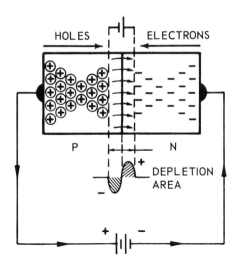

Fig. 12-4. The diode is forward biased. Conduction in P crystal is by holes; conduction in N crystal by electrons.

crystal; negative battery to N crystal. In this case, electrons from the P crystal are attracted to the positive source. For each electron leaving the P crystal, a HOLE is INJECTED into the P crystal. The holes drift toward the junction where they are filled with electrons from the N crystal. Electrons are supplied to the N crystal from the negative battery terminal. The depletion area decreases in width. The potential hill is considerably reduced by the opposing source voltage. Current flows through the junction diode as indicated by the arrows in the external circuit.

DIODE RESISTANCE

It is important to remember that a diode presents a HIGH resistance when connected in a reverse direction and a relatively LOW resistance in the forward direction. This characteristic is used to check the quality of an unknown diode.

POINT CONTACT DIODE

A point contact diode is another type of diode used in the detection of radio signals. It consists of a very small piece of N type germanium against which a fine phosphor bronze wire (catwhisker) is pressed. During manufacture, a high current is run through this

combination from wire to crystal. This forms a P type region around the contact point in the germanium crystal. In Fig. 12-5, a crystal diode and its symbol are illustrated. You will have an occasion to use several of these in the construction of projects in this text.

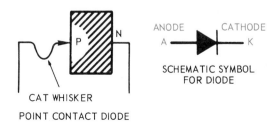

Fig. 12-5. A point contact crystal diode and the universal schematic symbol used for all diodes.

DIODE VOLT-AMPERE CHARACTERISTICS

The typical operation of a diode is displayed in the graph of Fig. 12-6. As forward bias voltage is applied to the diode, the barrier potential must first be overcome. Then, the diode starts conduction and there is an approximate linear relationship between voltage and current.

For practical purposes: forward voltage drop across a germanium diode approaches .2 volts; for a silicon diode, it is .7 volts. This will be

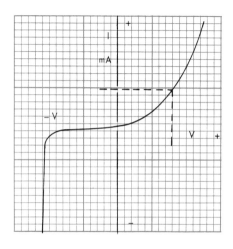

Fig. 12-6. The volt-ampere characteristic curve of a diode.

found useful in designing transistor circuits later in this text.

As a reverse bias voltage is applied to the diode, a minute current (a few microamperes) will flow due to minority carriers. This current is frequently called the REVERSE SATURA-TION CURRENT. The reverse current is the result of LEAKAGE RESISTANCE, which is another way of saying that only a very small current will flow.

At a certain reverse bias voltage, conduction suddenly increases. This reverse voltage is called BREAKDOWN, ZENER VOLTAGE or AVA-LANCHE voltage. The breakdown of the diode appears to be the result of ionization due to the strong electric field and collisions between current carriers.

In the ZENER region, you will note a very peculiar phenomenon. Current will increase with a very small increase in reverse voltage. This is the result of the high reverse electric field actually pulling carriers out of the atomic structure. The fact that the diode has a varying current range with a rather constant voltage makes it very useful in precise voltage regulation. The zener diode manufactured for this purpose will be discussed later in this chapter.

The dc resistance of a diode can be computed from the volt-ampere curves by using Ohm's Law:

$$R_{dc} = \frac{V}{I}$$

Select a point on the linear portion of the forward current curve as indicated by the dashed line in Fig. 12-6.

SPECIFICATION OF DIODES

In order to select a diode for a specific application, you must be aware of the many specifications supplied by the manufacturer. These specifications usually include the abso-

lute maximum ratings of the diode, the typical operating conditions, mechanical data for leads and mounting and set of curves showing typical operation.

1. PEAK INVERSE VOLTAGE (PIV). This is the maximum voltage that can be applied to the diode in a reverse direction without destruction.
2. AVERAGE RECTIFIED FORWARD CUR-RENT. The current the device is required to carry under normal operating range.
3. PEAK RECTIFIED CURRENT. The very maximum current the device can conduct for a partial cycle of operation.
4. SURGE CURRENT. The current that the device can conduct for a second.
5. FORWARD VOLTAGE. The forward oper-ating voltage, usually at a definite forward current.
6. REVERSE CURRENT. The maximum re-verse current at maximum reverse voltage.
7. POWER DISSIPATION. The power rating at 25 deg. C (77 deg. F). Derating specifications and curves are also given. In order to use these, refer to Fig. 12-7.

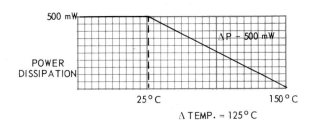

Fig. 12-7. Type of curves used in derating a diode due to an increase in temperature.

This curve shows that power dissipation of a certain diode is 500 mW, and is derated linearly to zero at 150 deg. C (302 deg. F). During a change in temperature to 125 deg. C (150 − 25 = 125 deg. C) or 257 deg. F (302 − 45 = 257 deg. F), power dissipation changes 500 milliwatts. The diode is derated at:

$$\frac{\Delta p}{\Delta C} = \frac{500 \text{ mW}}{125 \text{ C}} = 4 \text{ mW/degrees C}$$

To prevent excessive derating of a diode due to temperature, HEAT SINKS (a mass of metal used to carry heat away from a component) are generally used. The diode is mounted in contact with a larger mass of metal, usually aluminum. Commercial heat sinks shown in Fig. 12-8 have fins for air circulation and greater heat dissipation. Almost all high power diodes are mounted in this fashion.

Fig. 12-8. Heat sink. Note fins for air circulation and heat dissipation.

INTERELEMENT CAPACITANCE

In the junction diode, each crystal acts as a plate of a capacitor. The junction appears as the dielectric. In the forward bias arrangement, the junction capacitance increases as voltage in-creases. This is the result of the reduction of the potential barrier and the decrease in the width of the depletion area. This is, in effect, a reduction of the thickness of the dielectric and therefore an increase in capacitance.

In the reverse bias connection, the opposite holds true. The reverse voltage increases the potential hill, increases the width of the depletion area and increases the effective thickness of the dielectric. Therefore, interelement capacitance decreases with reverse bias. When diodes are used in the higher frequency ranges, the capacitive reactance of the interelement capacitance will be an important consideration.

DIODE CASE OUTLINES

Diodes come in many different cases depending on the type. Fig. 12-9 shows some typical outlines for diodes. Fig. 12-10 shows some typical sizes.

POWER SUPPLIES

A power supply may be defined as an electronic circuit that is designed to provide various ac and/or dc voltages for equipment operation. Sometimes such devices as batteries or generators also may be considered as power supplies. This book will discuss the electronic circuit that may be used as a power supply.

The circuit of a power supply can contain:

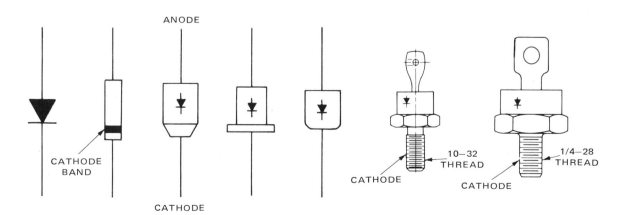

Fig. 12-9. Typical diode case outlines.

Fig. 12-10. A comparison of diode sizes and shapes.

1. A transformer.
2. A rectifier.
3. A filter.
4. A regulator.

Transformers and regulators are not found in all power supplies. See Fig. 12-11.

Most electronic circuits have built-in power supplies. For troubleshooting or experiments, laboratory power supplies can be used to provide the proper voltages and currents. See Fig. 12-12.

TRANSFORMERS

In Chapter 8, you studied the transformer as a device that operated from the principle of mutual induction. In power supplies, transformers may be used to step up or step down voltage or current.

Fig. 12-12. Laboratory power supply.

Another reason for using a transformer in a power supply is for ISOLATION. In a regular transformer, there is no physical connection between the primary winding and the secondary winding. Thus, the secondary is isolated from the primary. This is a safety factor to look for in power supplies.

Transformers are also used to provide two voltages, one 180 deg. out of phase with the other. These voltages are provided with the use of a center-tapped transformer. This concept will be discussed later in this chapter in the section on "full wave rectifiers."

HALF-WAVE RECTIFICATION

In the circuit of Fig. 12-13, the operation of the diode may be observed. Note that the letter symbol used for the diode is D_1.

The input generator supplies an ac voltage. When the cycle is positive, D_1 conducts and a

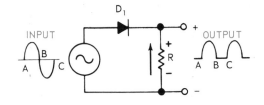

Fig. 12-13. A half-wave rectifier circuit, using a diode.

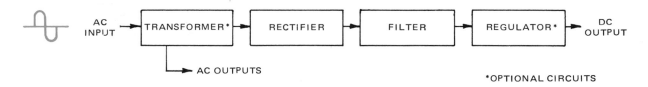

Fig. 12-11. Block diagram of a typical power supply.

voltage (I x R) appears across the load resistor R. When input voltage is negative, D_1 is reverse biased and does not conduct. This is illustrated by a straight line in the output wave form across R (points B to C). Output is a pulsating dc wave with a frequency the same as the input frequency. Since only one-half of input voltage is used, it is called a HALF-WAVE RECTIFIER.

Rectification is the process of changing alternating current to direct current. The peak output voltage will be the same as the peak input voltage, less any voltage drop across the diode. The average output voltage is one-half the average full-wave value, because the diode only conducts on half-cycles.

$$V_{av} = .318 \ V_{peak}$$

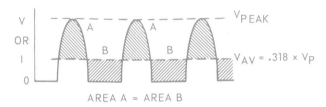

AREA A = AREA B

Fig. 12-14. Curves illustrate the average wave value of a half-wave rectifier output.

This is shown graphically in Fig. 12-14. The area A is equal to area B. Both the output voltage and the output wave forms may be observed by an oscilloscope connected across R.

FULL-WAVE RECTIFIER

The basic full-wave rectifier circuit is shown in Fig. 12-15, using a center-tapped transformer. The center tap is taken as a voltage reference point of zero. The total usable secondary voltage is cut in half to provide the required voltage wave for the rectifier.

When A in Fig. 12-15 is positive in respect to center tap, D_1 is forward biased and conducts as indicated by the arrows. A voltage appears across R equal to I x R. The output voltage is approximately one-half of the total transformer secondary voltage.

During the next half-cycle of the input voltage, B in Fig. 12-15 is positive and A is negative in respect to the center tap. Diode D_2 conducts (follow arrows) and produces an output voltage across R. BOTH half-cycles of input voltage produce a dc output voltage. It is called a FULL-WAVE RECTIFIER. The peak

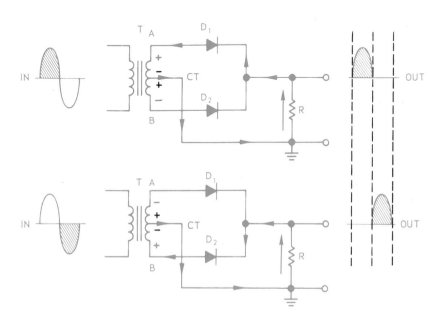

Fig. 12-15. A full-wave rectifier showing input and output wave forms.

output voltage is similar to one-half the total secondary voltage and the frequency of the output wave is twice the frequency of the input wave.

In Fig. 12-16, the output wave and average value is shown. This is a pulsating dc output

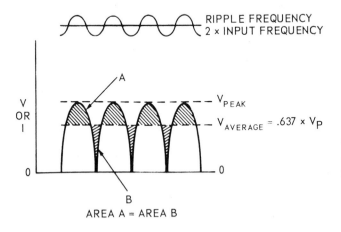

Fig. 12-16. Output wave form and average value of a full-wave rectifier.

wave which has a RIPPLE frequency twice the input wave frequency.

BRIDGE RECTIFICATION

The bridge rectifier circuit in Fig. 12-17 has the advantage of permitting the use of the total secondary voltage. Yet, it requires four diodes for full-wave rectification. To understand this circuit, you must realize that forward bias of a diode can be accomplished by making the anode positive in respect to the cathode or the cathode negative in respect to the anode. These two statements mean exactly the same thing.

When point A on transformer secondary is positive, both D_1 and D_2 are forward biased and both conduct (follow arrows). Voltage across R is result of this current. When point B is positive and A negative, diodes D_3 and D_4 conduct and produce voltage across R. Since both half-cycles of input voltage produce a dc wave, it is a full-wave rectifier. Output is pulsating dc with a frequency twice input frequency.

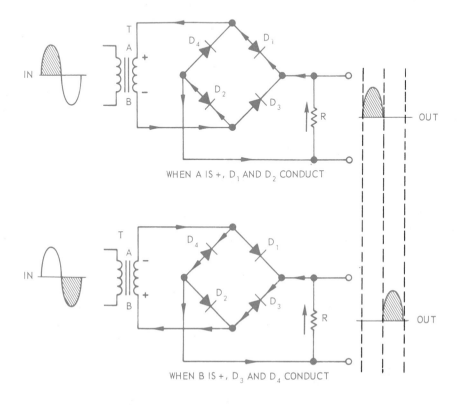

Fig. 12-17. Theory of operation of a full-wave bridge rectifier circuit.

Many times bridge rectifiers come as four separate diodes. It also is possible to purchase a full-wave bridge rectifier in a molded assembly with all four diodes in one unit. See Fig. 12-18 for a drawing of this unit.

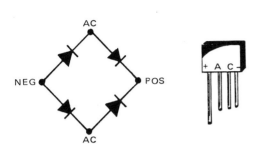

Fig. 12-18. Bridge rectifier assembly.

VOLTAGE DOUBLERS

It is possible to obtain a higher dc output voltage from a rectifier by "stacking up" or "cascading" voltages. In Figs. 12-19 and 12-20, a half-wave voltage doubler is shown. During the first half-cycle of input, D_1 will conduct and charge C_1 to about the full peak voltage.

D_2 is CUTOFF. During the second half-cycle, D_1 is cutoff and D_2 conducts. The voltage across C_2 is the PEAK VALUE OF THE INPUT WAVE PLUS THE CHARGE ON CAPACITOR

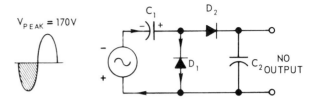

Fig. 12-19. A half-wave voltage doubler. During the first half-cycle, C is charged to peak voltage.

C_1 or twice the input voltage. This is shown in Fig. 12-20. Necessary filter circuits are added to this rectifier doubler.

Another type of voltage doubler circuit is illustrated in Fig. 12-21. During the first half-cycle of input voltage, D_1 conducts and charges C_1 to full peak voltage. During the second half-cycle, Fig. 12-22, D_2 conducts and charges C_2 to full peak voltage. The charges on C_1 and C_2 are in series and C_3 charges to the sum of these two series voltages. The output across C_3 is equal to two times the input peak voltage.

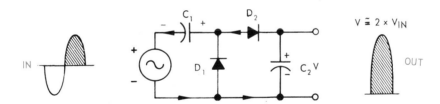

Fig. 12-20. Half-wave voltage doubler. On second half-cycle, D_2 conducts. Voltage output is equal to input voltage plus the charge on C_1.

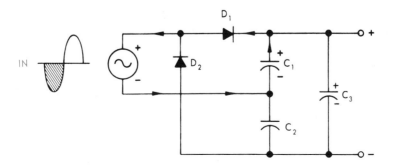

Fig. 12-21. Full-wave voltage doubler. During first half-cycle, D_1 conducts and charges C_1 to peak voltage.

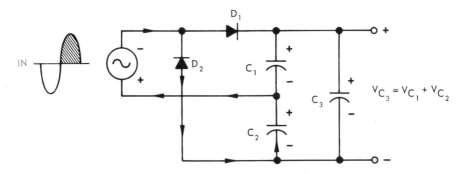

Fig. 12-22. During the next half-cycle, D_2 conducts and C_2 charges to peak voltage. Output is taken from across C_1 and C_2 in series.

RIPPLE VOLTAGE

Examine the output wave of a rectifier in Fig. 12-23. This pulsating dc wave is made up of the combination of a pure value of dc voltage which is the average voltage and an ac wave or ripple.

It is extremely important when working with transistor power supplies that the PERCENTAGE OF RIPPLE be kept at a very low value (3 percent or less). The percent of ripple may be measured and computed by the formula:

Percent of ripple =

$$\frac{V_{rms} \text{ of ripple voltage}}{V_{average}} \times 100$$

PROBLEM: The peak value of the output of a full-wave rectifier is 9 volts. The ripple is measured as .1 volt peak-to-peak. What is the percent of ripple?

1. Convert peak value to average value.

   9 volts x .637 = 5.73 volts

2. Find peak value of ripple voltage.

   .1 ÷ 2 = .05 volt

3. Find rms value of peak voltage.

   .707 x .05 volt = .035 volt

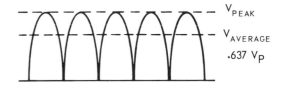

Fig. 12-23. The output wave consists of a dc component equal to average value of wave plus a ripple voltage.

4. Find the percent of ripple.

Percent of ripple =

$$\frac{.035}{5.73} \times 100 = .0061 \times 100 = .61 \text{ percent}$$

Since the power supply is common to most all amplifiers, oscillators and other circuits in a piece of equipment, an excessive ripple will introduce hum and distortion in the desired output.

PERCENT OF REGULATION

When a power supply is designed to produce a definite voltage output, it is desirable that the output voltage remains relatively constant when the load is applied to the supply. The percentage of regulation is a specification of quality of a power supply. It is expressed as the relationship between the difference in voltage at no-load to the voltage at full-load divided by the full-load voltage.

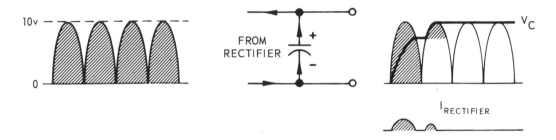

Fig. 12-24. The capacitor charges to full rectifier voltage and
remains charged.

Percent of regulation =

$$\frac{V_{no\text{-}load} - V_{full\text{-}load}}{V_{full\text{-}load}} \times 100$$

PROBLEM: A power supply has a no-load voltage of 12 volts which drops to 11.5 volts when connected to a load. What is percent of regulation?

$$\text{Percent of regulation} = \frac{12V - 11.5V}{11.5V} \times 100$$

$$= \frac{50}{11.5} = 4.3 \text{ percent}$$

FILTERING BY CAPACITANCE

Since ripple and poor regulation are undesirable, what can be done about it? In Fig. 12-24, a capacitor only is connected to a rectifier.

During the first half-cycle, capacitor C charges up to some voltage and during the next half-cycle C becomes fully charged, AND REMAINS CHARGED AT POLARITY SHOWN. (If resistance of the rectifier circuit were low, C might charge during one-half cycle.) The wave directly below the voltage output in Fig. 12-24 is the current wave supplied by the rectifier. Current only flows during the charging interval of C. When C is charged, no more current is required from the rectifier.

In Fig. 12-25, a load has been connected across the capacitor. Now C can continue to discharge through R and will continually need current from the rectifier to maintain its voltage. The voltage and current wave forms are shown with the diagram in Fig. 12-25. The capacitor has certainly raised the average level of the output as well as removed the ripple to some extent.

The same circuit is drawn in Fig. 12-26, but a load which requires more current is connected across C. Note the greater discharge of C. Note the increased current from the rectifier. The

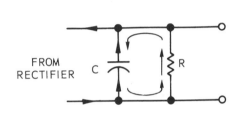

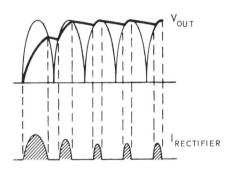

Fig. 12-25. The charge on C tends to keep the output voltage
at a constant value. Rectifier current recharges C as needed.

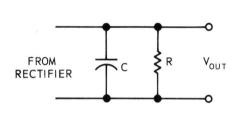 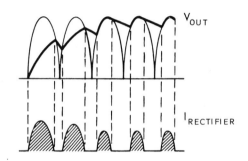

Fig. 12-26. If load R draws more current, the rectifier must supply more current. The ripple is more severe and the average voltage drops.

average dc voltage has dropped and the ripple is more severe.

Before making further improvements on this filter, consider the initial surge of current required to charge C in Fig. 12-27. A discharged capacitor, for an instant, appears like a short circuit across the rectifier. This surge of current could damage a diode if it had an insufficient surge current rating. Consult the specifications. Frequently, a resistor of a few ohms is placed in

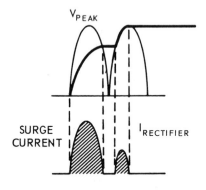

Fig. 12-27. The initial pulse of current to charge a filter capacitor must not exceed diode specifications.

series with the diodes to limt the initial surge of current. After C has become charged, then the current drawn by R should not exceed the average rectifier current specification.

FILTERING BY INDUCTANCE

The ability of a coil to resist a change in current has already been studied. In the case of the inductor, energy is stored in a magnetic field. In Fig. 12-28, a rectifier is connected to an inductor in series with the load R. This reduces the current variation a great deal, but some sacrifice is made in voltage. Nevertheless, the regulation is better in circuits which require larger currents. Inductors used in this manner are called CHOKES. See Fig. 12-29.

The combination of L and C in a filter circuit combines the advantages of each. This is diagrammed schematically in Fig. 12-30.

It is called an L type choke input filter. Its name comes from the inductor-capacitor con-

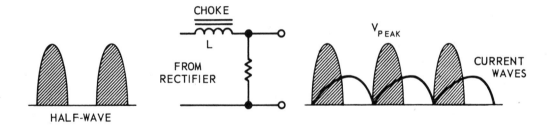

Fig. 12-28. A choke limits a change in current and improves the filtering.

Fig. 12-29. A typical filter for a power supply.

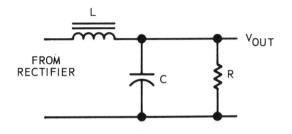

Fig. 12-30. An L filter consisting of choke and capacitor. Its diagram resembles an inverted L.

nected as an inverted L. The output of this filter approaches a dc voltage with only a little ripple and with good regulation.

An investigation into the theory of the LC filter requires a review of series RL circuits.

In Fig. 12-31, the L filter is redrawn. Make a comparison of both circuits and assure yourself

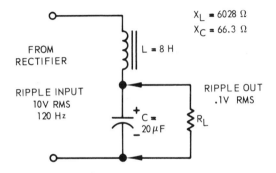

Fig. 12-31. Compare this circuit to Fig. 12-30. It is the same. This circuit is used for text problems.

they are the same. At a ripple frequency of 120 Hz, the reactance of X_L = 6028 ohms and X_c = 66.3 ohms. The impedance of X_L and X_c in series is Z = 6028 ohms − 66.3 ohms $\cong$ 5962 ohms. When a ripple voltage with an effective value of 10 volts is fed to the circuit, a current flows.

$$I = \frac{V}{Z} = \frac{10 \text{ volts}}{5962 \ \Omega} \cong .0017 \text{ amp}$$

The ripple voltage across C and the load equals:

$$IX_c = .0017 \times 66.3 = .113 \text{ volts rms}.$$

This is a fine improvement in the percent of ripple. By the filter network, the ripple voltage across the load has been reduced from 10 volts rms to a little over .1 volt rms.

Further improvement may be made by connecting a second capacitor to make a pi filter (π). From the rectifier, the first component is a capacitor. The filter is called a capacitor-input π filter. The input capacitor is selected to have a low reactance to the ripple frequency. Consequently, a major part of the filtering is accomplished by C_1. Most of the remaining ripple is filtered out by means of the LC network. Examine Fig. 12-32.

Although the filtering is better with the π filter, C_1 is still connected directly across the rectifier, and high pulses of current are necessary to keep C_1 charged, if high currents are needed for the load. These high currents may damage the rectifier diodes. It is, therefore,

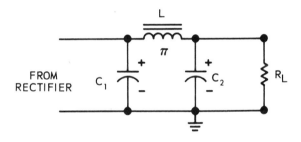

Fig. 12-32. The circuit is a capacitor input pi filter. The circuit resembles the Greek letter π.

only used with low current equipment. The capacitor-input filter does produce a higher output voltage, but its regulation is inferior to the L type filter.

RC FILTERS

In any of the previously discussed filters, a resistance may be substitued for the choke to save expense. The filtering action will be impaired, and there will be some loss of dc voltage because of the voltage drop across R. When using a choke filter, the only dc loss is due to the resistance of the choke windings, which is relatively low. A schematic of the RC filter is found in Fig. 12-33. The resistor R is usually a power resistor with sufficient wattage rating to carry the load current.

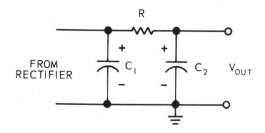

Fig. 12-33. A power resistor is used in this filter in place of a choke. It is cheaper and less effective.

BLEEDER RESISTOR

Very frequently, a resistor is placed across the power supply output. See Fig. 12-34. Several advantages are realized:

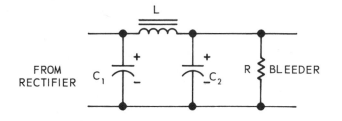

Fig. 12-34. A bleeder resistor across a filter circuit has several advantages.

1. This preload of the power supply causes an initial voltage drop. When the equipment or real load is connected, there is only a small amount of additional voltage drop. Since the difference between no-load and full-load voltage has been decreased, the regulation of the power supply output is much improved.

2. The resistor provides a path to DISCHARGE the filter capacitors after the power supply is turned off. From this advantage, it is called a BLEEDER RESISTOR. Without a bleeder, a power supply may retain, for some time, a high voltage which can be dangerous to personnel working on the equipment.

3. A bleeder maintains a current through the filter choke and stores energy in its magnetic field. The filter action of the choke is greatly improved. The resistance selected for this application should conduct ten percent of the total load current. It should have sufficient wattage rating to carry the current.

VOLTAGE DIVIDERS

The fixed resistance across a power supply may be a single resistor, or a resistor with taps, or a slider adjustment so that voltages other than the full voltage may be secured. The fixed

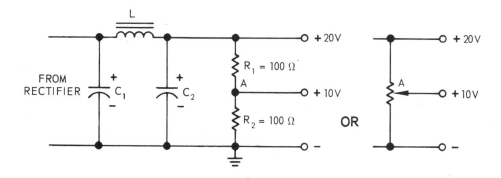

Fig. 12-35. The bleeder resistor may be used as a voltage divider.

resistance may be two or more resistors in series. This is called a VOLTAGE DIVIDER. See Fig. 12-35.

Using two 100 ohm resistors in a series voltage divider across the output of a 20 volt supply, the total divider resistance is:

$$R_1 + R_2 = R_T = 200 \text{ ohms}$$

The divider current is:

$$I = \frac{20V}{200\,\Omega} = .1 \text{ amp}$$

The voltage at point A is:

$$V_{R_1} = IR_1 = .1 \times 100 = 10 \text{ volts}$$

This seems simple enough, and now two ouput voltages are available: 10 volts at the tap; full supply voltage of 20 volts.

However, if a 100 ohm load is connected between ground and the tap, the voltage division is upset. Refer to Fig. 12-36.

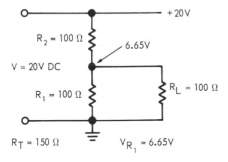

Fig. 12-36. This circuit shows the effect of loading the divider. Corrections must be made.

The load is in parallel with R_1 of the divider. The total resistance of the circuit is:

$$R_T = R_2 + (R_1 \text{ in parallel with } R_L)$$

$$R_T = 100\,\Omega + 50\,\Omega = 150 \text{ ohms}$$

The total current is:

$$I_T = \frac{20V}{150\,\Omega} = .133 \text{ amp}$$

The voltage at the tap now becomes:

$$V_{R_1} = IR = .133 \times 50 = 6.65 \text{ volts}$$

This remarkable change is caused by LOADING the divider circuit. Changes of this magnitude are not acceptable.

If 10 volts are required at the tap under load, the original divider resistances must be designed to provide this voltage under load.

VOLTAGE REGULATION

The output from a power supply filter is diagrammed in Fig. 12-37. The output voltage is designated as 25 volts. To this has been connected a load R_L and a variable series resistance R. The total voltage is divided between V_R and V_{R_L}, 5 volts and 20 volts respectively.

CASE 1. If the source voltage happened to increase to 26 volts, the voltage drop across R would have to increase to 6 volts to maintain a voltage across R_L of 20 volts. This would mean a change in resistance of R.

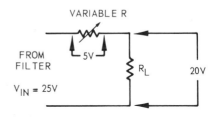

Fig. 12-37. The output voltage can be held constant under varying voltage inputs by varying regulation resistor R.

CASE 2. A decrease in the resistance of R_L would increase circuit current and increase voltage drop across R. It would be necessary to decrease the value of R to maintain 20 volts across R_L.

CASE 3. An increase in R_L would cause a decrease in current and R would have to be increased to hold voltage across R_L at 20 volts.

CONCLUSION: By varying regulator resistance R, a constant voltage can be maintained across

R_L. However, it would be impossible to vary this resistance manually and follow any sudden changes in load current. It can be done, but it must be done electronically.

ZENER DIODE REGULATION

Zener voltage occurs when a diode reaches a certain point in reverse bias. This phenomenon has been discussed earlier in this chapter under DIODE VOLT-AMPERE CHARACTERISTICS. Also see Fig. 12-6.

Special diodes called zener diodes are manufactured purposely to operate in a reverse direction. They operate with desired reverse breakover voltages from about 3 volts to 250 volts and with a minimum impedance. Also, there are power ranges available from 150 milliwatts to 50 watts.

Fig.12-38 refers to the zener diode as a voltage regulator. Note the component symbol used for the zener diode (D_1).

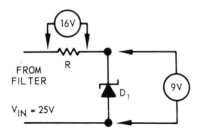

Fig. 12-38. Basic zener diode voltage regulation circuit.

The diode in this illustration is selected to have an operating voltage of 9 volts at 7.5 milliamps. Any variation of input voltage within

operating limits of D_1 will appear across R, and the voltage across D_1 will remain 9 volts. For example, if V_{in} is increased to 26 volts, V_R will increase to 17 volts.

In Fig. 12-39, a 5 kilohm load is attached to the circuit. To compute the value of regulator resistor R for this load, follow this procedure:

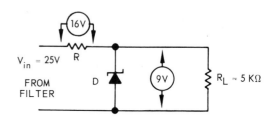

Fig. 12-39. Circuit for problem in computing regulation resistor R. Refer to text.

Compute load current:

$$I_L = \frac{9V}{5\,K\Omega} = .0018\ amp$$

Compute total current:

$$I_L + I_D = .0018 + .0075 = .0093\ amp$$

Compute R:

$$R = \frac{16V}{.0093A} \cong 1.7\ kilohms$$

If the load resistance and load current should change, the current through D_1 will also change. It will change in a direction that will hold the total current at a constant value and hold voltage V_R at 16 volts. If load current decreases, zener current increases.

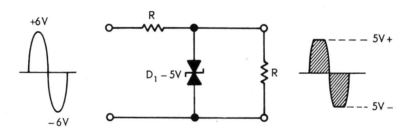

Fig. 12-40. Back-to-back zener diodes are used to limit voltage in both positive and negative directions.

In some applications, it is necessary to limit a voltage in both a positive and negative direction. Two zener diodes made back-to-back are called a double anode "symmetrical zener diode," Fig. 12-40. A 5 volt zener is used. A 6 volt peak ac wave is clipped off at 5 volts in both positive and negative directions.

SILICON CONTROLLED RECTIFIERS

A silicon controlled rectifier (SCR) diode is made by a process where two diodes are formed so that their junctions appear as in Fig. 12-41.

Note that junction 1 is forward biased, junction 2 is reverse biased and junction 3 is forward biased. Also, a third connection is made to the diode called the GATE. Regardless of the direction in which this SCR is connected in a circuit, it will not conduct, since one of the diode junctions will be reverse biased. However, the SCR will conduct if a voltage or current pulse is applied to the gate in a direction to forward bias junction 2. Also, it will remain in conduction until its anode voltage drops to zero or changes polarity, or its cathode voltage changes polarity.

In order to conduct a minimum holding current, I_H must be maintained. When the SCR is "triggered" or "fired," a maximum current will flow, depending upon power supply voltage and external circuit component values.

When the forward voltage (V_F) is gradually decreased, a point will be reached when V_F will not sustain I_H. When I_H is not maintained, the SCR will stop conducting.

In the circuit of Fig. 12-42, the SCR action can be observed.

In its present state with S_1 open, the SCR does not conduct and the lamp is out. When switch S_1 is momentarily closed, a positive voltage applied to the gate forward biases the center PN junction. The SCR is pulsed into conduction and remains so until forward voltage is removed or reversed.

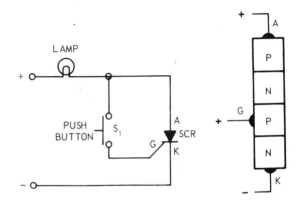

Fig. 12-42. A simple SCR switching circuit.

In recent years, hundreds of devices have been designed around the SCR. These devices include motor controls, light dimmers, automo-

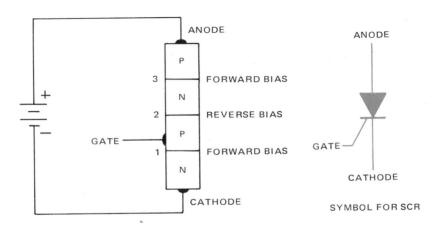

Fig. 12-41. A sketch and symbol for a silicon controlled rectifier.

tive ignition systems, battery chargers and in numerous devices which require a proportional power control. They are made in current ranges from a few milliamperes to a 150 amperes and more. See a typical example in Fig. 12-43.

Fig. 12-43. A high power SCR with ratings to 1200 peak inverse volts and a current rating of 275 amperes rms. (IRC, Inc. Semiconductor Div.)

TRIACS

Both the SCR and the TRIAC are members of the THYRISTOR family. A thyristor is a semiconductor switch whose bistable action depends on P-N-P-N regenerative feedback.

The SCR is a unidirectional device, which means that current flows in only one direction (from anode to cathode). Triacs are bidirectional devices which can conduct current in either direction.

Triacs, like SCRs, have three leads or electrodes. Refer to Fig. 12-43. Note that the names of the three leads are: Terminal No.1, Gate and Terminal No.2. Fig. 12-44 also shows the triac junction diagram.

A comparison can be drawn from a triac and two SCRs connected in parallel and turned in opposite directions. See Fig. 12-45.

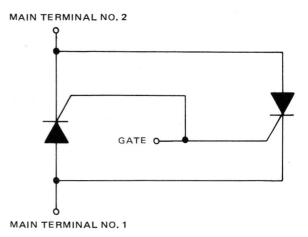

Fig. 12-45. Triac comparison as two SCRs are connected in parallel.

The output wave forms of the SCR and the triac are shown in Fig. 12-46. Note that the SCR produces a clipped-off dc wave form while the triac produces an ac wave form that only conducts partly during the full sinc wave.

LIGHT EMITTING DIODES

Light emitting diodes (LEDs) have, in just a few short years, become a major indicator device in transistor electronics. In chapter 7, a LED tester was presented as an electronic

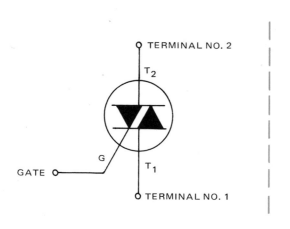

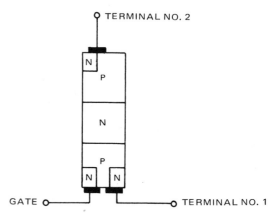

Fig. 12-44. Triac symbol and junction diagram.

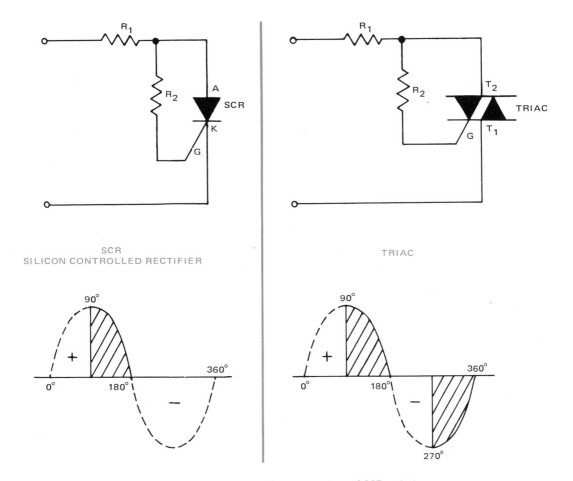

Fig. 12-46. Output wave form comparison of SCR and triac.

project. Maybe you built it. It included a light emitting diode as the indicator.

A LED is a type of diode that is made up of a P-N junction. Like regular light diodes, these devices have a low forward voltage threshold. When the threshold is overcome, the junction has a low opposition, and current flows easily. Usually, an external resistor in the circuit limits this current. The symbol for a LED is shown in Fig. 12-47.

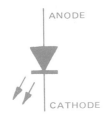

Fig. 12-47. Symbol for light emitting diode.

The LED comes in various colors such as red, yellow or green. They are smaller in size than a regular incandescent lamp. LEDs consume very little power and have an extremely long life (100,000 hours or more).

POWER SUPPLY PROJECT

Build a power supply that produces a 9 volt dc output. This project can be used to power any electronic device including a transistor radio that operates from 9 volt batteries. This circuit uses many of the concepts that you have learned in this chapter.

A printed circuit layout for the 9 volt power supply is shown in Fig. 12-48. The schematic and parts list is given in Fig. 12-49. Fig. 12-50 illustrates the chassis layout for this power supply project.

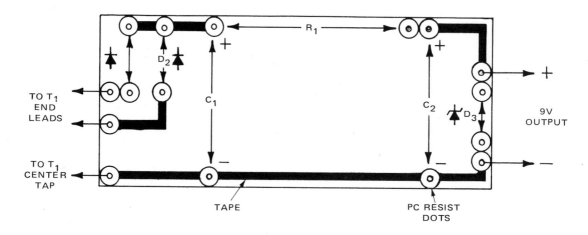

Fig. 12-48. Printed circuit layout for 9 volt power supply (actual size).

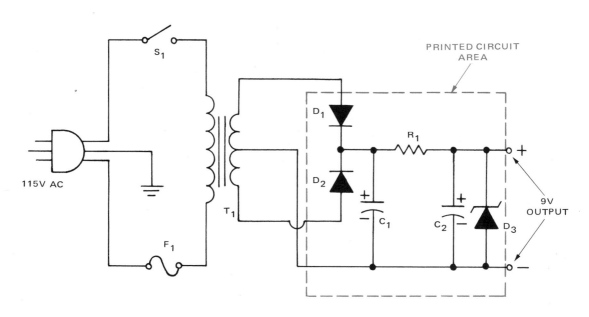

PARTS LIST FOR 9 VOLT POWER SUPPLY

C_1 — 2000 μF, 25V electrolytic capacitor
C_2 — 500 μF, 25V electrolytic capacitor
D_1, D_2 — 50 PIV @ 1A silicon diodes
     (HEP R 0050 or RCA SK 3030)
D_3 — 9.1V, 5W zener diode (HEP Z 2513 or IN 5346A)
F_1 — 1/4A fuse
R_1 — 10 Ω, 10W resistor

S_1 — SPST switch
T_1 — 25.2V CT transformer (Stancor P-8180 or
     Radio Shack 273-1512)
Misc. — chassis, printed circuit board, grounded
     line cord, fuse holder, grommets, solder,
     hardware, wire, etc.

Fig. 12-49. Schematic and parts list for 9 volt power supply.

Fig. 12-50. A 9 Volt Power Supply.

TEST YOUR KNOWLEDGE

1. Why must some forward voltage be applied to a diode before conduction starts and increases with voltage?
2. _____ are majority carriers in a N type crystal.
3. _____ are the minority carriers in a P type crystal.
4. What is meant by the depletion area and space charge distribution of a diode?
5. In a forward bias condition, the _____ terminal of the source is connected to the _____ type crystal and the _____ type crystal of the diode.
6. What is meant by the PIV of a diode?
7. Why is a heat sink used with diode rectification?
8. Rectification is the process of changing _____ to _____.
9. The output frequency of a full-wave rectifier is four times input frequency. True or False?
10. An inductor used in series with a rectifier and load is called a _____.
    a. Choke.
    b. Triac.
    c. Voltage divider.
11. What is the purpose of placing a small resistor in series with a diode and the input filter capacitor?
12. Why is the voltage output of a capacitor input filter higher than a choke input filter?
13. A 12 volt, 10 milliamp zener diode is used in a regulation circuit. The power supply has a 25 volt output. Compute the value of the regulation resistance when using a 5 kilohm load.
14. What power must be dissipated in both zener and regulation resistor?
15. Can an SCR be used as a light dimmer? Explain.
16. What is a triac?
17. A power supply has what purpose?
18. In power supplies, _____ may be used to step-up or step-down voltages.
19. A light emitting diode (LED) is made up of a _____ junction.
    a. NPN.
    b. PNP.
    c. PN.
20. A bleeder resistor provides a path to discharge the filter capacitors after the power supply is turned off. True or False?

Chapter 13

THE TRANSISTOR

The transistor has revolutionized the electronics industry. The transistor was discovered or born in the Bell Laboratories in 1948. Its advantages have opened many new possibilities for design engineers. It has almost replaced the vacuum tube as a basic amplifier.

In this chapter, you will:

1. Become familiar with the theory and action of a transistor.
2. Study the characteristics of transistors.
3. Identify the common configurations of transistor circuits and their phase relationships.

Some advantages of the transistor are:

1. Extremely long life, which contributes substantially to the dependability and maximum life of electronic equipment. In fact, the actual life of the transistor is yet to be accurately determined.

2. No power is required to "heat it up." The transistor is ready to go to work at once. This saves electrical energy, and cooler operation is assured. You might remember waiting for the old vacuum tube radio to warm up before it was ready to use.

3. The transistor operates on low voltages, which is a distinct advantage. Since the current drain is also low, small portable equipment may be satisfactorily supplied by dry cells, similar to flashlight cells. The need for high voltage transformers and rectifiers is considerably reduced.

4. The small physical size of the transistor has been a major contributing factor in the development of light and portable equipment. It has made possible the almost infinite variety of electronic devices used in our space programs and orbiting satellites. It has decreased the size of a sophisticated computer, which previously filled a large room, to a single console about the size of an office desk. Now, transistors are found in microelectronics, many of them so small that they must be observed under a powerful magnifying glass. Each day, new purposes and applications are found for this marvelous transistor.

5. The transistor is a rugged component. It will withstand excessive vibration and shock. Contrast this to the glass vacuum tube filled with delicate, precisely spaced elements.

6. Vacuum tubes cost considerably more than transistors.

Is it any wonder that the transistor has assumed such a place of importance in all industries concerned with the design and fabrication of electronic gear? A major disadvantage, as you will discover, is the inability of the transistor to operate satisfactorily in high surrounding temperatures. The characteristics of a transistor are very sensitive to changes in ambient temperatures.

THE TRANSISTOR

The transistor is a three element device made up of semiconductor materials. One of the

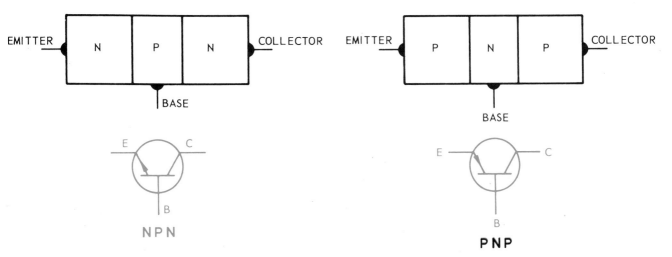

Fig. 13-1. Transistors are manufactured in two basic types. Notice the symbols used in schematic diagrams for each type.

popular methods of manufacturing is by growing a crystal. Transistors are made and the junctions between elements are formed by methods similar to the making of a PN diode. (Review Chapter 12.) This process produces a sandwich of semiconductors of specific types. See Fig. 13-1.

The transistor may be a PNP or an NPN type. The symbols used in electronic schematics are also displayed in Fig. 13-1. Note that each element of the transistor is named: EMITTER,

BASE and COLLECTOR. These terminals will be assigned the letter symbols of E, B and C respectively.

After the transistor crystals are formed and leads attached, the entire assembly is welded in a metal enclosure which protects the delicate transistor from heat, light, dust and moisture which might change its characteristics. The outline forms of enclosures and lead placement will be found in the JEDEC diagrams in any transistor manual.

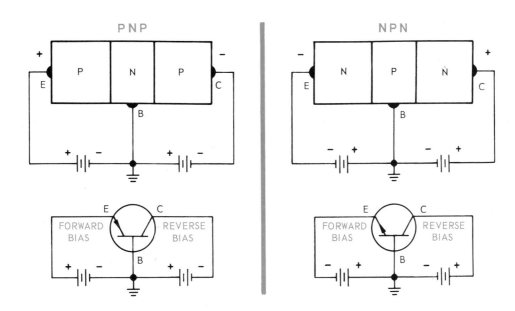

Fig. 13-2. These circuits show the forward and reverse bias of transistor junctions.

TRANSISTOR BIASING

Review the theory of PN junction diodes before you advance to the transistor. In Fig. 13-2, both the PNP and NPN transistor block diagrams and schematics are shown.

It is of particular importance to note that the emitter-base junction of each transistor is connected in a FORWARD BIAS manner. Therefore, the emitter-base junction is a low resistance or, more accurately, a low impedance circuit, and current could flow rather freely in this circuit. The collector-base junction, however, is connected in a REVERSE BIAS manner. It is a high impedance circuit.

Currents flowing in or out of the emitter are termed I_E. Currents in the base circuit are I_B. Collector currents are I_C. Capital letters are used to describe these currents because they are the dc values of currents. Later, we will consider ac signal currents which will be designated by lower case letters.

TRANSISTOR POTENTIAL HILLS

The potential hills and depletion areas of the two junctions of a transistor are illustrated in

Fig. 13-3. The establishment of these areas is a result of diffusion and recombination of electron-hole pairs along both the EB and CB junctions. Generally, the collector-base field is larger than the EB field due to the amount of dopants used in the collector crystal. Without external biasing voltage, these potential hills remain in a static or equilibrium state.

The charge distribution across the total transistor is also shown in Fig. 13-3. The NET CHARGE of the transistor is ZERO.

Biasing voltages are now applied in the proper directions in Fig. 13-4.

Consider the potential barriers and depletion areas. Like the semiconductor diode, a forward bias reduces the width of the depletion area and opposes the potential barrier. Conduction will

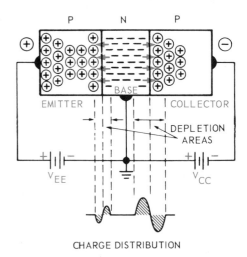

CHARGE DISTRIBUTION

Fig. 13-4. The forward bias of the EB junction decreases the potential hill and the depletion area. The reverse bias of the CB junction increases the depletion area and potential hill.

occur as the holes in the emitter drift toward the EB junction and combine with the electrons in the base section.

In case of the reverse biased CB junction, very little current will flow from C to B. However a larger amount will flow between C and E.

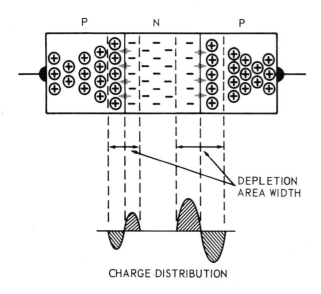

CHARGE DISTRIBUTION

Fig. 13-3. Like the diode, potential hills and depletion areas are formed at the EB and CB junctions of a transistor.

TRANSISTOR CURRENT PATHS

Study the circuit of the PNP transistor in Fig. 13-5. Current flows through the EB junction since it is forward biased. Conduction in the P emitter is by holes; conduction in the N base is by electrons. The holes and electrons combine by diffusion across the EB junction. This current is marked as I_B. But the base section is very thin and most of the holes injected into the base section DO NOT recombine to form a base current. These injected holes, which are MINORITY CARRIERS in the N type base, come under the influence of the

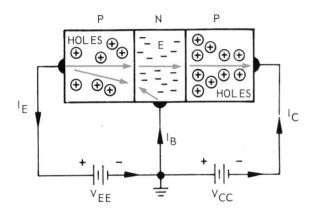

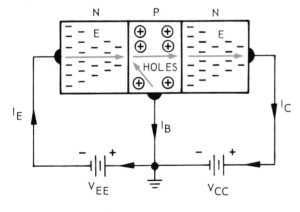

Fig. 13-5. Arrows indicate the current paths of the majority carriers in both types of transistors.

strong electric field of the CB junction and cross with little effort into the collector section. Electrons from the collector supply (V_{CC}) fill the holes in the P collector section, and this constitutes the collector current I_C. Note that

regardless of direction of conduction in the transistor, the electron flow in the external circuit is from negative to positive.

The action within the NPN transistor is also described in Fig. 13-5. The polarities of the supply voltages, V_{EE} and V_{CC} have been reversed for proper biasing. Electrons from V_{EE} flow through the N type emitter and combine with holes in the P type base. This constitutes a base current I_B. However, the base section is very thin and electrons in a P crystal are minority carriers. As a result, most of the electrons are influenced by the strong CB field and flow easily across the CB junction into the collector. Electrons in the collector are attracted to the positive terminal of V_{CC}, and this flow of electrons is the collector current I_C.

SATURATION CURRENT

Up to this point, we have only discussed the current in a transistor by majority carriers in P and N crystals. There are always some minority carriers in each crystal due to thermal or light energy producing electron-hole pairs. In Fig. 13-6, the collector-base junction is reverse biased when considering majority carriers, but forward biased for minority carriers.

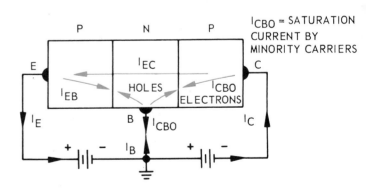

Fig. 13-6. Diagram shows direction of I_{CBO} saturation current in a PNP transistor.

So a few microamperes of reverse current flow in the collector-base circuit. This current is designated as I_{CBO} or the reverse SATURATION CURRENT. The total base current I_B is equal to I_{EB} minus I_{CBO}.

The designation I_{CBO} means: Current from C to B with emitter circuit open (O). Notice that I_c is equal to the sum of I_{EC} and I_{CBO} since both are flowing in the same direction.

In a similar diagram for the NPN transistor, Fig. 13-7, only the bias battery polarities have been reversed. Minority carriers in the N collector are holes; minority carriers in the P base are electrons. Compare the current paths and carriers in the NPN and PNP transistor to satisfy yourself that the same equations apply.

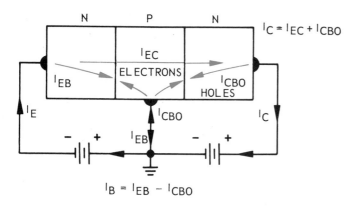

Fig. 13-7. This diagram shows the direction of I_{CBO} saturation current in a NPN transistor.

TRANSISTOR CURRENT GAIN

In our previous discussion, it was explained that only a small portion of the emitter current I_E flowed to the base connection. Most of the current carriers continued on directly to the collector. Generally, about 95 to 99 percent of emitter current I_E flows directly to the collector and constitutes collector current I_c. As a result, I_c is slightly less than I_E and the CURRENT GAIN IS LESS THAN ONE.

Current gain in this circuit configuration may be expressed as:

$$\text{Current amplification factor } (\propto \text{alpha}) = \frac{I_c}{I_E}$$

The amplification factor $\propto$ is also expressed as h_{FB}. The subscript letter F means "forward direction," and B refers to a common base circuit. Now these simple equations may be set up. Be certain that you understand each of them. Terms are placed in their appropriate places in Fig. 13-8.

1. I_E is the sum of base current I_B and collector current I_C.

$$I_E = I_B + I_C$$

2. Since I_C is equal to $\propto I_E$, then:

$$I_E = I_B + \propto I_E$$

3. I_B is the difference between emitter current I_E and collector current I_C.

$$I_B = I_E - I_C$$

4. And since I_C is equal to $\propto I_E$, then;

$$I_B = I_E - \propto I_E$$

5. I_C equals the product of I_E and the current amplification factor.

$$I_C = \propto I_E$$

6. Also, I_C equals the difference between I_E and I_B.

$$I_C = I_E - I_B$$

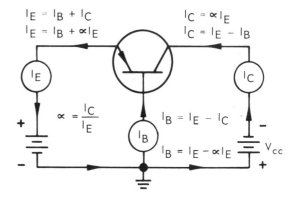

Fig. 13-8. The relationship between I_C, I_B and I_E are illustrated in this diagram.

THE TRANSISTOR AMPLIFIER

If the forward current amplification ∝ is less than one, how can a transistor amplify a signal? It is possible due to a difference in the resistance of the input circuit to the resistance of the output circuit. This is called the RESISTANCE RATIO. This is how the transistor was named. It is a "transfer of resistance," and the term was shortened to "transistor." It has already been explained that due to forward bias, the emitter-base resistance is low. Due to reverse bias, the collector-base resistance is high. Study Fig. 13-9 in which an approximate equivalent circuit of the transistor is shown.

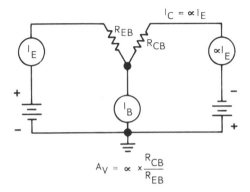

$$A_V = \propto \times \frac{R_{CB}}{R_{EB}}$$

Fig. 13-9. Voltage gain in a transistor is realized by a high resistance ratio.

The current I_E in the emitter circuit is relatively high due to its low resistance R_{EB}. However, current $\propto I_E$ also flows in the high resistance collector base circuit. The approximate voltage gain may be computed by comparing the voltage drops across the resistances.

Voltage gain,

$$A_V = \frac{\propto I_E R_{CB}}{I_E R_{EB}} \quad \begin{array}{l}\text{(voltage across } R_{CB}) \\ \text{(voltage across } R_{EB})\end{array}$$

The term I_E cancels out, and the equation is rewritten to:

$$A_V = \propto \times \frac{R_{CB}}{R_{EB}} \quad \text{or} \quad \propto \times \text{ resistance ratio}$$

EXAMPLE: Assume an emitter-base input resistance of 50 ohms and the collector-base output resistance as 500 kilohms. When ∝ equals .99, compute voltage gain:

$$A_V = .99 \times \frac{5 \times 10^5}{5 \times 10}$$

$$= .99 \times 10,000 = 9900 \text{ volts}$$

The computed voltage gain will be reduced considerably by other external resistors needed in a practical circuit. This explanation purposely omits several contributing factors to voltage gain in order to explain the resistance ratio concept of transistor amplification.

NOTE: Always disconnect the power source from a transistor circuit before making any disconnects or changes in a circuit. Transient voltages and current surges may be sufficient to destroy the transistors.

THE COMMON BASE CIRCUIT

When a transistor is connected so that the BASE is common to both the input and the output circuits, it is called the COMMON BASE or CB circuit. The CB circuit is illustrated in Fig. 13-10.

The input signal is applied across the EB junction. The output signal is taken from the CB junction. The base is common to both signals and usually is grounded. The circuit is sometimes called the "grounded-base circuit."

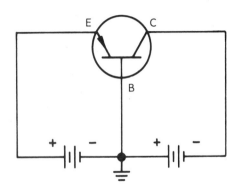

Fig. 13-10. The common base circuit (CB).

The circuit in Fig. 13-11 is used to measure the static collector characteristics of the transistor connected in this configuration. Setting the emitter current at a fixed value, the collector voltage V_{CB} will be set in steps of one volt to determine its effect on collector current I_c. For each fixed value of emitter current, the measurements are repeated. From the information obtained, a curve may be plotted, Fig. 13-12, which shows graphically the effect collector voltage has on collector current for each constant value of emitter current. At any fixed value of voltage V_{CB}, the current amplification factor may be found by:

$$\propto \; = \; \frac{I_c}{I_E}$$

The fact that the curves in Fig. 13-12 are very flat indicates a relatively high output

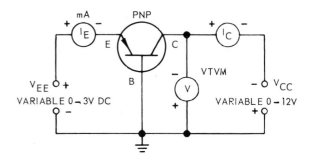

Fig. 13-11. This circuit is used to determine the static collector characteristics of a common base connected transistor.

impedance. The uniform spacing between each emitter current curve would indicate a linear relationship between input and output currents.

For small applied signals, a dynamic value of alpha may be found by the equation:

$$\propto \; = \; \frac{\Delta i_c}{\Delta i_e} \; \bigg| \quad V_{CB} \; \text{constant}$$

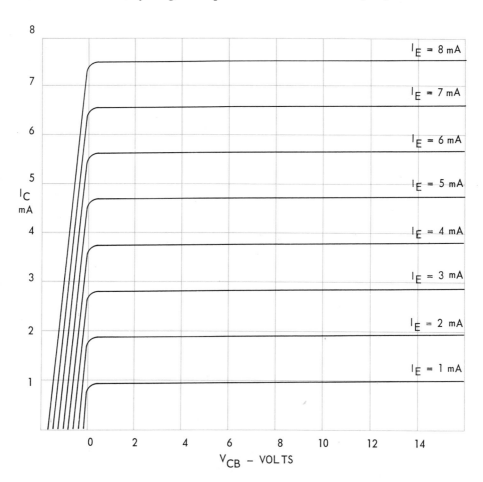

Fig. 13-12. The static characteristic curves of a transistor connected in a CB circuit.

In more advanced studies, the short circuit current gain is expressed as h_{fb}.

THE COMMON EMITTER CIRCUIT

The COMMON EMITTER or CE circuit is the most popular method of connecting a transistor. As an amplifier, Fig. 13-13, it has decided advantages over other configurations. In the CE circuit, the signal is applied across the base-emitter junction, and the output is taken from across the collector-emitter output circuits. It is frequently called the "grounded emitter circuit."

The circuit in Fig. 13-14 is used to measure the static collector characteristics of the common emitter circuit. Resistor R is a current limiting resistor used to protect the transistor from excessive currents while the test is being made. The base current is initially set at zero by adjusting V_{EB} to zero. The collector voltage V_{CE} is increased in steps of one volt, and the collector current I_c is recorded for each value

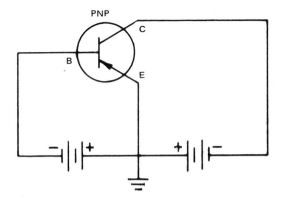

Fig. 13-13. The common emitter circuit.

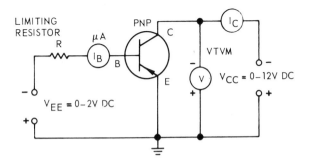

Fig. 13-14. This circuit is used to determine the static collector characteristics of the common emitter connected transistor.

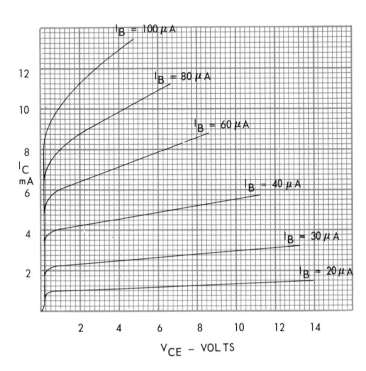

Fig. 13-15. The static characteristic curves of a transistor connected in the CE circuit.

of V_{CE}. Then, the base current is adjusted to another fixed value, and the collector voltages and currents are recorded in steps as before. A complete set of curves is exhibited in Fig. 13-15. Note that base current is in microamperes. There is a separate curve for each fixed value of base current.

Next, we will examine these static characteristic curves and see what information they contain.

1. At a certain value of V_{CE} and I_B, a collector current I_C may be found. By increasing V_{CE}, and holding the value of I_B fixed, very little increase in I_C is realized.

   CONCLUSION: A variation in collector voltage has only a minimal effect on collector current.

2. If, at a certain value of V_{CE}, the value of I_B is increased up to the next higher curve, there is a large increase in collector current.

   CONCLUSION: A small variation in base current will produce a large change in collector current.

The forward current amplification factor of the common emitter circuit is called BETA (β). Expressed in equation:

$$\text{(BETA)}\ \beta_{dc} = \frac{I_C}{I_B}\ \bigg|\ V_{CE}\ \text{constant}$$

Usually, transistor manufacturers will supply you with specifications and common-emitter characteristic curves for any specified transistor.

The dynamic value of beta for small signals also can be found from the curves.

$$\beta_{ac} = \frac{\Delta i_c}{\Delta i_b}\ \bigg|\ V_{CE}\ \text{constant}$$

The symbol for beta also will be expressed as h_{fe}.

From the equation $I_C = I_E - I_B$ and $I_C = \propto I_E$, the following relationship can be found:

$$\beta_{dc} = \frac{\propto I_E}{I_E - \propto I_E} = \frac{\propto}{1 - \propto}\quad \beta = \frac{\propto}{1 - \propto}$$

and by complementary relationships:

$$\propto = \frac{\beta}{\beta + 1}$$

THE COMMON COLLECTOR CIRCUIT

The COMMON COLLECTOR or CC circuit is diagrammed schematically in Fig. 13-16. The value of R is quite large and effectively isolates the transistor emitter from ground. The value of C is selected for a very low reactance path for signal frequencies to ground. The collector is at ground potential for signal frequencies. Review Chapter 9, Fig. 9-18 and text.

The input to the circuit is applied between base and ground. The output is taken from across R or between emitter and ground. Since the collector is at ground potential, the collector is common to both input and output signals.

This particular circuit has a high input impedance and a low output impedance. It is used as an impedance matching circuit. Its voltage gain must be less than one.

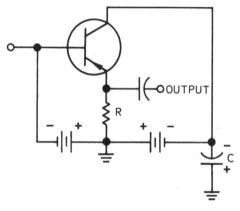

Fig. 13-16. The common collector circuit.

TYPE OF CIRCUIT	R_{IN}	R_{OUT}	A_V	A_I	A_P
COMMON BASE – CB	LOW 50–150 Ω	HIGH 300 K–500 KΩ	HIGH 500–1500	LESS THAN ONE	MEDIUM 20–30 db
COMMON EMITTER – CE	MEDIUM 500 Ω–1.5 KΩ	MEDIUM 30 K–50 KΩ	MEDIUM 300–1000	MEDIUM 25–50	HIGH 25–40 db
COMMON COLLECTOR – CC	HIGH 20 K–500 KΩ	LOW 50 Ω–1 KΩ	LESS THAN ONE	MEDIUM 25–50	MEDIUM 10–20 db

Fig. 13-17. Typical values of input and output impedances of a transistor in three configurations, CB, CE and CC.

POWER GAIN

In previous discussions, the voltage gain A_V and the current gain A_I have been mentioned. By the power law, $P = I \times V$, the power gain A_P can be found by the product of the current and voltage gains:

$$A_P = A_V A_I \quad \text{and} \quad A_P = \frac{P_{out}}{P_{in}}$$

Since $\propto$ is current gain in the CB circuit, the power gain may be approximated by:

$$A_P = \propto \times A_V$$

The same applies to the CE circuit since β is the current gain:

$$A_P = \beta \times A_V$$

Consideration of the resistance ratio may be used to compute a satisfactory power gain since

$P = I^2 R$. For the CB circuit:

$$\frac{P_{out}}{P_{in}} = \frac{(\propto I_E)^2 \times R_{CB}}{I_E^2 \times R_{EB}} = (\propto)^2 \frac{R_{CB}}{R_{EB}}$$

Typical values of input and output impedances of a transistor in the three configurations CB, CE and CC are found in the table in Fig. 13-17. You should memorize this table.

These approximations are for the transistor only and do not consider external circuit components.

VOLTAGE MEASUREMENTS IN TRANSISTOR CIRCUITS

The polarity of voltages has been discussed as it applies to reverse and forward bias of transistor junctions. These measurements are summarized in Fig. 13-18.

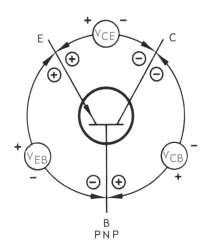
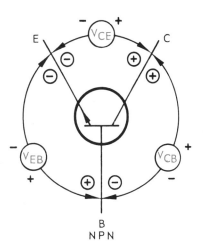

Fig. 13-18. These diagrams summarize the polarities of voltages at transistor terminals and indicate correct meter connections.

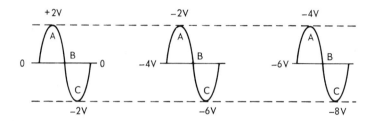

Fig. 13-19. Waves show the voltage variation of a 2 volt peak signal around three different reference levels. A is always more positive than B. C is always more negative than B.

It is extremely important to understand that two elements of a transistor may be positive in respect to ground, yet one may be more positive or more negative than the other. In measuring transistor voltages, it is recommended that you use a Vacuum Tube Voltmeter (VTVM) or Digital Voltmeter. The high internal impedance will not load the circuit and produce inaccurate readings. Voltage waves with three different reference levels are drawn in Fig. 13-19.

PHASE RELATIONSHIP IN COMMON BASE AMPLIFIER

In the illustrations showing phase relationship of input and output signals, certain resistors are added to complete the amplifier circuit. In Fig. 13-20, the input signal appears across R_E and the output appears across collector load resistor R_C.

First, consider the transistor as cutoff and not conducting. Since there is no collector current, the voltage drop across R_C equals zero,

and the collector voltage V_C is maximum and equal to V_{cc}.

Next, consider the transistor in full conduction and maximum collector current. The voltage across R_C approaches V_{cc}, $I_c \times R_c = V_{cc}$, and the collector voltage V_c approaches ZERO.

In Fig. 13-20 for the PNP transistor, an operating current I_c is considered to be flowing, which produces some voltage drop across R_C. The voltage at C is less negative than the supply voltage V_{cc} by the amount of the voltage drop across R_C. Now that a quiescent (no signal) voltage is established, a signal is applied to the emitter base circuit. Follow the action:

1. As the input signal moves in a positive direction, the forward bias of the EB junction is increased. This increases the emitter current AND the collector current.

2. An increase in I_c produces a larger voltage drop across R_C.

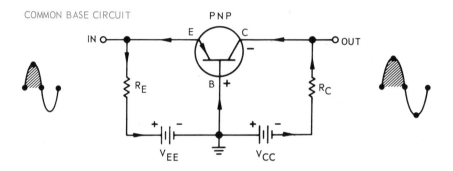

Fig. 13-20. The common base circuit does not invert the signal. Input and output are in phase.

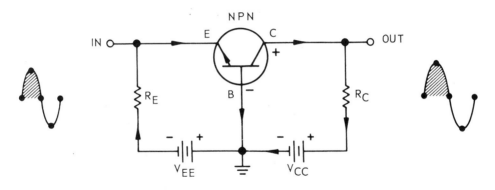

Fig. 13-21. A positive input decreases forward bias of EB, and decreases I_C and the voltage drop across R_C. Therefore, C becomes more positive.

3. The voltage at C must become less negative or moves in a POSITIVE DIRECTION.

4. As the input signal moves in a negative direction, the forward bias of the EB junction is reduced. This will cause I_c to decrease.

5. A decrease in I_c will produce a lesser voltage drop across R_c.

6. The voltage at C must become more negative than it was at its quiescent state.

CONCLUSION: The CB amplifier does not cause a phase reversal of the signal. The input and output signals are in phase.

In Fig. 13-21, the same circuit is drawn for the NPN transistor. All polarities are reversed, yet the same theory of operation will apply. Study the circuit and be able to explain the in-phase signal relationship with either type of transistor.

PHASE RELATIONSHIP IN COMMON EMITTER AMPLIFIER

In the circuits of Figs. 13-22 and 13-23, the CE amplifier is diagrammed for the PNP and NPN transistor. In the PNP circuit, once again assume a quiescent operating current and voltage drop across R_c. V_c is still negative, but less than V_{cc} by the amount of voltage drop across R_c.

1. A positive going signal applied to the base-emitter junction will make the base less negative and reduce the forward bias of the EB junction. The voltage at B is the algebraic sum of fixed bias voltage and the signal voltage.

2. A reduced forward bias will cause a decrease in I_c and a lesser voltage drop across R_c.

3. The voltage at C must become more negative and approaches the value of V_{cc}.

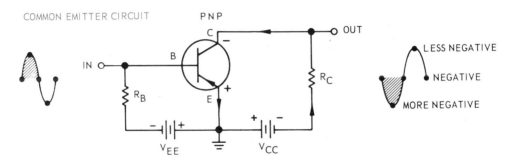

Fig. 13-22. The common emitter circuit causes a 180 deg. phase inversion of the signal.

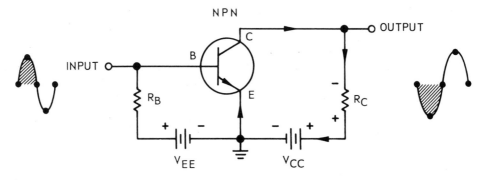

Fig. 13-23. Using a NPN transistor. A positive input signal makes B more positive and increases V_{EB}. As a result I_C increases, causing an increased voltage drop across R_C. Then C becomes less positive or more negative. Signal is inverted.

4. When a negative going signal is applied to B, then B becomes even more negative in respect to E and forward bias is increased.

5. Consequently I_C will increase and the voltage drop across R_C will increase.

6. The voltage at C must become less negative or more in a positive direction.

CONCLUSION: The CE amplifier causes a 180 deg. phase shift of the signal. It inverts the signal.

In Fig. 13-23, the same circuit is drawn using the NPN transistor. Study circuit and understand how phase inversion is accomplished.

PHASE RELATIONSHIP IN COMMON COLLECTOR AMPLIFIER

In the CC circuit, Fig. 13-24, the output signal is taken from the emitter end of R_E.

Capacitor C is a low reactance path to ground for the signal. The collector is at signal ground potential. Again, a definite quiescent current is assumed which produces a voltage drop across R_E. Then E is more negative than ground by the amount of the voltage drop.

1. A positive going signal applied to the base of the transistor will make the base more positive or less negative in respect to V_E. This decreases the forward bias of the emitter-base junction.

2. This will cause a decrease in I_C and also I_E since they are both the same current. The voltage drop across R_E will reduce, and the emitter end of R_E will become less negative or will move in a more positive direction.

3. A negative going input signal will make the base more negative and increase the forward bias. I_C and I_E will both increase.

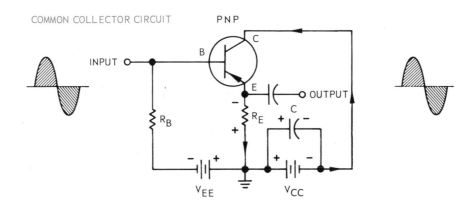

Fig. 13-24. The common collector circuit does not invert the signal.

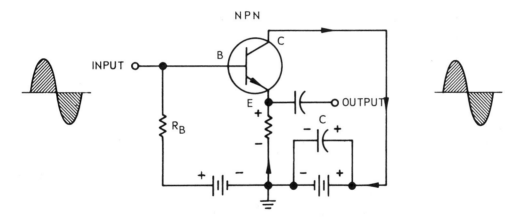

Fig. 13-25. An increase in I_E causes the E end of R_E to become more positive. The signal is not inverted.

4. The voltage across R_E and E will become more negative.

CONCLUSION: The input and output signals of a CC amplifier are IN PHASE.

Follow the diagram of the NPN transistor in Fig. 13-25 and assure yourself that this conclusion holds true.

The relationships of the input and output waveforms are summarized in Fig. 13-26.

TRANSISTOR TESTER PROJECT

This chapter dealt with the fundamental principles of the transistor. A transistor tester is an excellent project that almost every experimenter needs.

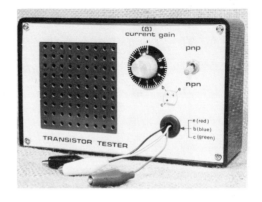

Fig. 13-27. Transistor Tester.

This inexpensive tester will test almost any type of conventional transistor. The circuit uses an oscillator, and the transistor to be tested becomes a part of the oscillator circuit. A "good" transistor will give an audio tone in the speaker. If the transistor is bad, no tone will be heard.

TYPE OF AMPLIFIER	INPUT WAVEFORM	OUTPUT WAVEFORM
COMMON BASE (CB)		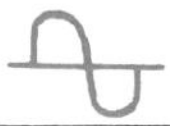
COMMON EMITTER (CE)		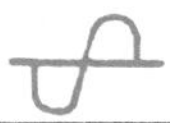
COMMON COLLECTOR (CC)		

Fig. 13-26. Phase relationships of waveforms in transistor amplifiers.

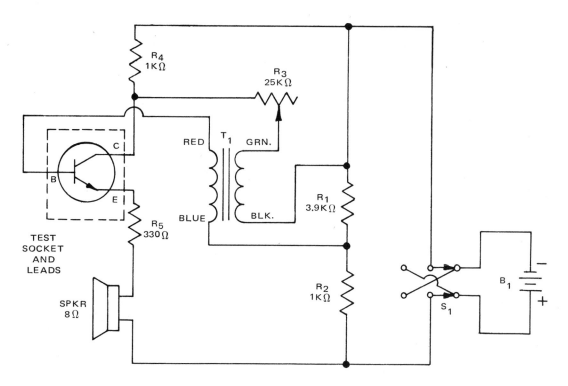

PARTS LIST FOR TRANSISTOR TESTER

R₁ — 3900 Ω, 1/2W resistor	T₁ — transistor output transformer (primary
R₂ — 1000 Ω, 1/2W resistor	10,000 Ω, secondary 2000 Ω)
R₃ — 25 K Ω, potentiometer with switch	STANCOR TA-35 or equivalent
(linear taper)	SPKR — 8 Ω miniature speaker
R₄ — 1000 Ω, 1/2W resistor	B₁ — 1 1/2V "D" cells (2)
R₅ — 330 Ω, 1/2W resistor	Misc. — plastic chassis, grille for speaker, knob,
S₁ — DPDT switch	alligator clips, wire, solder, grommets, etc.

Fig. 13-28. Schematic and parts list for Transistor Tester.

The current gain (beta) is controlled by the variable resistor R_3. The transistor may be plugged into socket, or the test leads may be used to connect to the transistor. The type of transistor (PNP or NPN) needs to be known. Switch (S_1) must be set to the proper position before operation. See Fig. 13-27.

A schematic and parts list are provided in Fig. 13-28.

TEST YOUR KNOWLEDGE

1. State four advantages of a transistor over a vacuum tube.

2. The three terminals on the transistor are called _____, _____ and _____.

3. What causes a potential hill?

4. In a transistor, the emitter-base junction is always _____ biased, and the collector-base junction is always _____ biased.

5. What is the cause of saturation or reverse leakage current?

6. A certain transistor has an I_E of 10 mA and I_B is 50 μA. What is its current gain?

7. A certain transistor has a alpha of .98 and the I_E is 10 mA. Then, I_C is _____.

8. A certain transistor has an alpha of .99 and a resistance ratio of 50/100,000 ohms. What is its voltage gain?

9. What is the power gain of the transistor used in question 8?

10. Draw a circuit for a common base transistor configuration.

11. Draw a circuit for a common emitter transistor configuration.

12. Draw a circuit for a common collector transistor configuration.

13. The designation I_{CBO} stands for current from C to B with _____.

14. Generally, about 95 to 99 percent of emitter current I_E flows directly to the collector and constitutes collector current I_C. True or False?

15. In measuring transistor voltages, it is recommended that you use a _____ or a _____.

16. What is the phase relationship between input and output of a common base (CB) amplifier circuit?

17. What is the phase relationship between input and output of a common emitter (CE) amplifier circuit?

18. In a common emitter (CE) transistor circuit, a change in I_B of 40 μA produces a change in I_C of 4 mA. What is the ac beta?

Chapter 14

TRANSISTOR AMPLIFIERS

A major function of the transistor is to amplify. In this chapter you will:

1. Describe various methods of biasing transistor circuits.
2. Learn how to graphically design a transistor amplifier by using load lines.
3. Apply maximum power dissipation specifications in the design of a transistor amplifier.
4. Distinguish between types of amplifiers.

GENERAL BIAS CIRCUITS

A general circuit for biasing a transistor is illustrated in Fig. 14-1. This circuit will be applied to each of the three circuit configurations (CB, CE and CC) by selecting the required input and output points. V_{cc} is the collector supply voltage that places a reverse bias on the collector-base junction. V_{EE} is the emitter supply voltage that biases the emitter-base junction in a forward direction. R_c is the

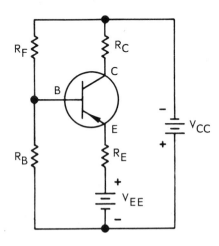

Fig. 14-1. The general bias circuit for transistors.

collector load resistor. R_E is the emitter resistor. R_B is the base resistor and R_F is the feedback resistor.

The **COMMON BASE** circuit is drawn in Fig. 14-2. In this circuit, the signal is applied

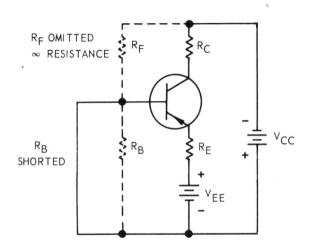

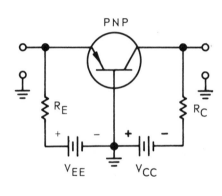

Fig. 14-2. General bias circuit converted to a CB amplifier. Compare these two schematics.

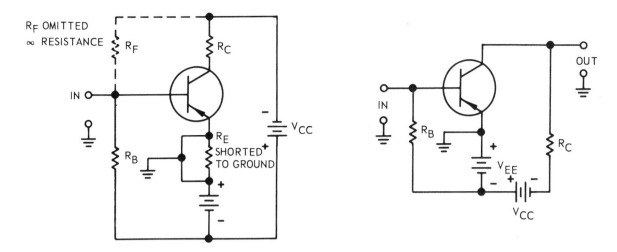

Fig. 14-3. General bias circuit converted to the CE amplifier.

between the emitter and base; the output is taken from the collector and base. The base is common to both input and output. This being the case, B is grounded and resistor R_B of the general bias circuit is short circuited. Resistor R_F is considered to have infinite resistance. Therefore, R_F is similar to an open circuit and is omitted.

The general bias circuit is converted to the **COMMON EMITTER** circuit in Fig. 14-3. The input signal is applied between base and emitter; the output is taken from the collector-emitter junction. The emitter is common to both input and output, and it is grounded. Therefore, R_E is short circuited and E is connected directly to ground. Again, R_F is considered to have infinite resistance and is represented as an open circuit.

In the **COMMON COLLECTOR** circuit in Fig. 14-4, the input signal is applied between the base and collector. The output signal is taken from the emitter and collector.

The collector is common to both input and output. Therefore, it is grounded. Since C is the common grounded terminal, R_c is short circuited. R_F is still considered as an infinite resistance and is represented as an open circuit. In practice, C is grounded for signal voltages only through a low reactance capacitor.

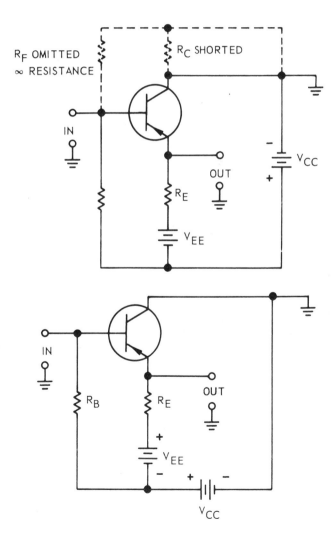

Fig. 14-4. General bias circuit converted to a CC amplifier. Compare these two schematics.

Up to this point R_F seems to always have been omitted or considered as an infinite resistance. In practical circuits, R_F will assume a value which will be explained later in this chapter.

SINGLE BATTERY CIRCUIT

For simplification, two voltage sources have been used in all circuits discussed. One source is used for the forward biasing of the EB junction; the other for the reverse biasing of the CB junction. Two batteries are unnecessary, and the CE amplifier in Fig. 14-5 is drawn and only one battery is used.

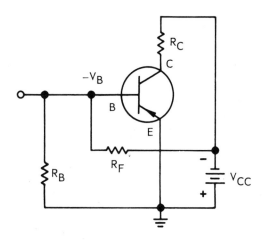

Fig. 14-5. This circuit converts the CE amplifier to a single power source.

There is no question about the reverse bias of the collector-base junction, because C is connected through R_C to the most negative point of the circuit which is the negative supply terminal. The most positive point in the circuit is the ground, which is connected directly to the positive terminal of V_{cc}.

R_F and R_B form a resistance voltage divider connected directly across V_{cc}. The voltage at B is less negative than the negative terminal of V_{cc} by the amount of the voltage drop across R_F. It is certainly negative in respect to E, which is at ground or the most positive point in the circuit. Now, you can see that the emitter-base junction is forward biased. By selecting the

proper values for R_F and R_B, the desired forward bias voltage and current can be established. The series combination of R_F and R_B must be large enough so that current drain from the supply battery will be small and long life will be assured.

METHODS OF BIAS

The FIXED BIAS method is illustrated in Fig. 14-6. Notice that R_B has been omitted. This circuit sets essentially a constant base current and is very sensitive to variations in the circuit. By the proper selection of R_F, the required forward bias voltages and base current may be set up.

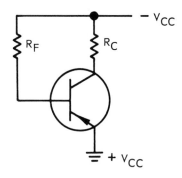

Fig. 14-6. The fixed bias method of connecting a transistor.

The SINGLE BATTERY BIAS scheme is one of the more popular methods of biasing transistors, and it enjoys some advantages. Fig. 14-7

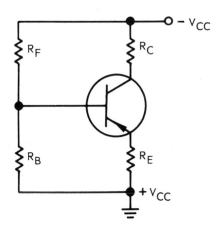

Fig. 14-7. Schematic of the single battery bias in popular use.

shows a typical circuit. Compare it to Fig. 14-5 and see that they are similar. The voltage divider $R_F R_B$ has already been discussed. It is called a "BIAS STICK."

EMITTER BIASING is a third method used to establish the forward bias of the EB junction. Refer to the circuit in Fig. 14-8.

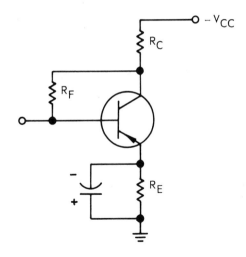

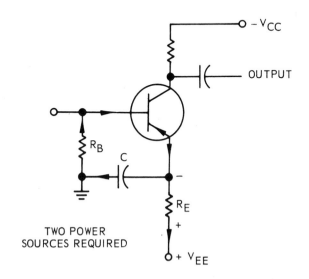

Fig. 14-8. Schematic circuit for emitter biasing.

Fig. 14-9. Schematic of transistor self-bias.

In this case, the forward bias V_{EE} will establish a constant emitter current, I_E, producing a voltage drop across R_E. R_E is selected to provide the proper forward bias. R_B is the "return" to complete the emitter-base circuit. A signal applied to the amplifier will produce an ac component in the collector current. However, this does not upset the emitter bias, because a low reactance path around R_E is provided by bypass capacitor C. This circuit almost guarantees a very stable operating point for the transistor. However, this advantage is overshadowed by the requirement of two power supplies or batteries.

The SELF BIAS schematic is shown in Fig. 14-9. This circuit differs from the fixed bias method in that bias resistor R_F is connected to the collector rather than V_{cc}. This method provides a more stable operating point than fixed biasing, and only one power source is needed.

If a fixed collector current is assumed at some selected operating point, the voltage V_c will be constant but lower in value that V_{cc} due to the drop across R_c. Any change in I_c will consequently change the value of V_c. Since the base is connected to V_c through R_F, a certain amount of degeneration will result.

What is meant by degeneration? A positive signal at the input of the CE amplifier (PNP) will make the base more positive and decrease its forward bias and decrease I_c. A reduction in I_c means a lesser voltage drop across R_c, and collector voltage V_c becomes more negative as it approaches the value of negative V_{cc}. This more negative voltage to the base through R_F tends to increase the forward bias. In other words, it opposes the increase caused by the input signal. A signal from the output of a device fed back to the input in a phase relationship which opposes the input is called DEGENERATION. The major disadvantage of the self-bias scheme is the loss of amplifier gain due to degeneration.

OPERATING POINT

The previously discussed bias schemes all tend to set up a constant value of forward bias V_{EB} which, in turn, sets up a constant collector current I_c when no signal is applied to the circuit. This fixed base current caused by V_{EB}

is the OPERATING POINT, and all variations caused by input signals should vary around this operating point.

The graph in Fig. 14-10 shows a selected operating point as 100 microamps of base current. At a constant collector voltage of 6 volts, the collector current is 3 milliamps. A signal input which causes a variation in I_B between 90 and 110 microamps will cause the collector current I_C to vary between 2 and 4 milliamps. This curve shows a linear realtionship between I_B and I_C.

This is not always so. Observe Fig. 14-11 in which part of the curve is nonlinear. If the operating point moved to the curved portion of the curve, the variation in I_C would become very distorted. If reliable operation of a transistor is to be expected, the operating point must be STABLE. Not only must V_{CE} and I_C remain stable, but also I_B at its quiescent point of operation. Therefore, the operating point is called the Q point. It is selected by reference to transistor specifications and the purpose of the amplifier.

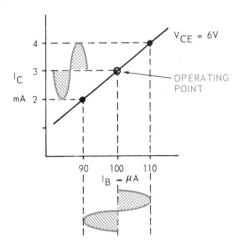

Fig. 14-10. The operating point is selected on the linear portion of the curve.

The worst enemy of a transistor is HEAT. Many nice little transistor radios have been destroyed by sunshine on the beach. Heat energy, when added to a semiconductor, raises the energy level of electrons and produces many electron-hole pairs. The N crystal is seriously

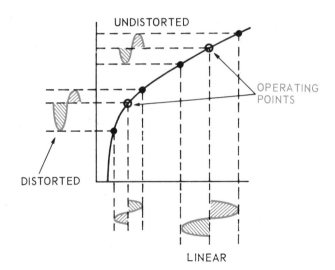

Fig. 14-11. These curves show distortion of output wave form, due to incorrect selection of bias and operating point.

affected because heat energy increases its number of holes or minority carriers. In the P crystal, many electrons or minority carriers are created.

Refer again to the equation:

$$I_C = I_{EC} + I_{CBO} = I_E + I_{CBO}$$

Now I_{CBO} represents the reverse saturation current due to minority carriers in the reverse biased CB junction. See Fig. 13-6 in Chapter 13. The value of alpha remains rather constant for a given transistor, and most variations are the result of the manufacturing process. There is little we can do about that. Therefore, we conclude that heat will raise the collector current and change the designed operating point of a transistor amplifier.

A block diagram of an N type transistor is shown in Fig. 14-12. The excessive minority carriers from the collector caused by a rise in temperature accumulate in the base section if the resistance of the base section is high. This makes the base MORE POSITIVE which, in effect, INCREASES the forward bias of the EB junction which also increases I_{EB}. The increase in current creates more heat, which again increases I_{CBO}, which again increases I_{EB} until the transistor destroys itself by thermal run-

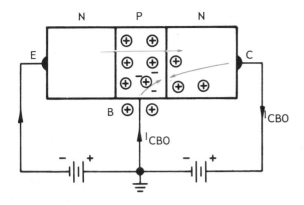

Fig. 14-12. An accumulation of holes in base section will increase forward bias of EB junction and will increase I_C.

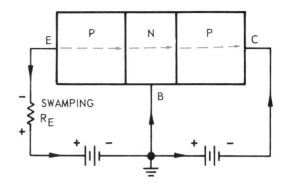

Fig. 14-14. An increase in I_E creates a greater voltage drop across R_E making the emitter end less positive. This reduces the forward bias of EB junction.

away. The accumulation of positive holes in the base section is the result of the high resistance in the base section and circuit.

A similar block diagram for the PNP transistor is found in Fig. 14-13, showing the accumulation of electrons in the base section by I_{CBO}, which makes the base more negative and starts the runaway cycles.

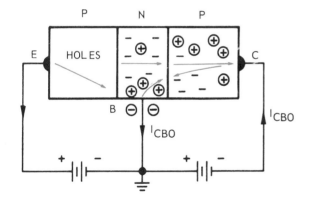

Fig. 14-13. An accumulation of electrons in base section will increase forward bias of EB junction and increase I_C.

EMITTER BASE RESISTANCE

A variation in temperature will also increase the emitter-base current. An increase in current means a decrease in resistance. So the EB junction resistance may be said to have a negative temperature coefficient. This decrease in resistance may be overcome by placing a relatively large "swamping resistor" R_E in the emitter lead as in Fig. 14-14.

Any increase in emitter current produces a voltage drop across R_E which decreases the forward bias of the EB junction and reduces the current flow. In other words, R_E counteracts the effect of a decrease in resistance of the EB junction caused by a rise in temperature.

STABILITY FACTOR

These previous discussions introduce a new term called the STABILITY FACTOR of a transistor circuit. It is assigned the letter S. The reasons that a circuit should be stable, both from the voltage and current points of view, have been illustrated.

The S of a transistor circuit is defined as the relationship between a change in emitter current to a change in saturation current.

In equation form:

$$S = \frac{\Delta I_E}{\Delta I_{CBO}}$$

S is a number only and has no units of measurement. The ideal circuit would have a stability factor of 0, which would mean that any change in I_{CBO} would have no effect whatsoever on emitter current.

$$S = \frac{0}{\Delta I_{CBO}} = 0$$

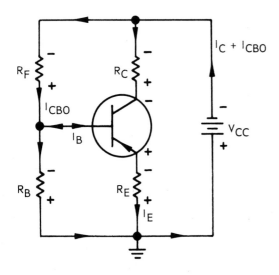

Fig. 14-15. The general bias circuit showing conductance paths for I_E, I_C, I_B and I_{CBO} and bias currents.

In practice, the design engineer attempts to keep the stability factor below 5, except in highly sophisticated types of circuits.

The general bias circuit is illustrated in Fig. 14-15 with resistors indicated. By computing the total conductance path for I_E and I_{CBO} and eliminating any effect they might have on the circuit, a very useful approximate equation for the value of S may be found.

$$S = \frac{R_F R_B}{R_E (R_F + R_B)} = +1$$

NOTE: The derivation of this formula is beyond the scope of this text.

If the value of R_F is ten times greater than R_B, then the equation may be further simplified to:

$$S = \frac{R_B}{R_E} + 1 \quad \text{or} \quad S\text{-}1 = \frac{R_B}{R_E}$$

This approximate equation will always give you a slightly worse stability factor than actually exists, so you may assume your circuit will be better.

LOAD LINES

The load line will represent graphically the performance of an amplifier circuit under load-

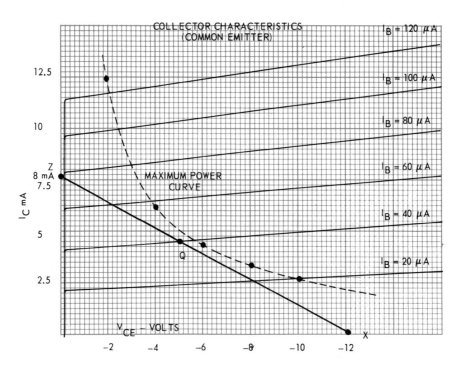

Fig. 14-16. The load line is drawn on the family of characteristic curves.

ed conditions. The family of static characteristic curves of a typical transistor is illustrated in Fig. 14-16 for a CE amplifier. A supply voltage of 12 volts will be used. $V_{cc} = -12$ volts.

1. Use Fig. 14-17 to follow computations. At the point of zero collector current, the voltage drop across R_c is equal to zero and the collector voltage V_c is close to V_{cc} or -12 volts. This point of $I_c = 0$ and V_{CE} is 12 volts is plotted in Fig. 14-16 as X.

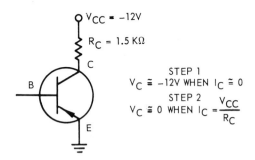

Fig. 14-17. The circuit used with characteristic curves to establish dc load line.

2. Using a load resistance $R_c = 1500$ ohms, the opposite end of the load line may be found when the maximum current I_c is flowing. The assumption is made that the transistor has zero resistance in full conduction, and the current is limited only by R_c. With maximum current the voltage drop across R_c equals V_{cc}, and collector voltage V_c equals ZERO. The current at which this occurs is:

$$I = \frac{V}{R} = \frac{12V}{1500\ \Omega} = .008 \text{ amp} = 8 \text{ milliamps}$$

The point Z is plotted on Fig. 14-16, using $V_{CE} = 0$ and $I_c = 8$ milliamps.

3. A straight line is drawn connecting the two points (X and Z). This is called the dc load line. The slope of the line is determined by the load resistance. The circuit current at any voltage between zero and 12 volts may quickly be determined.

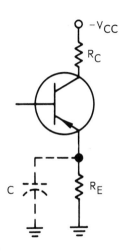

Fig. 14-18. R_E is not considered part of circuit resistance for ac if bypassed by C.

In Fig. 14-18, the circuit is redrawn to include emitter resistor R_E. Now the total resistance would be used to plot the dc load line. If an ac load line were plotted, R_E would not be a part of the total resistance if it is bypassed by a low reactance capacitor, shown as C in Fig. 14-18.

4. An operating point is selected well within the linear portion of the curves. In this example, the point is selected where the load line crosses the base current curve $I_B = 40$ microamps. A perpendicular line dropped from this point will show a collector voltage V_c and a horizontal line over to the current scale will show the quiescent current at this Q point. $V_{CE} = -5.2$ volts and $I_c = -4.5$ milliamps.

USING CHARACTERISTIC CURVES

By continuation of the previous problem, it is assumed that an input signal will cause the base current to vary between 30 and 50 microamps. These points are marked A and B in Fig. 14-19.

By projecting down to voltage scale, V_{CE} at point A is 3.6 volts and at point B, V_{CE} is 6.8 volts. The change in collector voltage due to signal input is:

$$\Delta V_{CE} = 6.8V - 3.6V = 3.2 \text{ volts}$$

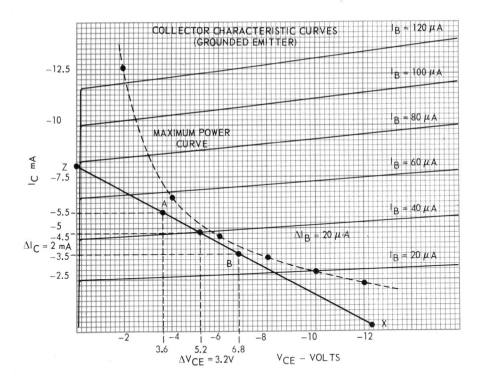

Fig. 14-19. Using the characteristic curves to find current and voltage excursions with signal input.

By projecting horizontally to the current scale, the current at point A is 5.5 milliamps and at point B is 3.5 mA. The change in collector current due to signal input is:

$$\Delta I_c = 5.5 \text{ mA} - 3.5 \text{ mA} = 2 \text{ milliamps}$$

Also, the input signal has caused a change in base current:

$$\Delta I_B = 50 \ \mu A - 30 \ \mu A = 20 \text{ microamps}$$

So a change in base current I_B of 20 microamps produced a change in collector current of 2 milliamps.

CURRENT GAIN

Now that we have found both input and output currents, the gain may be calculated.

$$A_I = \frac{\Delta i_c}{\Delta i_b} = \frac{2 \text{ mA}}{20 \ \mu A} = \frac{2 \times 10^{-3}}{2 \times 10^{-5}}$$

$$= 1 \times 10^2 = 100$$

VOLTAGE GAIN

Assuming a typical input impedance of 500 ohms, the change in voltage required to produce a change of base current can be found by Ohm's Law.

$$\Delta V_{EB} = \Delta i_b \times R_E$$

$$= 20 \ \mu A \times 500 \ \Omega = .01 \text{ volt}$$

An input voltage change of .01 volt will produce a V_{CE} change of 3.2 volts. See curves in Fig. 14-19. Voltage gain can be stated as:

$$A_V = \frac{\Delta V_{out}}{\Delta V_{in}} \quad \text{and} \quad A_V = \frac{3.2 \text{ volts}}{.01 \text{ volts}} = 320$$

POWER GAIN

The power formula states that:

$$P = I \times V \text{ and}$$

$$A_P = \frac{P_{out}}{P_{in}} \text{ or } A_I \times A_V = A_P$$

Using the previous figures:

$$A_P = 100 \times 320 = 32,000$$

CONVERTING GAIN TO DECIBELS

Gain in a circuit, whether it is current, voltage or power, is a logarithmic relationship between output and input expressed in decibels (db).

$$A_P \text{ (db)} = 10 \log \frac{P_{out}}{P_{in}} = 10 \log A_P$$

so,

$$A_P = 10 \log 32,000 = 10 \times 4.5 = 45 \text{ db gain}$$

MAXIMUM POWER CURVE

All transistors are rated according to their ability to dissipate power at a given ambient temperature. If this maximum power rating is exceeded, the transistor may be destroyed. Remember, too, that if a transistor is to be operated above normal temperatures of 25 deg. Celsius (77 deg. F), its ability to dissipate power must be derated according to a scale supplied by the manufacturer.

The power dissipation may be calculated at any given point on the transistor curves in Fig. 14-19 by the formula: $P = I \times V$ or $P = I_c \times V_{CE}$. To plot the maximum dissipation curve, it is assumed that this particular transistor is rated at 25 mW.

To find the permissible current for each value of V_{CE}, the power equation is transposed to:

$$I_c \text{ (mA)} = \frac{25 \text{ mW}}{V_{CE}}$$

These computations are made and plotted on the curves in Fig. 14-19:

$$\begin{array}{llll}
\text{For } V_{CE} = 12V, & I_c = & 2.08 \text{ mA} \\
V_{CE} = 10V, & I_c = & 2.5 \text{ mA} \\
V_{CE} = 8V, & I_c = & 3.1 \text{ mA} \\
V_{CE} = 6V, & I_c = & 4.1 \text{ mA} \\
V_{CE} = 4V, & I_c = & 6.2 \text{ mA} \\
V_{CE} = 2V, & I_c = & 12.5 \text{ mA}
\end{array}$$

A smooth curve is drawn through these points of maximum power dissipation. The transistor can be damaged if the product of the current I_c and the voltage V_{CE} falls at a point above this line.

If maximum power is to be realized from a transistor, the load line should be tangent to the maximum dissipation curve. The resistance load which will produce this line is found by drawing a load line from the $V_{CC} = 12$ volts tangent to the curve. It will intersect the current scale at some value of I_c. The load resistance will be:

$$R_{load} = \frac{12V}{I_c}$$

In our example, we could have used a load not lower than:

$$R_L = \frac{12V}{9 \text{ mA}} = 1330 \text{ ohms}$$

SINGLE BATTERY BIASING

The previous characteristic curves and computations will be used to determine the value of the bias voltage divider circuit with a stability factor of 5.

Examine Fig. 14-20 and be sure you understand all voltages around the circuit.

In the collector circuit at our selected Q point, $I_c = 4.5$ mA, the load has been divided between R_C and R_E. We need R_E to assure better stability.

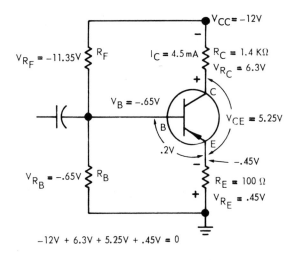

$$-12V + 6.3V + 5.25V + .45V = 0$$

Fig. 14-20. This diagram is marked with voltages and polarities as computed in the text analysis.

By Kirchoff's Voltage Law:

$$V_{RE} + V_{CE} + V_{RC} = -12 \text{ volts}$$

E is more negative than ground by the amount of voltage drop across R_E, which is .45 volt. The emitter-base junction must be forward biased, and the voltage drop across the EB junction for a germanium transistor may be considered at .2 volt (for silicon, it is .7 volt). Therefore, B must be .2 volt more negative than E, or at a potential of .65 volt negative in respect to ground. The voltage drop across R_B may be calculated:

$$-V_{R_B} = V_{R_E} + V_{B_E}$$

or

$$-V_{R_B} = .45V + .2V = .65 \text{ volt}$$

and

$$V_{R_F} = V_{cc} - V_{R_B} = 11.35 \text{ volts}$$

NOTE: Current I_B is considered as zero.

It is convenient now to find the voltage ratio between V_{R_F} and V_{R_B}.

$$\frac{V_{R_F}}{V_{R_B}} = \frac{11.35}{.65} = 17.46$$

Applying the stability factor equation to find the resistance of R_B:

$$S - 1 = \frac{R_B}{R_E} \qquad 4 = \frac{R_B}{100} \qquad R_B = 400 \text{ ohms}$$

and R_F is 17.46 times greater than R_B.

$$R_F = 17.46 \times 400$$

$$= 6984 \text{ ohms or } 6900 \text{ ohms (approx.)}$$

NOTE: 6984 ohms changed to standard value resistor of 6900 ohms.

To compute further proof of the voltage division:

$$V_{R_F} = \frac{R_F}{R_B + R_F} \times V_{cc}$$

$$= \frac{6900}{7300} \times 12 = .95 \times 12 = 11.4 \text{ volts}$$

$$V_{R_B} = \frac{R_B}{R_B + R_F} \times V_{cc}$$

$$= \frac{400}{7300} \times 12 = .055 \times 12 = .66 \text{ volt}$$

Small errors are caused by rounding off figures.

CLASSES OF AMPLIFIERS

Amplifiers may be classified according to their bias current, which determines the output wave form compared to the input wave form.

CLASS A: The class A amplifier is biased so that an output signal is produced for 360 deg. of the input signal cycle. See Fig. 14-21. Current I_c flows at all times. The output of a

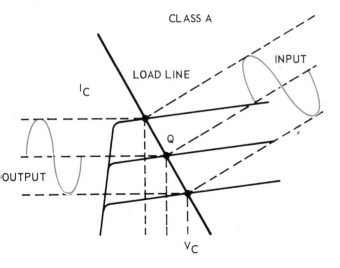

Fig. 14-21. A Class A amplifier has a 360 deg. signal output for a 360 deg. input signal.

class A amplifier is a faithful, but amplified, reproduction of the input signal. Its efficiency is low. This class is widely used in high fidelity sound systems.

CLASS B: A class B amplifier is biased so that an output signal is reproduced during 180 DEG. OF INPUT SIGNAL. The class B amplifier is biased close to cutoff. See Fig. 14-22. Since collector current flows only one-half of the time, this amplifier has medium efficiency. It is similar to a half-wave rectifier. Its output is

distorted. Two amplifiers operated in the push-pull configuration can restore both halves of the signal in the output. The push-pull circuit can enjoy the increased efficieny and greater power output. It will be used in many audio amplifiers as the final output stage. Power amplifiers will be studied in the next chapter.

CLASS C: A class C amplifier is biased so that LESS THAN 180 DEG. OF OUTPUT SIGNAL is produced during 360 deg. of input signal. This class finds limited use in transistor circuits and will not be discussed.

Of greater importance than amplification in our studies will be the switching characteristics of transistors. Switching will be discussed in detail in later chapters. A transistor may operate in the ON with full conduction or OFF with zero conduction conditions. This switching action can occur with extreme rapidity. Such circuits are found in our modern computers.

FIELD EFFECT TRANSISTORS (FET)

As mentioned earlier, the major difference between a transistor and a vacuum tube is that a vacuum tube is controlled by voltage and the transistor is controlled by current. All transistors are current amplifiers with the exception of the FIELD EFFECT TRANSISTOR. This special component is a voltage controlled amplifier. Its theory now may be studied. In Fig. 14-23, the construction of the device is shown.

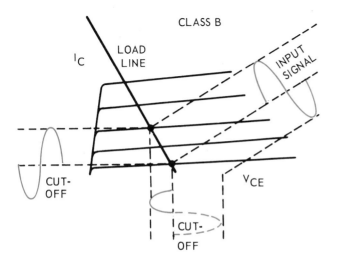

Fig. 14-22. A Class B amplifier has 180 deg. signal output for a 360 deg. input signal.

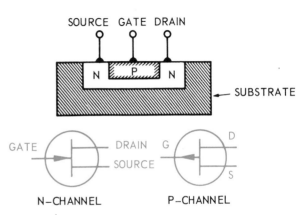

Fig. 14-23. Physical construction of a FET and its schematic symbol.

An N type material is diffused into a P type substrate material to form a relatively thin channel. A gate of P material is established between the source and drain. The FET depends on majority carriers for conduction purposes. In this case, electrons will be the major current carriers between the source and drain. This current is indicated by arrows in Fig. 14-24.

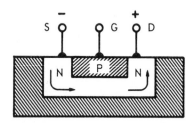

Fig. 14-24. Maximum current flows from source to drain. The channel is unrestricted.

A review of Chapter 12 on SEMICONDUCTOR DIODES will reveal that a PN junction will have a depletion area, void of current carriers, due to diffusion of holes and electrons across the junction. Refer back to Fig. 12-2. When the PN junction diode is REVERSE BIASED, the depletion area INCREASES IN WIDTH. This is shown graphically in Fig. 12-3.

The same effect occurs when the gate junction of the FET is reverse biased in respect to the source and drain. This is illustrated in Fig. 14-25. The conduction channel between the source and drain is now restricted and the drain current is substantially reduced. A negative voltage can be applied to the gate which will cut off the current between source and drain. This is called "PINCH OFF" voltage. In normal operations, the gate is not forward biased.

Because a relatively small voltage change at the gate will produce a large change in drain current, the device is able to amplify.

The major advantage of the FET over the typical transistor is its high input impedance. A comparison of the output characteristic curves

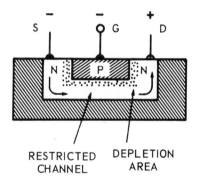

Fig. 14-25. Increased depletion area due to reverse bias of drain-gate junction will restrict the channel and reduce current flow.

of the field effect transistor to the vacuum tube pentode will show many similarities. In Fig. 14-26, the curves of a typical FET show the variation of drain current I_D because of an increase in drain-source voltage V_{DS} for fixed values of gate bias voltage.

A circuit for a simple amplifier employing an FET is schematically drawn in Fig. 14-27. It may be compared to the common-emitter transistor circuit or the common cathode tube circuit. In this circuit; R_1 is the gate resistor; R_2 is the load resistor; bias is established by R_3 with bypass capacitor C_2. Several projects in this text will use FETs as amplifiers and oscillators.

UNIJUNCTION TRANSISTORS

A unijunction transistor is a type of controlled diode with some amplifying capabilities. It may be best understood by reference to Fig. 14-28.

A small bar of N type crystal has a P emitter section alloyed to it. The N type material is uniformly doped and its resistance is also uniform. Therefore, the voltage at any point between B_1 and B_2 is proportional to the length of the section of crystal involved. For example, if 12 volts are applied between B_1 and B_2, the voltage at E, if at the center, would be +6 volts. If a bias voltage of +6 volts were now applied to the emitter, the forward bias of the B_1 to E junction would be zero.

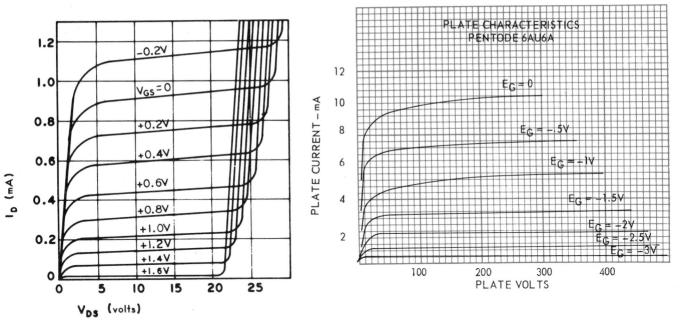

Fig. 14-26. A comparison of FET and Vacuum Tube Pentode characteristic curves.

Then, if bias voltage is increased to +7 volts, E would be one volt more positive and the B_1 to E junction would be forward biased. Under these conditions, holes are injected into the P emitter which unite with the electrons in the base section and current will increase between B_1 and B_2 due to the higher conductivity between B_1 and E.

The unijunction transistor is classified as a negative resistance device. Why is this so? Under the previous circumstances, where the base resistance was effectively reduced between B_1 and E because of forward bias, now the voltage division is upset. The voltage division at E becomes less, which further increases the forward bias. Actually, the emitter current can INCREASE even though the applied emitter voltage is decreasing. This is negative resistance.

The symbol for the unijunction transistor is shown in Fig. 14-29. A voltage applied to the emitter of this component can gate or trigger the current between base 1 and base 2.

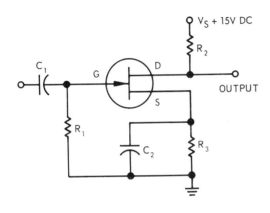

Fig. 14-27. Basic amplifier circuit using a FET.

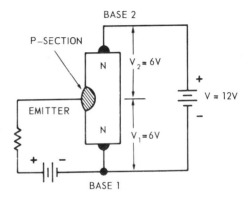

Fig. 14-28. This circuit is used to explain the theory of a unijunction transistor.

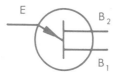

Fig. 14-29. Symbol for the unijunction transistor.

APPLAUSE METER PROJECT

You probably have seen applause meters used on television. This meter actually measures the loudness of noise or applause. Your school assembly programs may wish to measure the acceptance of a performer by measuring the applause received. The circuit and parts list for an applause meter project are shown in Fig. 14-30.

In this circuit, the speaker is used as a microphone which is matched to the amplifier circuit by transformer T_1. Then the signal is amplified by three stages of amplification. The output of Q_3 is rectified by the bridge rectifier,

and its output registers on meter A. The deflection of the meter will increase by the loudness of the applause. The circuit is not critical and is easy to build. You may use printed circuit techniques or build it on vector board.

The major cause of circuit failure in student built transistor circuits is using incorrect voltage or polarity, or both. Check and recheck the polarity of your voltage source before connect-

Fig. 14-31. Basic Transistor Amplifier.

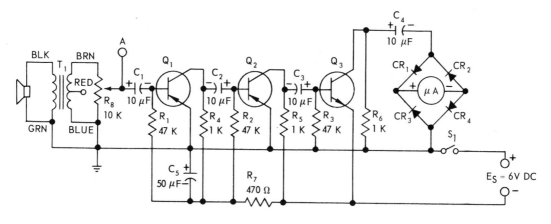

PARTS FOR APPLAUSE METER

*A — milliammeter with 0-.1 mA, 0-1 mA, 0-10 mA ranges (VOM)

Q_1, Q_2 — 2N408 transistor or RCA SK 3003
Q_3 — 2N649 transistor or RCA SK 3010
R_1, R_2, R_3 — 47 KΩ, 1/2W resistor
R_4, R_5, R_6 — 1 KΩ, 1/2W resistor
R_7 — 470 Ω, 1/2W resistor
R_8 — 10 KΩ, 1/2W potentiometer
CR_1, CR_2 — silicon diode or RCA SK 3091

CR_3, CR_4 — silicon diode or RCA SK 3091
C_1, C_2 — 10 μF electrolytic capacitor
C_3, C_4 — 10 μF electrolytic capacitor
C_5 — 50 μF electrolytic capacitor
T_1 — 500 Ω CT/3.2 Ω output transformer
S_1 — SPST switch
SPKR — 4 Ω speaker
E_S — 0-6V power source or four D cells
* — use VOM

Fig. 14-30. Schematic diagram and parts list for Applause Meter.

ing to the circuit. A little flashlight cell can destroy a transistor if connected with the wrong polarity.

BASIC TRANSISTOR AMPLIFIER PROJECT

A transistor amplifier can be an excellent piece of equipment for the hobbyist or experimenter. This amplifier has a simple, yet straightforward design, and the output is approximately 1.5 watts. Fig. 14-31 shows a photo of the amplifier. The schematic and parts list are shown in Fig. 14-32. The layout for the printed circuit board is shown in Figs. 14-33 and 14-34.

TEST YOUR KNOWLEDGE

1. In a common _____ amplifier circuit, E is isolated from ground by R_E. C is at signal ground potential by low reactance of the capacitor.

2. Name four bias methods used in transistor circuits.

3. What is meant by the Q point?

4. What considerations must be studied for correct selection of the Q point?

5. What is the normal operating temperature of a transistor?
   a. 25 C (77 deg. F).
   b. 35 C (95 deg. F).
   c. 45 C (113 deg. F).

6. Why does the CE amplifier require more stabilization than a CB amplifier?

7. What two heat effects cause thermal runaway?

8. What is emitter "swamping?"

9. The unijunction transistor is classified as a _____ resistance device.
   a. Positive.
   b. Negative.

10. A unijunction transistor is a type of controlled diode with some amplifying capabil-

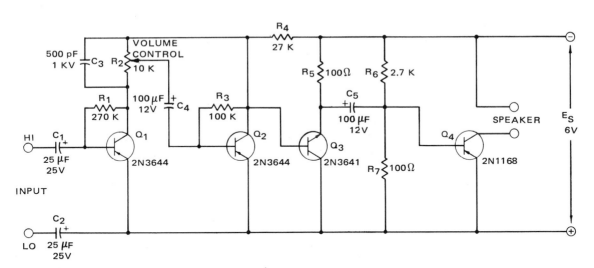

PARTS LIST FOR TRANSISTOR AMPLIFIER

R_1 — 270 KΩ , 1/2W resistor (10 percent)
R_2 — 10 KΩ , 1/2W potentiometer
R_3 — 100 KΩ , 1/2W resistor (10 percent)
R_4 — 27 KΩ , 1/2W resistor (10 percent)
R_5 — 100 Ω , 1/2W resistor (10 percent)
R_6 — 2.7 KΩ , 1/2W resistor (10 percent)
C_1, C_2 — 25 μF, 25V electrolytic capacitors
C_3 — 500 pF, 100V disc capacitor

C_4 — 100 μF, 12V electrolytic capacitor
Q_1, Q_2 — 2N3644 or RCA SK 3114 transistors
Q_3 — 2N3641 or RCA SK 3018 transistor
Q_4 — 2N1168 or RCA SK 3009 transistor
E_S — 6V power supply or battery
Misc. — heat sink for Q_4 chassis, knob, input jack, speaker jack, jacks for power input, printed circuit materials, decals

Fig. 14-32. Schematic and parts list for Basic Transistor Amplifier.   (Hickok Teaching Systems)

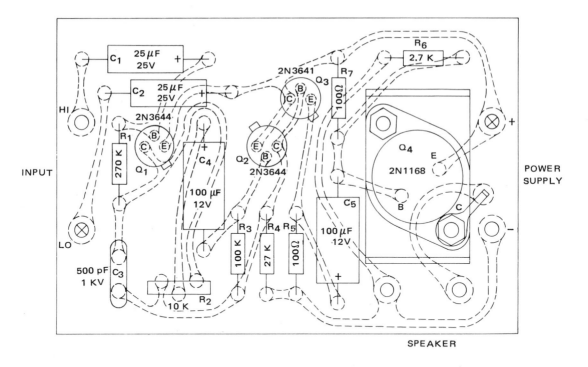

Fig. 14-33. Printed circuit parts layout for amplifier. (Hickok Teaching Systems)

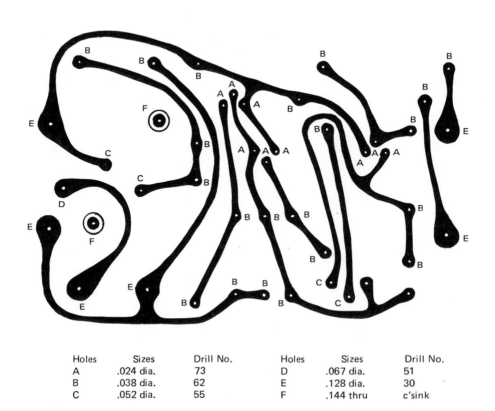

Holes	Sizes	Drill No.	Holes	Sizes	Drill No.
A	.024 dia.	73	D	.067 dia.	51
B	.038 dia.	62	E	.128 dia.	30
C	.052 dia.	55	F	.144 thru	c'sink

Fig. 14-34. Layout and hole sizes for printed circuit for transistor amplifier.

ities. True or False?

11. A field type transistor (FET) is considered to be a voltage controlled _____.

12. The main advantage of the FET over the typical transistor is its high input _____.
    a. Voltage.
    b. Current.
    c. Impedance.

Study the following equations and mark them TRUE (T) or FALSE (F):

13. $I_C = I_E - I_B$

14. $\propto = \dfrac{I_C}{I_B}$

15. $\beta I_B = \dfrac{I_C}{I_B}$

16. $I_B = I_E + I_C$

17. $\propto I_E = I_C$

18. $\beta I_B = I_C$

19. $I_E = I_C - I_B$

20. $A_p = A_i A_e$

21. $A_p = (\propto)^2 \dfrac{Z_{out}}{Z_{in}}$

22. $A_p = (\beta)^2 \dfrac{Z_{out}}{Z_{in}}$

23. $\propto = \dfrac{1}{\beta + 1}$

24. $\beta = \dfrac{\propto}{1 - \propto}$

25. $P = I_C \times V_{CE}$

Chapter 15

TRANSISTOR POWER AMPLIFIERS

Chapter 15 involves a detailed study of transistor power amplifiers. In this chapter, you will:

1. Investigate the several methods of coupling transistor stages together in cascade.
2. Examine the major difference between small signal and large signal amplifiers.
3. Study the theory and operation of phase inversion circuits to drive a push-pull amplifier.
4. Learn how to select component values and predict the performance of a push-pull amplifier.
5. Explore complementary symmetry circuits to be used for push-pull amplifiers.
6. Investigate the construction and advantages of selected types of speakers.

In truth, all transistors are current amplifiers. There are major differences, however, between a tiny transistor whose purpose is to increase the voltage level of a microsignal and the larger power transistor used to drive a speaker or output device.

Power transistors are designed with larger collector surfaces and with provisions for heat dissipation. More often than not, they are mounted on heat sinks with cooling fins. A power transistor is shown in Fig. 15-1.

This explanation suggests that more than a single transistor is required to amplify a signal to a useful value; and this is correct. Transistor

Fig. 15-1. A typical power transistor. The case is the collector terminal. When mounted in a heat sink, the collector will have an increased cooling surface. (Westinghouse)

stages are coupled together in CASCADE to produce the desired output.

RC COUPLING

A review of RC networks in Chapter 9 may be necessary before proceeding further. Note that a typical RC circuit may have two stages of transistor amplifiers coupled with a transformer. If the transistor in one stage has an output impedance of 20 kilohms, and the transistor in the other stage has an input impedance of 1 kilohm, a severe mismatch and loss of gain would result.

The step-down transformer is a convenient device to match these impedances. A low secondary voltage means a higher secondary current, which is just fine for transistors because they are current operated. A simple method of coupling transistor amplifier stages is by using resistor-capacitor (RC) coupling. A step-by-step explanation is given in Figs. 15-2 through 15-5.

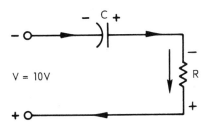

Fig. 15-2.

In this circuit, capacitor C will charge to V equals 10 volts. Only during the charging of C will a current cause a voltage to appear across R. After C is charged, V_c equals 10 volts and V_R equals zero.

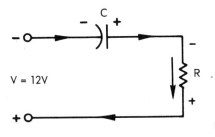

Fig. 15-3.

If the source voltage is changed to 12 volts, C will increase its charge to 12 volts also. The charging current will produce a momentary two volt pulse of voltage across R in the polarity shown.

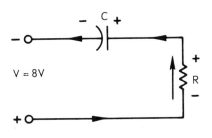

Fig. 15-4.

If source voltage is changed to 8 volts, C will discharge to 8 volts. The discharge current will produce a momentary pulse of voltage across R in the polarity shown.

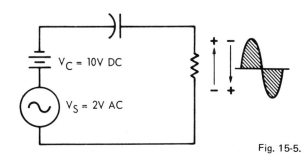

Fig. 15-5.

Here both a dc and an ac voltage are connected to the RC circuit. The ac signal causes the total voltage to vary between 8 and 12 volts. Therefore, C will charge and discharge at the frequency of the ac generator.

The voltage appearing across R will rise and fall at the same frequency as the generator voltage. However, take particular note of Fig. 15-6.

Where the input signal varied around a dc level of 10 volts, the output signal varies around the zero level. The dc component has been removed. A capacitor blocks dc. In this respect, it is called a BLOCKING CAPACITOR.

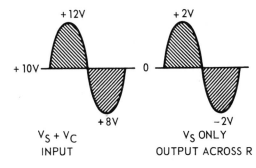

Fig. 15-6. The capacitor blocks the dc voltage, but permits the ac signal to pass.

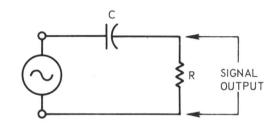

Now, examine this circuit from a mathematical point of view. C and R in series form an ac voltage divider. Refer to Fig. 15-7.

The voltage division will depend upon the reactance of X_c at the frequency of the signal. The voltage output across R is the important consideration, and we want as much of the ac voltage to appear across R as possible. Assuming a frequency of 1000 hertz and a value of C equal to .01 microfarad, then:

$$X_c = \frac{1}{2\pi fC} = \frac{1}{6.28 \times 10^3 \times 1 \times 10^{-8}}$$

$$\cong .16 \times 10^5 = 16,000 \text{ ohms}$$

If the value of R is ten or more times greater than X_c, then most of the voltage will appear across R. Assuming R equals 160 kilohms, then:

Fig. 15-7. R and C form a voltage divider for the ac signal.

$$V_R = \frac{R}{R + X_c} \times V$$

$$= \frac{1.6 \times 10^5}{(1.6 \times 10^5) + (1.6 \times 10^4)} \times V$$

$$= \frac{1.6 \times 10^5}{17.6 \times 10^4} \times V = .91 \times 10V$$

$$= 9.1 \text{ volts}$$

(From the practical point of view, phase shift is disregarded.)

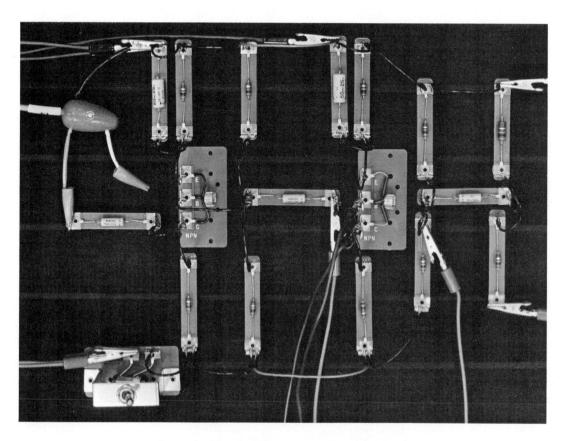

Fig. 15-8. RC coupling is used in this breadboard circuit.
It is a popular method of coupling transistor stages.

It seems that almost all the voltage does appear across R. Less than a volt has been lost as signal output. If R were made larger, then even more of the total output would be developed across R. In vacuum tube circuits, the tube has a high input impedance and is considered voltage controlled. This circuit then would be quite satisfactory. But look at the two stage amplifier using transistors in Figs. 15-8 and 15-9.

These CE amplifiers have a fairly low input impedance. Refer to Fig. 13-17 in Chapter 13. Since the input impedance of the second stage is in series with coupling capacitor C, then the reactance of C must be quite low so that it will not rob a large part of the signal from Q_2.

Also, consider the signal voltage at the collector of Q_1. It has a choice of paths to go and will take the easiest path. It can go through R_c or through the coupling network which may be considered in parallel. If the impedance of the network is higher than R_c then signal currents will go through R_c instead of to the next stage. So, therefore, the coupling capacitors used in transistor circuits must have values in the vicinity of 8 to 10 microfarads.

To prove the point, assume the input impedance of stage Q_2 is 500 ohms, then X_c would need to be 50 ohms or less. At 1000 hertz frequency:

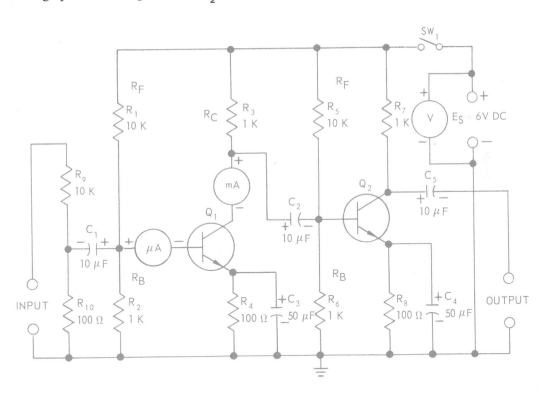

PARTS LIST FOR RC COUPLING

V — 0-25V dc voltmeter	R_4, R_8 — 100 Ω, 1/2W resistors
μA — 0-100 μA dc ammeter	R_{10} — 100 Ω, 1W resistor
mA — 0-10 mA dc ammeter	C_1, C_2, C_5 — 10 μF electrolytic capacitors
Q_1, Q_2 — 2N649 transistor or RCA SK 3010	C_3, C_4 — 50 μF electrolytic capacitors
R_1, R_5, R_9 — 10 KΩ, 1/2W resistors	C_6 — .01 μF capacitor
R_2, R_3, R_6, R_7 — 1 KΩ, 1/2W resistors	S_1 — SPST switch

Fig. 15-9. Schematic and parts list for RC Coupling.

$$C = \frac{1}{2 \pi f X_c} = \frac{1}{6.28 \times 10^3 \times 5 \times 10}$$

$$\cong 3.2 \text{ microfarads}$$

At lower frequencies, which the amplifier would also amplify, X_c would be higher and a capacitor of 8 to 10 microfarads would be required to prevent loss of amplifier gain.

Note also in Fig. 15-9 that R_F and R_B are in parallel with the emitter-base circuit. It is desirable that the values of these resistors be sufficiently high so that the signal will not be bypassed around the EB junction.

The values of R_B and R_F were determined by the required bias and stability of the circuit. Also, the current drain from the source must be considered.

Circuit design with transistors can become quite complex, and it frequently is a matter of give and take. Compare the output and input impedances of the transistors in the CE configuration. The input can be in the range of 500 ohms to 1.5 kilohms, and the output impedance in the range of 30 to 50 kilohms. This is a severe mismatch. With the RC coupling, the mismatch must be tolerated with its accompanying loss of power gain. However, when cost is a factor, it may be cheaper to add another transistor stage to offset the loss due to mismatch rather than purchase a transformer for interstage matching.

TRANSFORMER COUPLING

The circuit in Fig. 15-10 shows two stages of transistor amplifiers coupled with a transformer. Assume that transistor Q_1 has an output impedance of 20 kilohms, and Q_2 has an input impedance of 1 kilohm. A severe mismatch and loss of gain would result.

The transformer is a convenient device to match these impedances. A step-down transformer is required. A low secondary voltage means a higher secondary current. This is just fine for transistors because they are current operated. Many special sub-ouncer and sub-sub-ouncer transformers have been developed for this purpose. Some of these are illustrated in Fig. 15-11.

The purpose of C_2 in the transformer coupling circuit of Fig. 15-10 is to block the dc bias voltage of the transistor from ground. Notice that with C_2 omitted, the transistor base would be grounded directly through the transformer secondary T_1.

The major disadvantage of transformer coupling, besides cost, is the poor frequency response of transformers. They tend to saturate at high audio frequencies. At radio frequencies,

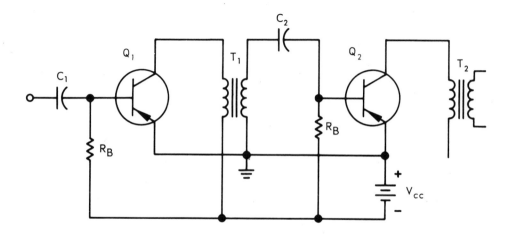

Fig. 15-10. A transformer coupled amplifier circuit.

Fig. 15-11. These tiny sub-ouncer and sub-sub-ouncer transformers are used in transistor circuitry. (Triad)

the inductance and winding capacitance will present problems.

A variation of transformer coupling you will see in many circuits uses a tapped transformer. These taps can be at medium, low and high impedance points. With this version of a transformer, good impedance matching can be attained as well as good coupling and gain. See Fig. 15-12. For radio frequency amplifier circuits, both the transformer primary and secondary windings can be tuned by variable capacitors for frequency selectivity. This will be discussed further in Chapter 17.

One other point about the transformer: its primary impedance acts as a collector load for the transistor. Since this impedance only appears under signal conditions, the load is X_L of primary. From the dc point of view, the only load is the ohmic resistance of the wire used to wind the transformer primary. You will use this information later in the design of a power amplifer, which will require an understanding of both dc and ac load lines.

DIRECT COUPLING

In many industrial circuits used today, it is necessary to amplify either very low frequency

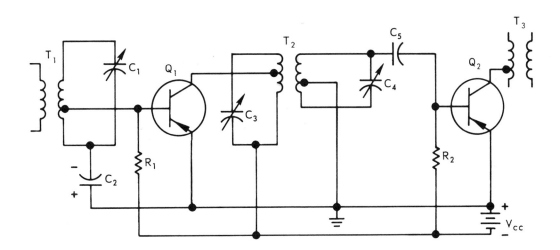

Fig. 15-12. In this circuit, the taps on the transformer primary and secondary windings provide a convenient matching point. The transformer can be designed for good overall gain.

signals, or it is required to retain the dc value as well as the ac value of a signal. Amplifier circuits using RC or transformer coupling will block out the dc component. The directly coupled amplifier shown in Fig. 15-13 is an answer.

In this circuit, the collector of Q_1 is connected directly to the base of Q_2. The collector load resistor R_c also acts as a bias resistor for Q_2. Any change of bias current is amplified by the directly coupled circuit. Therefore, it is very sensitive to temperature changes. This disadvantage can be overcome with stabilizing circuits. Another disadvantage appears when one realizes that each stage requires a different bias voltage for proper operation.

SINGLE-ENDED POWER AMPLIFIER

A final amplifier stage designed to drive a speaker or some output device is classified as a POWER AMPLIFIER. Voltage and current gain are not important considerations. However, the transistor must be capable of carrying relatively large currents and provisions must be made for heat dissipation. This LARGE SIGNAL AMPLIFIER usually is mounted on a HEAT SINK.

Methods of mounting can be examined in the exploded views in Fig. 15-14. Note that the

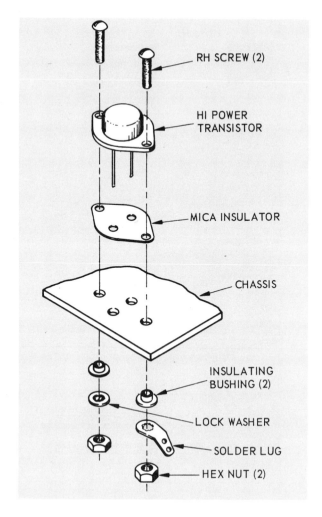

Fig. 15-14. A special mounting kit is used when mounting a power transistor on a heat sink. The kit contains the required washer, insulating sleeves and hardware.

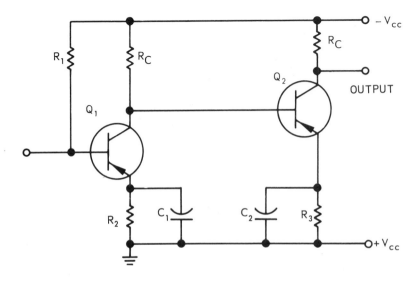

Fig. 15-13. The circuit of a directly coupled amplifier.

234

transistor case is the collector terminal of the power transistor. This arrangement provides greater surface area for heat dissipation in the collector circuit.

A single-ended power amplifier circuit appears in Fig. 15-15. Before analyzing this circuit, study the function of each component: T_1 serves as a matching transformer between the high output impedance of the transistor and the low impedance of the speaker voice coil. Z_p is also the collector load.

R_E is the emitter resistance selected to improve stabilization. C_E is the emitter resistor bypass for signal currents to maintain E at a constant voltage and prevent degeneration.

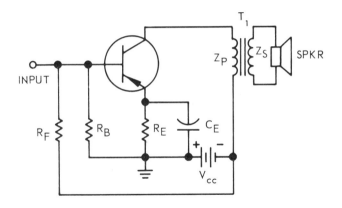

Fig. 15-15. Circuit of single-ended power amplifier.

R_B and R_F form a voltage divider selected at the correct proportion for correct bias voltage of B.

It might be wise, at this point, to review some of the earlier lessons on Ohm's Law and power. For emphasis, these equations will be repeated.

1. The impedance ratio of a transformer varies as the square of its turns ratio (Chapter 8).

$$\frac{Z_p}{Z_s} = \left(\frac{N_p}{N_s}\right)^2$$

2. Efficiency of a device is the ratio of the power out to the power in.

$$\text{EFFICIENCY} = \frac{P_{out}}{P_{in}}$$

3. The effective values (rms) of a peak-to-peak voltage or current wave is equal to:

$$\frac{V_{peak\text{-}to\text{-}peak}}{2} \times .707 = V_{effective}$$

Since $.707 = \frac{1}{1.414}$ or $\frac{1}{\sqrt{2}}$, then:

$$\frac{V_{p\text{-}p}}{2\sqrt{2}} = V_{eff}$$

4. Power equals I times V. If effective power is desired, then:

$$P_{eff} = I_{eff} \times V_{eff}$$

Then effective power will be equal to:

$$\frac{V_{p\text{-}p}}{2\sqrt{2}} \times \frac{I_{p\text{-}p}}{2\sqrt{2}} = \frac{P_{p\text{-}p}}{8} = P_{eff}$$

Also power equals $\frac{V^2}{P}$ and in ac circuits:

$$P = \frac{V^2}{Z} \quad \text{or} \quad V^2 = P \times Z$$

PROBLEM: A power amplifier is desired which will supply 50 milliwatts of power to a 4 ohm speaker. A matching transformer with an efficiency of 85 percent will be used. A battery of 9 volts is available. A 30 percent collector efficiency is desired with maximum signal applied. Characteristic curves for the transistor are shown in Fig. 15-16. The circuit is diagrammed in Fig. 15-17.

First, we need to further examine the term COLLECTOR EFFICIENCY. By formula:

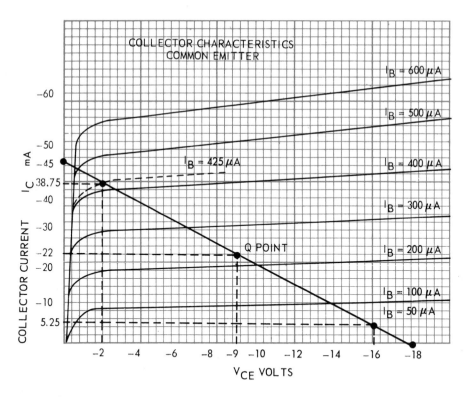

Fig. 15-16. The characteristic curves of the power transistor used in text problem.

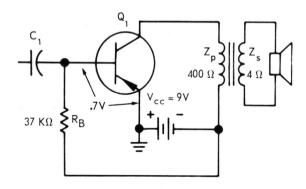

Fig. 15-17. The circuit used for designing the power amplifier in the text.

$$\text{Eff} = \frac{P_{out}}{P_{in}} \quad \begin{array}{l} \text{(ac power output)} \\ \text{(dc power to collector circuit)} \end{array}$$

With a collector voltage V_{cc} of 9 volts, the very maximum swing of a signal could not exceed 18 volts, which would cover the entire length of any load line. See Fig. 15-16. Under these conditions, the efficiency would be 50 percent, which is the maximum possible. But the curves are unevenly spaced and distortion of

the signal would occur at the extreme ends of the load line (saturation region and cutoff region). It is advisable to design the amplifier at an efficiency of something less than 50 percent. For our problem, we will try to attain a 30 percent collector efficiency.

PROBLEM SOLUTION:

1. The output transformer has an efficiency of 85 percent, and 50 milliwatts of power are required at the speaker. To achieve this, the $P_{primary}$ (power input) must be:

$$.85 = \frac{50 \text{ mW}}{P_{primary}} \quad \text{or}$$

$$P_{primary} = \frac{50 \text{ mW}}{.85} \doteq 58.7 \text{ milliwatts}$$

2. With a collector efficiency of 30 percent, the dc power input to the collector must be:

$$.3 = \frac{58.7 \text{ mW}}{P_{dc}} \quad \text{and}$$

$$P_{dc} = \frac{58.7}{.3} \cong 195 \text{ milliwatts}$$

3. With a power source V_{cc} = 9 volts, compute the current required for this power:

$$P = I \times V \quad \text{or} \quad I = \frac{P}{V}$$

then:

$$I = \frac{195 \text{ mW}}{9V} \cong 22 \text{ milliamps}$$

These figures establish the QUIESCENT POINT or NO SIGNAL POINT on the load line marked Q. Now is the time to check the absolute maximum power dissipation of the transistor in use. If the specifications state a higher figure, then we are safe and ready to continue.

4. With a maximum signal excursion about the Q point, the limits would occur at saturation and cutoff at equal and opposite distances along the load line. One end of the load line can be established at 18 volts (2 x 9V = 18V). The load line must pass through Q, so the current at zero voltage can be found graphically. This is approximately 45 milliamps.

5. The slope of the load line represents impedance Z. This should be the PRIMARY LOAD impedance.

$$Z = \frac{V}{I}$$

then:

$$Z_{primary} = \frac{18V}{45 \text{ mA}} = 400 \text{ ohms}$$

6. This impedance is matched to the 4 ohm secondary, and the transformer turns ratio must be:

$$\frac{Z_p}{Z_s} = \left(\frac{N_p}{N_s}\right)^2$$

$$\frac{N_p}{N_s} = \sqrt{\frac{Z_p}{Z_s}} = \sqrt{\frac{400}{4}} = \frac{10}{1}$$

7. Assuming a base-emitter voltage drop for a silicon transistor is .7 volts, the base-resistor may be calculated by Ohm's Law.

$$R_B = \frac{V_{cc} - V_{EB}}{I_B} =$$

$$\frac{9V - .7V}{225 \times 10^{-6}} = 37 \text{ kilohms}$$

Note that the Q point falls just above the 200 microamp I_B curve at about 225 microamps.

8. We have already observed that at a full signal excursion between 0 and 18 volts, distortion would occur at both the saturation and cutoff ends of the load line because of the irregular curves. So we plan to operate at only 30 percent efficiency. What will be the voltage swing for this efficiency?

$$P_{primary} = \frac{(V_c)^2}{Z} \quad \text{or}$$

$$P_{pri} \times Z_{pri} = (V_c)^2$$

and the rms or effective value of V_c is $\dfrac{V_c}{2\sqrt{2}}$

By substitution:

$$P_{pri} \times Z_{pri} = \left(\frac{V_c}{2\sqrt{2}}\right)^2$$

and: $V_c{}^2 = \left(2\sqrt{2}\right)^2 \times P_{pri} \times Z_{pri}$

or $\quad \Delta V_c{}^2 = 8 P_{pri} \times Z_{pri}$

but, $Z_{pri} = \dfrac{V_{cc}}{I_c}$ (see step 5)

then: $\Delta V_c^2 = 8 P_{pri} \times \dfrac{V_{cc}}{I_c}$

and: $P_{pri} = V_{cc} \times I_c$ (see step 3)

then: $\Delta V_c^2 = 8(V_{cc} \times I_c) \times \dfrac{V_{cc}}{I_c}$

But we desire a collector efficiency of 30 percent, so the equation must be rewritten to:

$$\Delta V_c^2 = 8(V_{cc} \times I_c \times Eff) \times \dfrac{V_{cc}}{I_c}$$

Combine terms:

$$\Delta V_c^2 = 8 V_{cc}^2 \times Eff$$

(The I_c cancels out.)

We are looking for the voltage of swing V_c which will produce an output at a specified collector efficiency, so:

$$\Delta V_c = \sqrt{8 V_{cc}^2 \times Eff}$$

$$\Delta V_c = \sqrt{2 \times 4 \times V_{cc}^2 \times Eff}$$

$$\Delta V_c = 2 V_{cc} \sqrt{2 \times Eff}$$

Applying this formula to the problem:

$$\Delta V_c = 2 \times 9V \sqrt{2 \times .3}$$

$$= 18V \times \sqrt{.6} \cong 14 \text{ volts}$$

One-half of this voltage swing will be above the Q point and the other half below.

$$V_c \text{ (min)} = 9V - \dfrac{14V}{2} = 2 \text{ volts}$$

$$V_c \text{ (max)} = 9V + \dfrac{14V}{2} = 16 \text{ volts}$$

These points are now located on the curves in Fig. 15-16, along the V_{CE} in volts. Perpendicular projections from these voltage points to the load line will also show the current swing ΔI_c.

9. Proof of the current swing plot can be found in these calculations:

$$P_{out} = \dfrac{\Delta V_c \Delta I_c}{8} \quad \text{and} \quad \Delta I_c = \dfrac{8 P_{out}}{\Delta V_c}$$

Substituting the values from previous steps:

$$\Delta I_c = \dfrac{8 \times 58.7 \text{ mW}}{14V} = 33.5 \text{ milliamps}$$

One-half of the current swing will be above the Q point, and the other half below. As a result:

$$I_{max} = 22mA + \dfrac{33.5}{2}$$

$$= 38.75mA, \text{ at } -2 \text{ volts}$$

$$I_{min} = 22mA - \dfrac{33.5}{2}$$

$$= 5.25mA, \text{ at } -16 \text{ volts}$$

These plotted points now represent the maximum peak-to-peak current and voltage swings with maximum signal.

10. Looking at Fig. 15-16 again, the change in base current ΔI_B may be approximated. Base current curves do not run through the maximum excursion points on the load line, but they can be quite accurately determined:

$$\Delta I_B = 425 \ \mu A - 50 \ \mu A = 375 \ \mu A$$

Refer now to the INPUT CHARACTERISTIC CURVES OF THE TRANSISTOR, Fig.

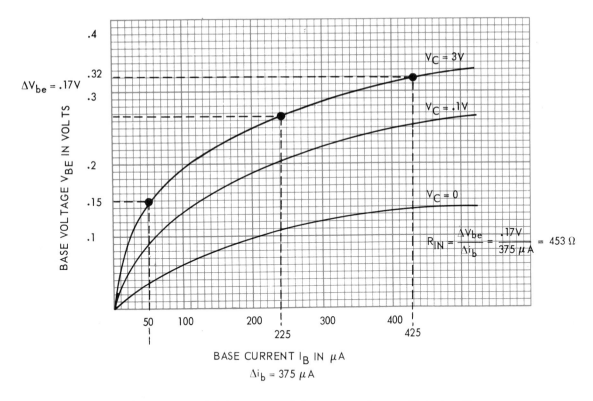

Fig. 15-18. The characteristic input curves of the power transistor used in text problem.

15-18. The curves represent the change of base-emitter voltage as a result of a change in base current at a constant collector voltage. Ordinarily, these would be plotted on $-9V$ V_C curve for our problem. However, all curves above $-3V$ will be found to be almost identical to the $-3V$ curve, so this one is conveniently used.

The following information is plotted on these curves.

Base Current I_B at Q point — 225 μA
Max excursion of I_B — 425 μA
Min excursion of I_B — 50 μA

The intersects with the $V_C = -3V$ curve are plotted, and the base voltage excursions determined.

V_{BE} at Q point — .257V
V_{BE} at I_B Minimum — .15V
V_{BE} at I_B Maximum — .32V

The INPUT RESISTANCE is found by Ohm's Law:

$$R_{in} = \frac{\Delta V_{BE}}{\Delta I_B} = \frac{.17V}{375 \ \mu A} = 453 \text{ ohms}$$

Using the power formula for effective power:

$$P_{in} = \frac{\Delta V_{BE} \times \Delta I_B}{8}$$

$$= \frac{.17V \times 375 \ \mu A}{8} = 7.97 \text{ microwatts}$$

The power gain A_p is calculated:

$$A_p = \frac{P_o}{P_{in}} = \frac{58.7 \text{ mW}}{7.97 \ \mu W} = 7360$$

and

$$Db = 10 \log 7360 \ \tilde{=} \ 38.6 \text{ db}$$

The voltage gain A_V is computed as:

$$A_V = \frac{\Delta V_c}{\Delta V_{BE}} = \frac{14V}{.17V} = 82.4$$

And the current gain A_I by the formula:

$$A_I = \frac{\Delta I_c}{\Delta I_B} = \frac{33.5mA}{375 \ \mu A}$$

$$= \frac{33.5 \times 10^{-3}}{375 \times 10^{-6}} = 89.3$$

This completes the practical analysis of this single-ended, Class A power amplifier. Do not proceed further until you have mastered the procedure. If it seems difficult, your trouble will be found in your inability to read characteristic curves or insufficient study of the chapter on Ohm's Law and the Power Law.

PUSH-PULL AMPLIFIERS

In order to attain the maximum power output and efficiency from a power amplifier, two transistors may be operated in the PUSH-PULL power amplifier circuit, Fig. 15-19.

Note that this basic circuit is operating at zero bias when no signal is applied. No voltage is applied across the EB junction, so the current I_B is zero. When point A of the input transformer becomes negative (as the result of a

signal) then B of Q_1 is more negative than E. This is a forward bias and Q_1 conducts. At the same time, the base of Q_2 is being driven positive which increases the reverse bias on Q_2. Q_2 does not conduct.

On the second half of the input signal, the reverse is true. Q_2 is driven into conduction when a negative signal is applied to its base. Q_1 is cut off as a reverse bias is applied by a positive signal to its base. So, first Q_1 conducts, then Q_2 conducts. On one half-cycle, current I_c flows in Q_1; on the second half-cycle, current I_c flows in Q_2. These transistors are operating Class B. One-half the time each transistor is resting and cooling. Is there any question that this circuit has increased efficiency? In fact, the maximum possible efficiency approaches 78.5 percent.

The two half signals, or waves of current from the transistors, are restored to their original input form by transformer action in the output transformer. This requires an explanation.

In Fig. 15-20, when Q_1 conducts, current flows as indicated by arrows and creates a magnetic field of one polarity. This indicates a half-wave form in the secondary of the transformer. When Q_2 conducts, the polarity of the primary reverses and induces the wave of opposite polarity in the secondary. Thus, the complete wave is restored.

CROSSOVER DISTORTION

Operation of the push-pull amplifier as Class B does cause some distortion of the wave form. In Fig. 15-21, the output curves of each tube are drawn together. It is customary to draw the curves in this manner.

In Fig. 15-22, a signal applied causes variations in base current, and the resulting collector current wave form is also shown. It is distorted, and the distortion occurs due to the change-over between Q_1 conducting and Q_2 con-

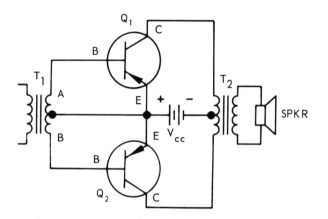

Fig. 15-19. The basic PUSH-PULL amplifier circuit.

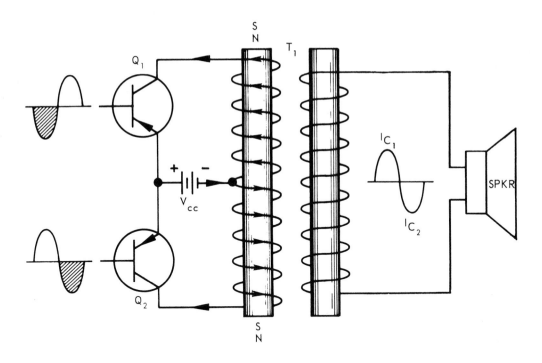

Fig. 15-20. Each half of the signal is joined in the output transformer to restore the original wave.

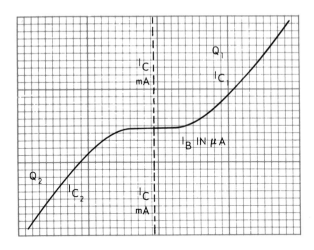

Fig. 15-21. The two curves of the push-pull operation are joined together.

ducting. This is referred to as CROSSOVER DISTORTION and should be eliminated.

To remedy this distortion, a small bias voltage is applied to each transistor, and the push-pull amplifier is operated Class AB. This is done in the usual manner with a voltage divider, Fig. 15-23, consisting of R_1 and R_2.

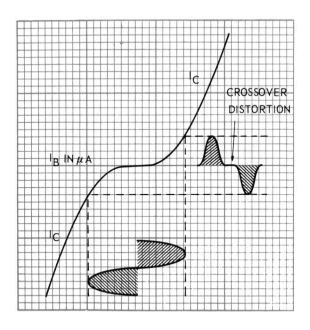

Fig. 15-22. Distortion results from crossover between conduction of transistors.

Current through the divider is indicated by arrows, and the polarity of the voltage drops is shown. Now the bases of both transistors are slightly negative in respect to their emitters.

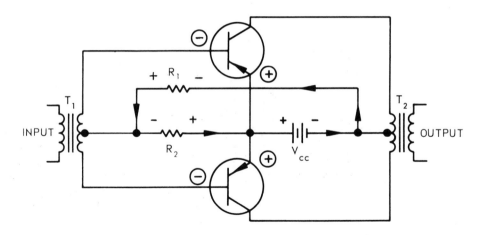

Fig. 15-23. A voltage divider provides bias for the transistors.

Therefore, both are forward biased. The effect of this biasing may be observed in Fig. 15-24.

The curves are matched at their bias points. This small bias effectively cancels out the crossover distortion.

COMPLEMENTARY SYMMETRY

As your refer back to earlier chapters in this text, you will remember there are two kinds of transistors, NPN and PNP. Currents in the NPN transistor are exactly opposite to currents in the PNP. This fact can be made quite useful in the push-pull amplifier.

Study Fig. 15-25 where two transistors Q_1 and Q_2 are fed with a single input signal. Q_1 and Q_2 are opposite types. A positive going signal causes the base of Q_1 to become positive and Q_1 will conduct. At the same time, the same signal causes the base of Q_2 to become positive, which cuts it off. The wave forms and their phase relationships are illustrated. The signals are joined in the output transformer T for the full-wave output.

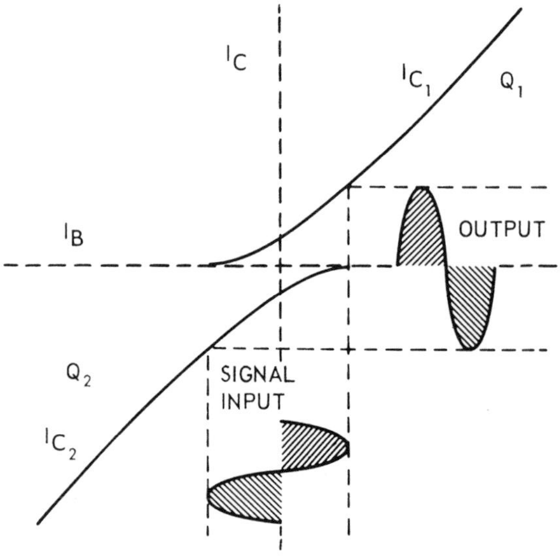

A disadvantage of this circuit is the requirement of two supply voltages. (A single battery can be used with a floating ground.) A major advantage is the lack of need for a transformer or inverter circuit to provide out-of-phase signals for the input.

PHASE SPLITTERS

In the previous discussion, a transformer was used to provide the 180 deg. out-of-phase signals for the push-pull amplifier. Transformers are expensive and have a rather poor frequency response. A simple circuit which will supply two outputs, 180 deg. out-of-phase, to drive the push-pull amplifier is drawn schematically in Fig. 15-26.

Fig. 15-24. Operation Class AB removes the crossover distortion.

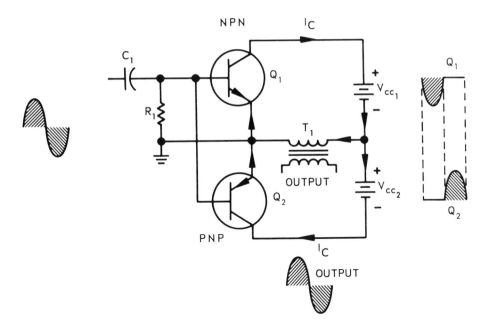

Fig. 15-25. The circuit of a complementary symmetry push-pull amplifier.

A review of transistor operation will tell you that $I_C = I_E - I_B$. An increase in I_E also produces an increase in I_C. If R_1 and R_2 are selected approximately equal, then the voltage drops across R_1 and R_2 will be equal. An increase in collector current will make C more

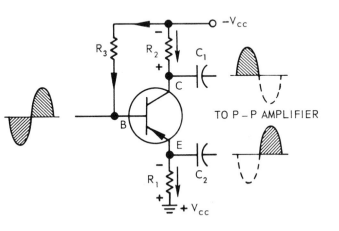

Fig. 15-26. A phase splitter circuit.

positive and, at the same time, make E more negative. The opposite is also true. Thus, equal out-of-phase signals are taken from different points in the circuit. This circuit does not provide any amplification of the signal.

PHASE INVERTERS

Another circuit which produces two out-of-phase signals and also amplifies the signal is illustrated in Fig. 15-27. It is a simple circuit to understand when you realize that a CE amplifier always inverts the signal. In this circuit, the output for one signal is taken from the collector of Q_1. A small part of this output is fed to stage Q_2 where it is amplified and INVERTED. The second output is taken from the collector of Q_2.

Note that the load for Q_1 consists of $R_3 + R_4$ in series. The reason for dividing this load resistance is to secure a small part of the output of stage Q_1 to drive stage Q_2. If R_3 is made variable, then the exact amount of signal may be obtained which, after amplifciation by Q_2, will produce the same amplitude signal as found at the output of Q_1.

In other words, the signals at points A and B should be equal but opposite in phase. Other resistors in the circuit include networks $R_1 R_2$ and $R_6 R_7$ which are voltage dividers to fix operating bias. R_5 and R_9 are emitter resistors.

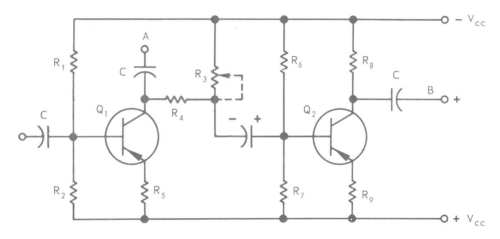

Fig. 15-27. The circuit of a typical phase inverter.

SPEAKERS

A speaker is a device which converts electrical energy into sound energy. It also converts an audio frequency wave into sound waves. The physicist describes a sound wave as alternate rarefactions and condensations of air which, when impressed upon one's ears, will produce the sensation of sound. A sound wave is

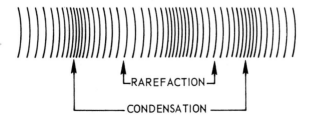

Fig. 15-28. A representation of sound waves as alternate rarefactions and condensations of the air.

illustrated in Fig. 15-28. These sound waves may be set in motion by a vibrating diaphragm or disc.

A PERMANENT MAGNET SPEAKER is sketched in Fig. 15-29. This is by far the most commonly used speaker in audio systems today. It consists of a permanent magnetic field. At the center is a coil suspended in the field and attached to the flexible paper speaker cone. A varying current through the voice coil will produce its own magnetic field which reacts

with the PM field and causes motion. This is similar to the MOTOR ACTION described in Chapter 11. A current of one polarity will cause the coil to move to the right; the opposite polarity causes the coil to move to the left. Consequently, the coil and cone move inward and outward at the same FREQUENCY and AMPLITUDE as the incoming signal fed to the voice coil. The cone movement sets up sound waves that the human ear will sense as sound.

The voice coil is wound with only a few turns of low resistance wire. The common types have an impedance of 3.2 ohms at signal frequencies. It is for this reason that an output transformer is used to match this low impedance. You will find speakers with an impedance of 8, 16, 40 and 100 ohms in use with audio

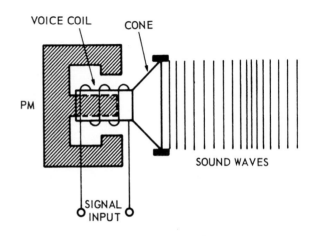

Fig. 15-29. A sketch of a PM or permanent magnet speaker.

equipment. The better quality audio amplifiers may use a universal output transformer with taps and terminals for 4, 8 and 16 ohms. Also line amplifiers may have impedance taps for 500, 333, 250, 125 and 50 ohms.

ELECTRODYNAMIC SPEAKER

The action of the electrodynamic speaker, Fig. 15-30, is the same as the PM speaker. The major difference is the use of an electromagnet rather than the permanent magnet for the speaker magnetic field. The electromagnetic field is established in a field coil and supplied by a source of dc taken from the equipment power supply.

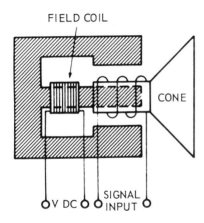

Fig. 15-30. A sketch of an electrodynamic speaker.

ELECTROSTATIC SPEAKER

An electrostatic speaker usually is used for high frequencies. It depends upon the characteristics of capacitance for its operation. In Fig. 15-31, a high voltage is applied to the two capacitor plates in the speaker. One of these plates will flex inward and outward slightly. The speaker cone is attached to this flexible plate. A signal now applied in series with the high voltage will cause a varying high voltage which, in turn, varies the electrostatic field between the capacitor plates. The flexible plate moves according to the applied signal. The cone converts the movement to sound waves.

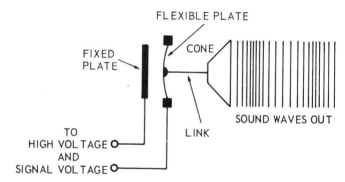

Fig. 15-31. Drawing shows the operation of a high frequency electrostatic speaker, usually called a TWEETER.

HIGH FIDELITY SYSTEMS

A speaker should be the proper size and design to best produce the audio sounds in a given range. Generally, the larger speakers are necessary to produce the low bass notes. These speakers are called WOOFERS. There are INTERMEDIATE RANGE speakers for the great majority of audio sounds. Speakers designed for only the high frequencies are called TWEETERS. See Fig. 15-32.

Fig. 15-32. A 3-way speaker system with a woofer, a mid-range and a tweeter. (Pioneer)

Some typical components usually found in a stereo system are shown in Figs. 15-33 through 15-36. These components include: an AM-FM stereo receiver, Fig. 15-33; a turntable, Fig.

Fig. 15-33. An AM-FM stereo receiver.　(Pioneer)

Fig. 15-34. A manual/multiple play turntable.　(B.I.C.)

Fig. 15-36. Stereo headphones.　(KOSS)

15-34; a pair of speaker enclosures, Fig. 15-35; and a set of stereo headphones, Fig. 15-36.

CROSSOVER NETWORKS

The various kinds of speakers suggest some circuitry which will direct a specified range of frequencies to the speaker which will best reproduce them. Such is the crossover network in Fig. 15-37.

All these networks are based on the characteristics of X_L and X_C. With higher frequencies, X_L increases in ohms. Therefore, it will pass the lower frequency currents better. With higher frequencies, X_C decreases in ohms and offers less impedance to higher frequencies.

Fig. 15-35. Speaker enclosures for a stereo system.　(BOSE)

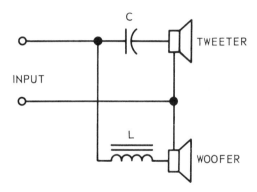

Fig. 15-37. Crossover networks.

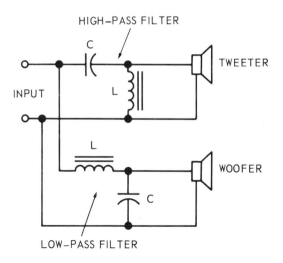

Fig. 15-38. A crossover network using filter circuits.

The more sophisticated circuit in Fig. 15-38 takes advantage of both changing values of X_C and X_L to form a HIGH PASS FILTER and a LOW PASS FILTER. Apply your previous knowledge of L and C circuits and be sure you understand the principles of these crossover circuits found in most high fidelity systems.

PUSH-PULL AMPLIFIER PROJECT

Here is a practical little amplifier which may be used for many purposes. You can use it to increase the volume of oscillators and sirens.

You can connect a record player or a microphone to its input. The complete push-pull amplifier is shown in Fig. 15-39. Its schematic and components list are given in Fig. 15-40.

The circuit board layout is presented for you in Fig. 15-41. However, it may be built on terminal strips or vector board.

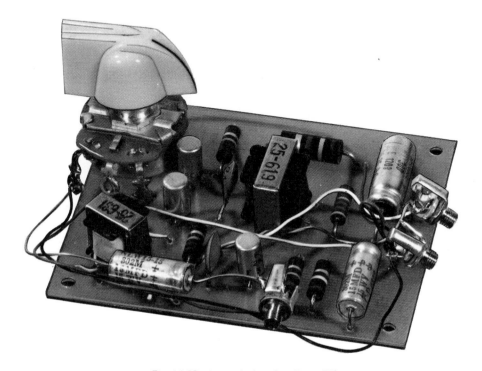

Fig. 15-39. A practical push-pull amplifier.

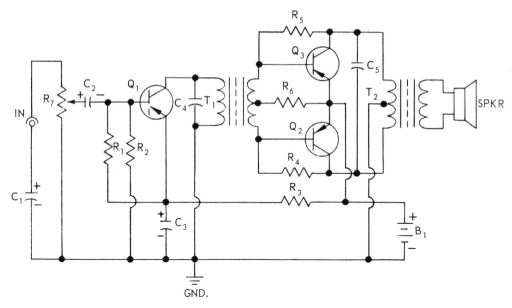

R_1 — 10 KΩ, 1/2W resistor
R_2 — 100 KΩ, 1/2W resistor
R_3 — 470 Ω, 1/2W resistor
R_4, R_5 — 22 KΩ, 1/2W resistor
R_6 — 270 Ω, 1/2W resistor
R_7 — 1000 Ω potentiometer
C_1, C_2 — 15 μF capacitors @ 15V dc
C_3 — 50 μF capacitor @ 15V dc
C_4 — .002 μF capacitor @ 50V dc

C_5 — .05 μF capacitor @ 50V dc
Q_1, Q_2, Q_3 — 2N408 transistors
T_1 — 10,000 Ω CT/1000 Ω transformer (Midland 25-619)
T_2 — 500 Ω CT/3.2 Ω transformer (Midland 25-631)
SPKR — 3.2 Ω speaker
B_1 — 9V battery or power supply

Fig. 15-40. Schematic and parts list for Push-Pull Amplifier.

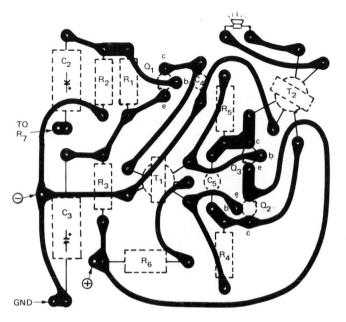

Fig. 15-41. A suggested circuit board layout and parts placement for amplifier project.

TEST YOUR KNOWLEDGE

1. Explain the purpose of a BLOCKING CAPACITOR.
2. A series RC coupling circuit has a value of R = 500 ohms. What value of C should be used at a frequency range of 400 hertz?
3. What method usually is used to overcome loss of gain due to the mismatch with RC coupling?
   a. Add another transistor stage.
   b. Remove a transistor stage.
   c. Bypass the capacitor.
4. What are the advantages of transformer coupling?
5. A _____ to 1 turns ratio is required for a transformer to match 1600 ohms to 4 ohms.
6. _____ is effective power of a signal with a

V_{p-p} of 10 volts and I_{p-p} of .4 amps.

7. What is the collector efficiency of a power transistor when its ac output power is 100 mW and its dc input power is 300 mW?

8. A small forward bias is applied to transistors in a push-pull amplifier to reduce crossover distortion. True or False?

9. Give an advantage and a disadvantage of using transistors in complementary symmetry for a power amplifier.

10. Why is a phase splitter required for the input to a push-pull amplifier?

11. What is the difference between a PM speaker and an electrodynamic speaker?

12. A "tweeter" is a _____ speaker.

13. A "woofer" is a _____ speaker.

14. Why are crossover networks used with speakers?

15. What is the purpose of a "heat sink?"
    a. To store heat.
    b. To dissipate heat.
    c. To preheat transistor.

Chapter 16
TRANSISTORIZED OSCILLATORS

An oscillator is an electronic circuit that produces a continuously variable alternating current (ac) or pulsating direct current (dc) at a desired frequency.

In this chapter you will:

1. Study the conditions which must exist in a circuit to create oscillation.
2. Become acquainted with some of the more common oscillator circuits.
3. Become familiar with special multivibrator circuits used in modern computers.
4. Study methods of converting ac to dc and dc to ac by power oscillators.

THE OSCILLATOR

A sound amplification system being used at a public gathering is one of the classic examples of conditions necessary to produce oscillation. If the microphone is close to the speaker system, it will pick up music, speech or just noise from the speaker. These sounds will be amplified and reproduced by the speaker, only to be picked up again and amplified; and around and around it goes, Fig. 16-1. A disagreeable howl is produced.

A more technical description of the condition would suggest that energy from the output of the system is fed back to the input and reamplified. An oscillator must always depend upon feedback to sustain oscillation. The feedback must reinforce the input signal or be REGENERATIVE. If the signal feedback opposed the input by being 180 deg. out of phase,

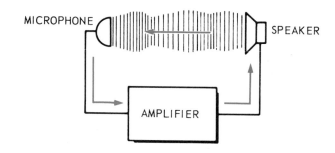

Fig. 16-1. The amplifier howls or oscillates due to regenerative feedback.

it would be DEGENERATIVE and oscillation would not be sustained.

In the case of the amplifier system, oscillation is undesirable. In many other types of equipment, circuits are purposely designed to produce an oscillating signal of a desired amplitude and frequency.

A review of tuned circuits in Chapter 10 may now be necessary. For emphasis, the resonant RCL circuit will be discussed again. In the circuit of Fig. 16-2, capacitor C is charged when switch S is closed. When switch S_1 is opened, C starts its discharge cycle, Fig. 16-3, and a magnetic field is produced around L. As C

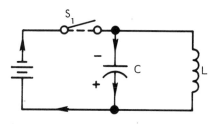

Fig. 16-2. When switch S_1 is closed, capacitor C charges.

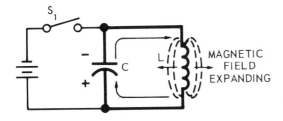

Fig. 16-3. C discharges through L and produces a magnetic field.

becomes discharged and the discharged current decreases, the magnetic field collapses and induces a current of such a polarity that C is charged to the opposite polarity, Fig. 16-4.

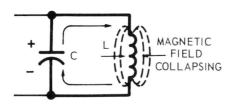

Fig. 16-4. The collapsing magnetic field will charge C to the opposite polarity.

Now C attempts to discharge in the opposite direction, as in Fig. 16-5. Again, the magnetic field is produced by the discharge current and, as the current decreases, the magnetic field again collapses and charges C to its original polarity. See Fig. 16-5.

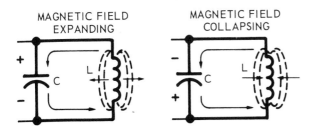

Fig. 16-5. C discharges through L. As field around L collapses, it charges C to the opposite polarity.

This current will continue to circulate until the energy originally supplied to charge C is dissipated or used up by the resistance of the circuit. The action of the current is described as FLYWHEEL ACTION. The tuned circuit is called a TANK CIRCUIT.

The circulating current or voltages across L or C can be plotted as shown in Fig. 16-6. Each successive wave is less in amplitude due to used up energy. The wave is called a DAMPED WAVE.

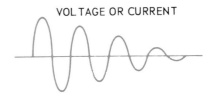

Fig. 16-6. A damped wave successively decreases in amplitude until it dies out.

The frequency of the wave output will depend upon the values of L and C. You will recall the resonant frequency formula as:

$$f_o = \frac{1}{2\pi \sqrt{LC}} \quad \text{or} \quad \frac{.159}{\sqrt{LC}}$$

One might conclude that if a little bit of energy were added to the tuned oscillating circuit at the proper time, it would no longer decay to zero, but would continue to oscillate. This is quite correct. Then, the output of the circuit would appear as a CONTINUOUS WAVE (CW), as illustrated in Fig. 16-7.

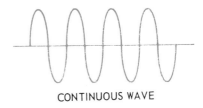

Fig. 16-7. A continuous wave of constant amplitude and frequency.

Originally, the circuit was activated by closing switch S_1 and charging capacitor C. This was a manual operation. Opening and closing a switch by hand even a few times per second would be physically impossible. Therefore, the transistor is prominently used as an extremely fast electronic switch. A description of this circuit will lay a firm foundation for more advanced oscillator circuits. Follow the action in Fig. 16-8.

In this circuit, transistor Q can be in one of two states: CUTOFF (similar to an open switch), or CONDUCTING (similar to a closed switch). Assuming proper bias voltages: When the circuit is turned on, collector current starts to flow. The increasing I_C through L_1 produces

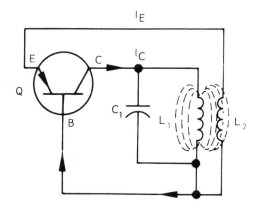

Fig. 16-8. This circuit explains the basic oscillator feedback system and transistor switching action. Voltage sources are omitted for simplification.

a magnetic field which is coupled to L_2 by transformer action. The increasing I_C produces a current in L_2 of such a polarity that I_E is increased, which also causes I_C to increase further.

This "feedback" action continues until transistor Q is conducting at SATURATION and cannot conduct any more. There is no further increase in I_C, so there is no further change in

I_C. As a result, the magnetic coupling between L_1 and L_2 drops to ZERO (coupling only exists when there is a CHANGE in current).

There is now no induced current to increase the forward bias of the EB junction, so the collector current starts to decrease. A decrease in I_C produces magnetic coupling again, but of such a polarity in L_2 that the EB junction becomes very much reverse biased, and the transistor is quickly driven to "cutoff." At cutoff, the current I_C again ceases to "change" and there is no magnetic coupling. The transistor returns to its normal bias state and the action is repeated and will continue to repeat itself.

The frequency of the output depends upon L and C of the tank circuit. Note that the tank circuit itself controls the point at which energy is added to the tank circuit. In the earlier example, energy was added to C from a battery source. In this circuit, energy is added again from a battery source, but is switched by a transistor which is controlled by magnetic feedback. Coil L_2 is frequently called a TICKLER COIL. The output wave form of this circuit is shown in Fig. 16-9.

The basic block diagram for an oscillator is shown in Fig. 16-10. Every oscillator circuit has a wave producing circuit, an amplifier and a feedback circuit. The feedback circuit sends a small electrical signal back to the wave producing circuit to keep the oscillations going or sustained.

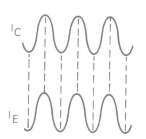

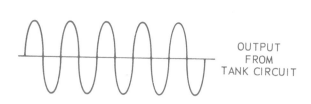

Fig. 16-9. These waves show the rise and decay of I_C and I_E as well as the oscillator continuous wave output.

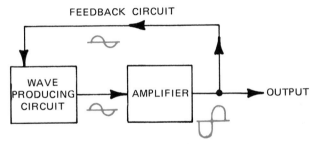

Fig. 16-10. Block diagram for a basic oscillator.

THE HARTLEY OSCILLATOR

Many of the same circuits used with vacuum tubes now find their counterparts in transistorized circuits. In either case, the transistor or tube acts as an amplifier and switch. The Hartley oscillator circuit, Fig. 16-11, depends upon magnetic coupling for its feedback. Its operation and theory is exactly the same as the tickler coil and transformer. The Hartley oscillator is identified by a tapped tank coil or autotransformer.

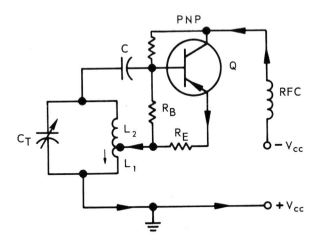

Fig. 16-11. A transistorized Hartley oscillator circuit.

The current I_E, which includes I_C, is indicated by arrows in Fig. 16-11. As the circuit is turned on, current flowing through L_1 induces a voltage at the top of L_2. This makes the base of Q more negative, which drives the transistor to saturation. At saturation, there is no longer a change of current, and the coupling between L_1 and L_2 falls to zero. The less negative voltage at

the base of Q causes the transistor to decrease in conduction. The decrease, in turn, induces a positive voltage at the top end of L_2. This is reverse bias for the transistor, and it is quickly driven to cutoff. Then, the cycle is repeated. The tank circuit is energized by pulses of current. The transistor alternates between saturation and cutoff at the same frequency.

The "at rest" bias condition of the transistor is established by R_B and R_E. The radio frequency choke (rfc) blocks the rf signal from the power source. In this circuit, note that the coil L_1 is in SERIES with the collector circuit of the transistor. It is called a SERIES-FED oscillator.

In Fig. 16-12, a SHUNT-FED oscillator is shown. The operation is the same. Note that the dc path for the emitter-collector current is not

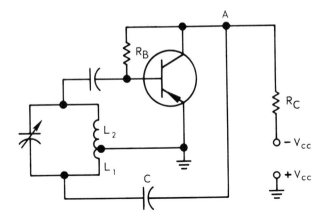

Fig. 16-12. A shunt-fed Hartley oscillator.

through the coil L_1. The ac signal path, however, is through C and L_1. At point A, the two current components are separated and required to take parallel paths. Both oscillators receive their feedback energy by means of magnetic coupling.

THE COLPITTS OSCILLATOR

Feedback to sustain oscillation may also be accomplished by means of an electrostatic field as developed in a capacitor. If the tapped coil from the Hartley oscillator is replaced with a split stator capacitor, a voltage of proper polarity may be fedback which causes the circuit to oscillate. This circuit has been named the COLPITTS OSCILLATOR, Fig. 16-13.

The Colpitts oscillator is similar to the Hartley oscillator in theory of operation. However, in the Colpitts setup, the signal is coupled

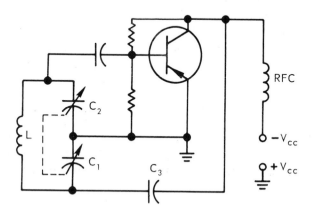

Fig. 16-13. Schematic diagram of a Colpitts oscillator.

back to C_1 of the tank circuit through coupling capacitor C_3. A changing voltage at the collector appears as a voltage across the tank circuit $LC_1 C_2$ in the proper phase to be a regenerative signal. The amount of feedback will depend upon the ratio of C_1 to C_2. This ratio usually is fixed, and both capacitors C_1 and C_2 are controlled by a single shaft (ganged capacitor). The natural frequency of the oscillator is determined in the usual manner.

The tuned tank consists of L and C_1 and C_2 in series. Note that the circuit is shunt fed. Series feed is impossible due to the blocking of dc by the capacitors.

CRYSTAL OSCILLATORS

The piezoelectric effect of crystals was studied early in Chapter 2. When the crystal energized or excited, it will vibrate at a frequency depending upon its physical dimensions. Crystals are cut from quartz as it is found in

nature. The characteristics of the crystal will depend on how it is cut.

Before applying a crystal to stabilize an oscillator, it is well to look at the ELECTRICAL EQUIVALENT of the crystal as shown in Fig. 16-14. A crystal is placed between two metallic holders. This forms a capacitor C_H with the crystal itslef as the dielectric. C_G represents the series capacitance between the metal holding plates and the air gap between

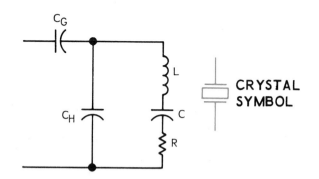

Fig. 16-14. The equivalent electrical circuit of a crystal. At left is its schematic symbol.

them as a dielectric. L, C and R represent the characteristics of the crystal. Of special importance to our studies is the similarity of the equivalent crystal circuit to a tuned circuit. It will have a resonant frequency.

A typical mounted crystal is displayed in Fig. 16-15. A crystal ground to vibrate at a frequency of one megahertz would be approximately one in. square and .1125 in. thick. To grind it thinner would increase its frequency.

Fig. 16-15. Typical crystal used to fix the transmitting frequency of a radio station. (Texas Crystals)

A transistorized crystal oscillator circuit is drawn in Fig. 16-16. Compare this circuit to Fig. 16-12. It is the same circuit with the crystal added to the FEEDBACK CIRCUIT. The crystal acts as a series resonance circuit and determines the frequency of the feedback currents. The tank circuit must be tuned to this frequency.

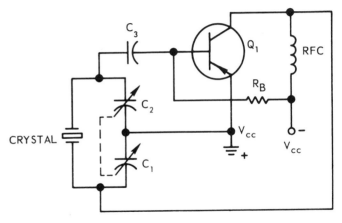

Fig. 16-17. Circuit of the crystal-controlled Pierce oscillator.

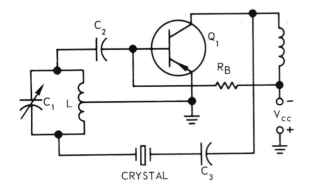

Fig. 16-16. A crystal controlled Hartley oscillator circuit.

In the circuit of Fig. 16-17, the crystal is used in place of the tuned circuit in a Colpitts oscillator. This version has been named the PIERCE OSCILLATOR. Compare the circuit to Fig. 16-13.

The amount of feedback to energize the crystal again depends upon the ratio of C_1 to C_2. These capacitors form a voltage divider

across the base-emitter of the transistor. The Pierce circuit is a very stable circuit under varying circuit conditions and changes.

RC OSCILLATORS

Remember, if part of the output of an amplifier is fedback to the input so that it is in phase or regenerative, one of the conditions of oscillation will be present. Refer to Fig. 16-18 in which two stages of CE amplifiers are coupled together.

First, consider the circuit without C_F, the coupling feedback capacitor. It is a CE amplifier. A positive going signal into Q_1 produces a negative going signal at C of Q_1. The signal is

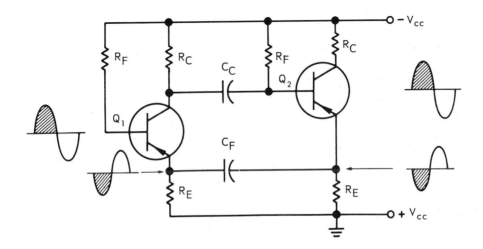

Fig. 16-18. A two-stage RC oscillator.

inverted. A negative going signal to Q_2 produces a positive going signal at output of Q_2. The signal now is in phase with the input to Q_1. A positive going signal at the collector will produce a negative going signal at the emitter of Q_2. This voltage is couple to E of Q_1, which decreases the forward bias of Q_1. This is exactly the same effect as a positive input to Q_1 so it is regenerative feedback. The circuit oscillates, depending on values of R and C, but it is not stable. It usually will operate over a range of frequencies. NOTE: More practical circuits can be designed.

In Fig. 16-19, a single transistor RC oscillator is diagrammed. This setup is called a PHASE SHIFT oscillator.

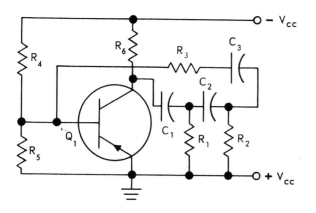

Fig. 16-19. A phase shift oscillator circuit.

This oscillator takes advantage of the fact that an RC circuit will produce a phase shift. That is, in a capacitive circuit, the current will lead the applied voltage by an angle of 90 deg. or less. In this circuit, $R_1 C_1$ shift the phase by 60 deg., $R_2 C_2$ shift it another 60 deg. and $R_3 C_3$ shift it a final 60 deg. which totals a 180 deg. phase shift. As a result, a positive going signal is inverted by the phase shift network to a negative going and regenerative signal for the transistor input. The conditions for oscillation are satisfied.

THE TUNNEL DIODE OSCILLATOR

Ordinarily, we would describe resistance in terms of Ohm's Law. An increase in resistance would produce a decrease in current; or an increase in voltage would produce an increase in current. There are some effects which seemingly contradict this law. An example of this contradiction is a device in which current would decrease as voltage is increased. It is technically called NEGATIVE RESISTANCE.

We have pretty well established the fact that a tuned circuit will not continuously oscillate unless energy is added to the circuit to overcome the resistance of the tank circuit. Why not add negative resistance to cancel the positive resistance? Then, the circuit would oscillate continuously.

A curve is drawn in Fig. 16-20 to help you to understand negative resistance. Among components which exhibit this peculiar characteristic is the TUNNEL DIODE or ESAKI DIODE, named after the inventor. In our studies in Chapter 12, we discovered that a PN junction would conduct when biased in a

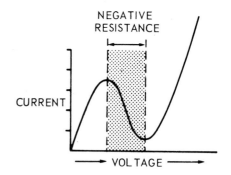

Fig. 16-20. In the negative resistance region, the current decreases as voltage is increased.

forward direction. At first application of a low voltage, conduction was not high because it was necessary for the bias voltage to overcome the potential hill. To express it another way, the carriers must gain sufficient energy to cross the depletion area. A typical diode curve is redrawn in Fig. 16-21.

With the discovery of the tunnel diode, certain classical theories were found to be incorrect. Current carriers, for example, might

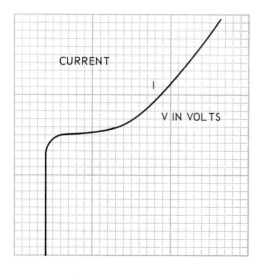

Fig. 16-21. The characteristic curve of a typical diode.

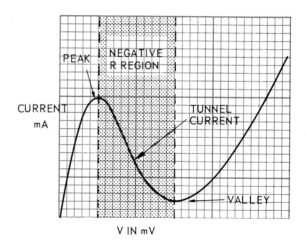

Fig. 16-22. The characteristic curve of a tunnel diode.

tunnel under the depletion area even though they have insufficient energy to cross over it. The necessary requirement is that the depletion area must be made very narrow. The tunnel diode is made of two very heavily doped crystals. The P crystal is rich in HOLES; the N crystal is rich with free electrons. Consequently, the depletion area is very narrow, and the carriers can tunnel under the area and cross the junction. This characteristic curve is shown in Fig. 16-22.

Note that when a small voltage is applied, a small current starts to flow. A further increase in voltage causes the current to decrease. At the point I_{valley}, the diode starts to appear as a positive resistance and current rises as would be expected from a conventional diode. From zero to peak, the current is a result of carriers

tunneling under the depletion area. The current after "valley" is the result of carriers crossing and joining at the junction. It is the area of negative resistance which interests us at the moment.

In some circuits, the negative resistance may be used in parallel with a tuned circuit or in series with a tuned circuit. In either case, its purpose is to cancel the positive resistance of the oscilliatory circuit. The tunnel diode will be biased near the center of the negative resistance portion of its curve.

Study the circuit in Fig. 16-23 and follow the action. R is selected to provide a voltage across D near the center of its negative resistance curve. When S_1 is closed, current rises to a value determined by the resistance of R, plus the resistance of D. The voltage divides across D and R according to the ratio of their resistances.

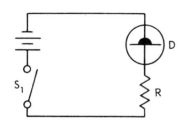

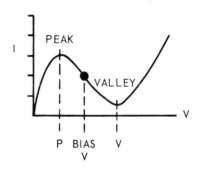

Fig. 16-23. Circuit is used to explain tunnel diode action.

However, as the voltage drop across D passes point P, it is in the negative area and resistance starts to increase. Therefore, a larger voltage appears across D, and its resistance continues to increase until V_D reaches point V on the curve. At this point, a further increase in voltage drives the diode into its positive resistance region. The resulting increase in current will increase the voltage across R, and the voltage across D will again decrease into its negative resistance region. This decrease in voltage across D decreases its resistance, and circuit current continues to rise until point P is again reached.

This describes one cycle of operation, and the circuit will continue to oscillate back and forth through the negative resistance region. Its output across R is similar to a sine wave.

A practical circuit is shown in Fig. 16-24. In this case, negative resistance is used in series with a tuned circuit consisting of C and the speaker voice coil.

At first, S_1 is closed, and the voltage is adjusted across divider R_1 and R_3 until the proper bias point is reached. This point is determined by a sine wave appearing on an oscilloscope connected across the speaker. An audible tone should be heard from the speaker. The oscillations of the tunnel diode are now energizing the tuned circuit, and a continuous wave is produced at its output. In this circuit,

R_3 sets the proper bias level for the diode and R_1, in parallel with the tank, sets the proper current level for the diode.

POWER OSCILLATORS AND CONVERTERS

Sometimes, it is desirable to convert dc to ac. Also, you may wish to convert dc to ac, transform it to a higher voltage and rectify it back to dc. A power oscillator may be used in this application. Fig. 16-25 shows a circuit of a converter or, in this case, a push-pull oscillator. The collector load of each transistor is the primary of the transformer. The ac output will be found at the secondary. A suitable turns ratio may be used if higher or lower voltages are desired.

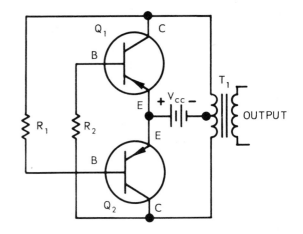

Fig. 16-25. The circuit of a power oscillator.

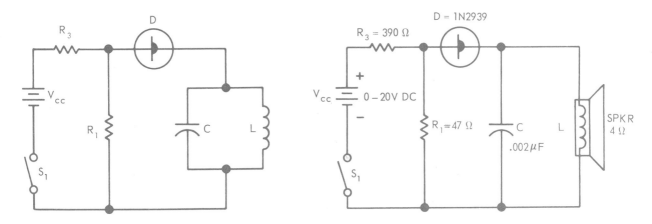

Fig. 16-24. A tunnel diode audio oscillator circuit.

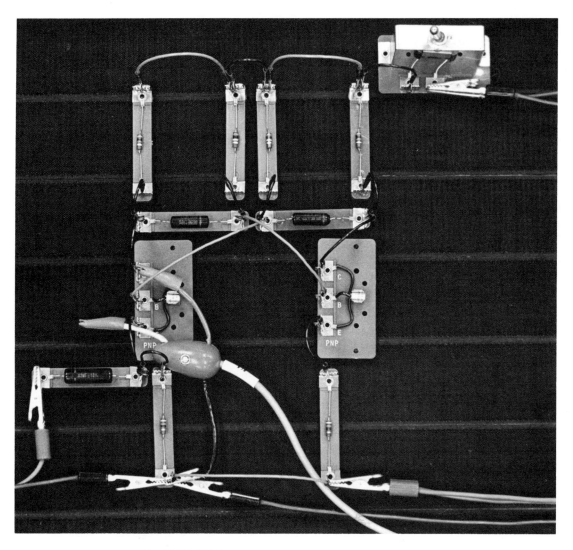

Fig. 16-26. This breadboard experiment demonstrates a typical transistorized multivibrator. (Lab Volt)

Only a slight imbalance in conductivity between Q_1 and Q_2 is required to start oscillation. This imbalance is always present due to components or temperature. The two transistors are either going toward saturation and cutoff respectively, or vice versa. Follow the action.

Assume Q_1 starts conducting. The voltage at C of Q_1 goes less negative, which makes the base of Q_2 less negative and drives Q_2 toward cutoff. A more negative voltage at C of Q_2 drives Q_1 base more negative and Q_1 reaches saturation.

At saturation, when there is no change in current, the reactance of the transformer primary drops to zero with a corresponding de-crease in collector voltage toward the value of V_{cc}. This more negative voltage coupled to the base of Q_2 through R_1 starts Q_2 toward conduction and saturation. The transistors conduct alternately, and output is combined into a complete cycle at transformer secondary output.

If the output from the transformer in Fig. 16-25 is connected to a rectifier and filter circuits, the output can again become dc.

ASTABLE MULTIVIBRATOR PROJECT

A free running multivibrator circuit is called ASTABLE. This circuit could be called an oscillator with a square wave output. A circuit for the oscillator, Fig. 16-26, is drawn schematically in Fig. 16-27. Examine its operation.

The multivibrator starts its oscillation due to a slight difference between the conduction of Q_1 and Q_2. Assume that Q_1 starts to conduct slightly more than Q_2. The conduction of Q_1 causes a larger voltage drop across R_{C_1} and consequently a less negative voltage at the collector of Q_1. This positive going signal is coupled through C to the base of Q_2. This decreases the forward bias of Q_2 and Q_2 is quickly driven to cutoff. A decreasing current in Q_2 causes its collector to become more negative. This negative going voltage coupled to the base of Q_1 quickly drives it into saturation.

At this time, C_2 charges to the collector voltage of Q_2, and C_1 discharges to the collector voltage of Q_1 according to the time constant of the circuit. After an interval, the base end of C_2 becomes positive, and the base end of C_1 becomes negative. This decreases the forward bias of Q_1 and decreases the conduction of Q_1. Also, the negative voltage at the base of

Q_2 increases the forward bias and Q_2 is driven into conduction. Action continues with Q_1 and Q_2 alternating conduction and a square wave is produced. The frequency depends upon the time constant of the circuit.

This circuit can be kept accurately "on frequency" by means of a synchronization pulse, called the "SYNC PULSE." To understand this action, assume a positive pulse voltage applied across R_{E_1}. This would drive Q_1 immediately into conduction by increasing the forward bias of its EB junction. When this occurs, even if Q_1 is not quite ready to conduct, it will be forced into conduction. The oscillator will lock-in on the pulse sync (oscillator must be designed to operate close to intended SYNC pulse frequency).

Pulse synchronization is used in your TV set to produce a stable picture. The "horizontal

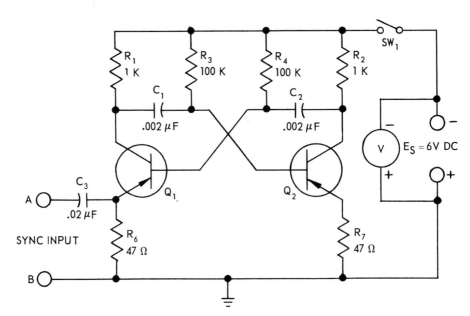

PARTS LIST FOR ASTABLE MULTIVIBRATOR

V — 0-25V dc voltmeter
Q_1, Q_2 — 2N408 transistors or RCA SK 3003
R_1, R_2 — 1 KΩ, 1/2W resistors
R_3, R_4 — 100 KΩ, 1/2W resistors
R_5 — 100 KΩ, 1/2W potentiometer
R_6, R_7 — 47 Ω, 1/2W resistors

C_1, C_2 — .002 μF capacitors
C_3 — .02 μF capacitor
C_4 — .01 μF capacitor
C_5 — 390 pF capacitor
S_1 — SPST switch
E_s — 0-6V dc power source

Fig. 16-27. Circuit and parts list for an Astable Multivibrator.

hold" and "vertical hold" controls on the TV set make slight adjustments to the horizontal and vertical oscillator frequencies so that they can lock-in on the sync pulse sent by the TV station.

Another method of designing an astable multivibrator is the use of a common emitter resistance for feedback, Fig. 16-28. When this circuit is energized, emitter currents of both transistors flow through R_E. Current flow will cause a voltage drop across collector load resistors R_{C_1} and R_{C_2}. The collector voltages will become less negative by the amount of the voltage drops. Which transistor will conduct first? Follow the action.

When the circuit is first turned on, C will instantaneously attempt to charge to the value of $-V_{CC}$. This charging current will make the base end of R_B more negative than the emitter of Q_2, which quickly drives Q_2 to saturation. The emitter end of R_E is driven quite negative due to the current of Q_2, and Q_1 is cut off.

Now, Q_2 conducts fully while C is charging. However, as C approaches full charge at an exponential rate, the current decays at the same rate. When it becomes zero, the base end of R_B approaches zero and Q_2 conducts less. A decrease in current causes the emitter end of R_E to become less negative. This starts Q_1 conducting. The collector of Q_1 becomes more positive, which is coupled back to the base of Q_2.

Q_2, then, is driven to cutoff, and C will discharge to the new value of Q_1 collector voltage. This discharge current makes the base of Q_2 even more positive. Q_1 remains conducting until C has discharged, then the reverse bias is removed from Q_2 and conduction is switched to Q_2. The oscillation continues. Its frequency is dependent upon the circuit time constants.

MONOSTABLE MULTIVIBRATOR

A monostable multivibrator circuit has one stable state. The circuit is frequently referred to as a ONE-SHOT MULTIVIBRATOR. A pulse from an external source will cause the transistors to switch conduction. After an interval of time, depending on the circuit time constant, it will switch back to its original state. See Fig. 16-29.

Assume that Q_1 is conducting at saturation and Q_2 is cutoff. This is true because the voltage dividers $R_F R_B$ set up a forward bias for Q_1 and Q_2 has no forward bias. The collector voltage of Q_1 is less negative due to the large drop across R_{C_1}.

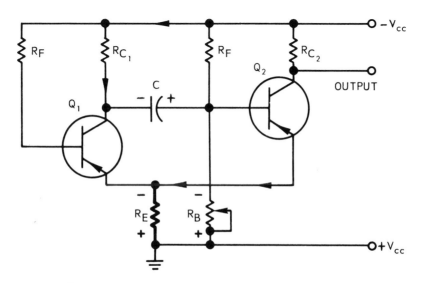

Fig. 16-28. An emitter coupled astable multivibrator circuit.

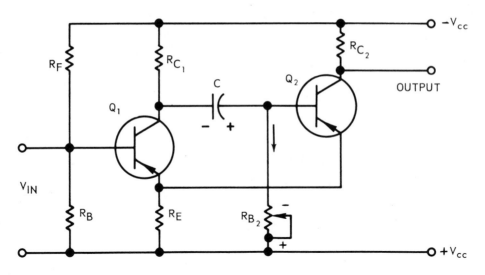

Fig. 16-29. A monostable multivibrator circuit.

When a positive pulse is applied to input across R_B, it places a reverse bias on Q_1 and cuts it off. The collector voltage of Q_1 approaches $-V_{cc}$ since there is no drop across R_{C_1}. Instantaneously, a large voltage appears across R_{B_2} in the polarity shown (due to charging current for C). Q_2 is driven into conduction and remains in conduction until C is charged to the value of $-V_{cc}$. At that time, the voltage across R_{B_2} disappears, Q_2 is cutoff and the circuit reverts to its original state with Q_1 conducting.

The duration of time that Q_2 will conduct will depend upon the time constant of the RC circuit. If the multivibrator is going to go through its entire cycle with one input pulse, the time required for Q_2 to turn on and off must be greater than the time duration of the triggering pulse.

FLIP-FLOP, BISTABLE MULTIVIBRATOR

Another variation of the multivibrator is a circuit which has two stable states. That is, either Q_1 conducting and Q_2 off or Q_1 off and Q_2 conducting. A pulse from an external source will make the stable state shift from one to the other. For this reason, it has been named the FLIP-FLOP.

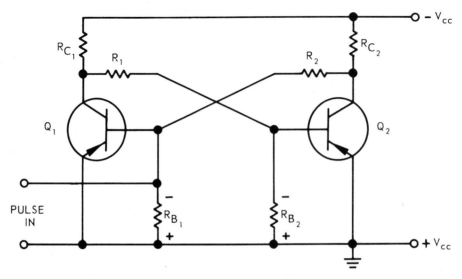

Fig. 16-30. The circuit of a flip-flop or bistable multivibrator.

The flip-flop, bistable multivibrator finds wide usage in computer circuitry. The circuit requires two pulses to return it to its original state. Remember, a transistor can be either ON or OFF in its stable state and can be switched from one to the other. Therefore, the ON and OFF can be related to the ONE and ZERO of the BINARY NUMBER SYSTEM.

Fig. 16-30 describes the circuit. Biasing resistors for each transistor are equally matched. The circuit is quite symmetrical.

Assume that Q_1 is ON and Q_2 is OFF. A positive pulse to the base of Q_1 will reverse bias the EB junction and Q_1 will be cut off. The collector voltage of Q_1 approaches $-V_{cc}$. Now R_1 and R_{B_2} form a bias voltage divider for Q_2. The base of Q_2 becomes more negative because of the increased divider current. Q_2 turns ON. The collector voltage V_c of Q_2 becomes more positive, and voltage at the base of Q_1 becomes less negative so the Q_1 is cut off.

These conditions are tabulated as follows:

Q_1	Q_2	V_{c_1}	V_{c_2}	V_{B_1}	V_{B_2}
ON	OFF	$\cong 0$	$-V_{cc}$	negative	positive
OFF	ON	$-V_{cc}$	$\cong 0$	positive	negative

You will observe that a positive pulse will switch from Q_1 to Q_2. In order to switch back to Q_1, a negative pulse must be used as a trigger if applied to Q_1. A positive pulse applied to Q_2 will switch the circuit to Q_1.

By using several of the flip-flops in cascade, a COUNTER CIRCUIT can be made with readouts in the binary number system, Fig. 16-31.

In each flip-flop, there is a lamp in the collector circuit of Q_1. When Q_1 is conducting, the lamp will glow. Assume that all transistors are OFF. See the Binary Pulse Table in Fig. 16-32.

In this table, note that:

FF1 switches with every pulse.
FF2 switches with every second pulse.
FF3 switches with every fourth pulse.
FF4 switches with every eighth pulse.

DIGITAL	PULSES IN	LIGHTS ON	BINARY
0		0000	0000
1		0000●	0001
2		000●0	0010
3		000●●	0011
4		00●00	0100
5		00●0●	0101
6		00●●0	0110
7		00●●●	0111
8		●000	1000

CAN CONTINUE TO 15 PULSES

Fig. 16-32. Binary Pulse Table.

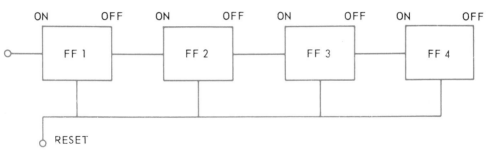

Fig. 16-31. Counter circuit made of flip-flops.

At 16 pulses, this counter returns to zero. All lights are out. Additional stages are added to count to any number desired.

At this time, you may wish to do some additional study in reference textbooks on binary number systems. A knowledge of numbers systems other than base 10 will be very useful to the electronic engineer or technician in today's industry.

CODE PRACTICE OSCILLATOR PROJECT

You may wish to build the exciting electronics project called the Code Practice Oscillator pictured in Fig. 16-33. The schematic and parts list is given in Fig. 16-34. By plugging in the key, Fig. 16-33, the circuit is activated. The tone of the audio signal is controlled by R_1.

This project is very simple, yet effective to operate.

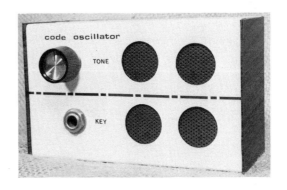

Fig. 16-33. Code Practice Oscillator.

TEST YOUR KNOWLEDGE

1. An oscillator is an electronic circuit that produces a continuously variable alternating current or pulsating direct current at a desired _____.
2. Energy from the output of a sound amplification system is fed back to the input and reamplified. This is called _____.
   a. Feedback.

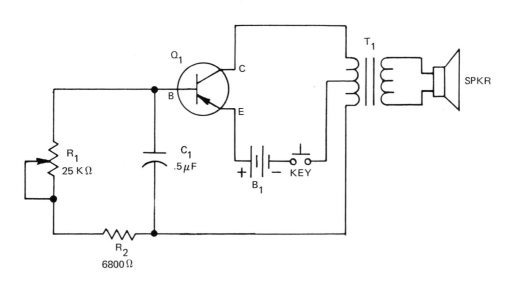

PARTS LIST FOR CODE PRACTICE OSCILLATOR

Q_1 — PNP transistor, RCA SK 3004 or Motorola HEP G0005

T_1 — 500 Ω CT/4 Ω transistor output transformer, Stancor TA-21 or equivalent

C_1 — .5 μF, 100V dc capacitor

R_1 — 25 KΩ, 1/2 to 2W potentiometer

R_2 — 6800 Ω, 1/2W resistor

SPKR — 4 Ω miniature speaker

Key — code key

B_1 — 9V battery or power supply

Misc. — chassis, knob, jack

Fig. 16-34. Schematic and parts list for Code Practice Oscillator.

b. Phase shift.

c. Flip-flop.

3. What conditions are necessary for a circuit to oscillate?

4. What method of feedback is used in the Hartley oscillator?

5. What method of feedback is used in the Colpitts oscillator?

6. Explain the meaning of "negative resistance."

7. An 800 KHz cut crystal has a negative temperature coefficient of 20 Hz/deg. C/MHz, and oscillates at 800 KHz at 50 C. What is its oscillating frequency at 60 C?

a. 7998.4 Hz.

b. 79,984 Hz.

c. 799,840 Hz.

8. Draw a dc-ac converter with full-wave rectifier and μ filter.

9. What would cause a wave in a tank circuit to dampen out?

10. A Hartley oscillator has a tank circuit consisting of a 100 μ H coil and a capacitor of 159 pF. What is its frequency?

a. 1.26 KHz.

b. 12.6 KHz.

c. 126 KHz.

11. A free running multivibrator is called _____.

12. A monostable multivibrator circuit has one stable state. A _____ multivibrator circuit has two stable states.

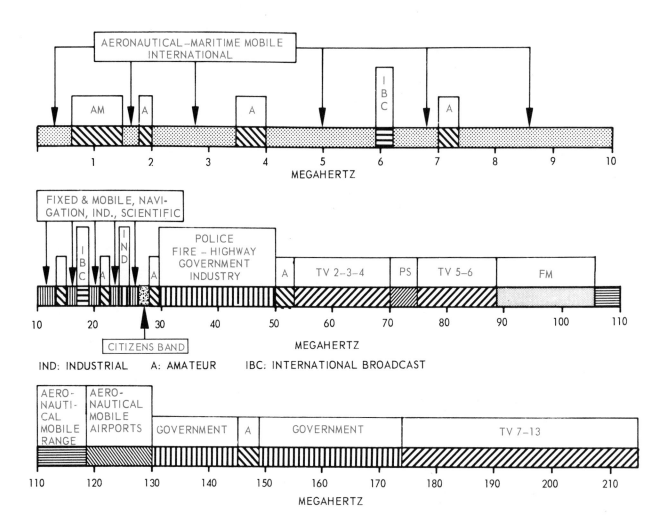

Fig. 17-1. The frequency spectrum and allocations.

Chapter 17

RADIO RECEPTION AND TRANSMISSION

In the field of electronics, the study of reception and transmission of waves is vital. In this chapter, the following major concepts will be explained:

1. The frequency spectrum and frequency location of various services.
2. The principles of amplitude modulation.
3. The principles of demodulation or detection.
4. The block diagram and assembly of selected units into a working radio receiver.
5. The principles of frequency modulation and detection.

THE FREQUENCY SPECTRUM

It is convenient to group certain bands of frequencies and label them for reference purposes. Common identification of these bands is listed in the following Table:

17-1. Major frequency allocations made by international agreements are indicated.

Imagine what would happen if all broadcasting companies were allowed to use as much of the frequency spectrum as they wished. There would be utter confusion, and the radio and TV signals would be badly distorted.

The following questions are for practice in using the frequency spectrum: What band of frequencies is used for AM broadcasting? In which band would TV Channel 4 be located? Where are FM stations located?

FREQUENCY AND WAVELENGTH

Any particular frequency can be identified by either HERTZ (cycles per second) or WAVELENGTH in METRES. The Greek letter lambda (λ) is used to designate wavelength.

	BAND	FREQUENCY RANGE
VLF	— Very Low Frequencies	3 KHz to 30 KHz
LF	— Low Frequencies	30 KHz to 300 KHz
MF	— Medium Frequencies	300 KHz to 3000 KHz
HF	— High Frequencies	3 MHz to 30 MHz
VHF	— Very High Frequencies	30 MHz to 300 MHz
UHF	— Ultra-High Frequencies	300 MHz to 3000 MHz
SHF	— Super-High Frequencies	3000 MHz to 30,000 MHz
EHF	— Extremely High Frequencies	Above 30,000 MHz

The FREQUENCY SPECTRUM is the entire range of frequencies of electromagnetic radiations. It is shown in abbreviated form in Fig.

Remember that radio waves travel at the speed of light, 3×10^8 metres/sec. or 186,000 miles/sec. Therefore, it is possible to compute a wavelength by the distance a wave will travel during one time period of its cycle.

$$\lambda \text{ in metres} = \frac{3 \times 10^8}{f \text{ in Hz}} \quad or \quad \frac{300}{f \text{ in MHz}}$$

$$or \ f \text{ in Hz} = \frac{3 \times 10^8}{\lambda \text{ in metres}}$$

PROBLEM: What is the wavelength of a 4 MHz signal?

$$\lambda = \frac{300}{4} = 75 \text{ metres}$$

What is the frequency of a 10 metre wave?

$$f = \frac{3 \times 10^8}{10} = 3 \times 10^7 = 30 \text{ MHz}$$

CW AND ICW

Oscillators studied in Chapter 16 were designed to produce a CONTINUOUS WAVE (CW). When these oscillators are connected to an antenna system, they will radiate waves into space. If the oscillator is keyed (turned on and off), it is possible to radiate or send out spurts of energy. These spurts of radio energy, when conforming to the MORSE CODE, provide one of the earliest and still quite popular methods of sending messages. Fig. 17-2 shows the continuous radio wave, then an INTERRUPTED CONTINUOUS WAVE (ICW).

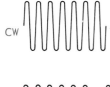

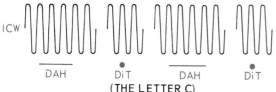

Fig. 17-2. An ICW can conform to the Morse Code for communications.

The Morse Code is illustrated in Fig. 17-3. One of the most exciting adventures in electronics is found in Amateur or HAM Radio. You must be licensed by the Federal Communications Commission before you can operate.

A ● —	K — ● —	U ● ● —
B — ● ● ●	L ● — ● ●	V ● ● ● —
C — ● — ●	M — —	W ● — —
D — ● ●	N — ●	X — ● ● —
E ●	O — — —	Y — ● — —
F ● ● — ●	P ● — — ●	Z — — ● ●
G — — ●	Q — — ● —	
H ● ● ● ●	R ● — ●	PERIOD ● — ● — ● —
I ● ●	S ● ● ●	COMMA — — ● ● — —
J ● — — —	T —	QUESTION ● ● — — ● ●
1 ● — — — —	5 ● ● ● ● ●	8 — — — ● ●
2 ● ● — — —	6 — ● ● ● ●	9 — — — — ●
3 ● ● ● — —	7 — — ● ● ●	0 — — — — —
4 ● ● ● ● —		

Fig. 17-3. The Morse Code.

You can obtain an operator's license by passing a written examination and by demonstrating your ability to send and receive Morse Code. To find out more about the process of obtaining a HAM license, write to:

American Radio Relay League
225 Main Street
Newington, Connecticut 06111

HETERODYNING

HETERODYNING is the process of mixing two signals together. When so mixed, four signals appear at the output of the MIXER. Namely, the original two signals, the sum signal and a difference signal. For example: A 1000 KHz signal is mixed with a 999 KHz signal. In the output, there is:

1000 KHz original signal — Incoming
999 KHz original signal — BFO
1999 KHz sum of two signals
1 KHz difference of two signals

At this point, you should notice that the 1 KHz signal is in the audio range, and a tone is produced which a human can hear. When a CW transmitter is sending spurts of energy in Morse Code, an ordinary receiver would not produce a sound. A CW receiver will have a BEAT FREQUENCY OSCILLATOR (BFO) built into it as an integral part of the receiver. The oscillator signal is mixed with the incoming signal and the audible difference signal is heard by earphones or speaker.

The beat frequency oscillator may have a variable panel control to produce an audible tone more pleasant to listen to or to distinguish it from other tones or interference. For instance, if a 400 Hz tone was desired, the BFO in the above example would be tuned to 999.6 Hz.

A block diagram of a transmitter and receiver are illustrated in Fig. 17-4.

ZERO BEATING

A practical method used to measure the frequency of an unknown signal will use a mixer and a known signal source. As the known frequency approaches the unknown frequency, a high audible tone is heard. As the difference becomes less, the pitch of the tone decreases to a point of no tone or "zero beat." At this point, both frequencies are the same. If the known source frequency passes beyond the unknown, the tone starts at a low pitch and increases in

pitch as the difference becomes greater. This is illustrated by the following examples:

UNKNOWN FREQUENCY	GENERATOR FREQUENCY	AUDIBLE TONE
1000 KHz	999 KHz	1000 Hz
1000 KHz	999.5 KHz	500 Hz
1000 KHz	1000 KHz	0
1000 KHz	1000.5 KHz	500 Hz
1000 KHz	1001 KHz	1000 Hz

As you advance in your studies of electronics, you will discover many applications of "zero beating."

AMPLITUDE MODULATION (AM)

Most of us enjoy music and speech from our radios. We cannot understand the Morse Code. Intelligence superimposed upon a radio frequen-

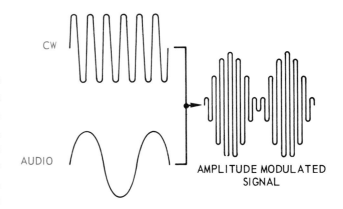

Fig. 17-5. A CW wave, an audio wave and the resultant amplitude modulated wave.

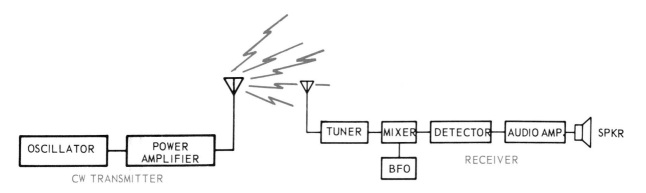

Fig. 17-4. A block diagram of a CW transmitter and receiver.

cy wave is called MODULATION. Assume a radio transmitter is operating on a frequency of 1000 KHz, and a musical tone of 1000 Hz is to be used for modulation. Referring to Fig. 17-5, the CW and the audio signal are illustrated. By a modulation circuit, the AMPLITUDE OF THE CARRIER WAVE IS MADE TO VARY AT THE AUDIO SIGNAL RATE.

Look at this modulation process another way. The mixing of a 1000 Hz wave with a 1000 KHz wave produces a sum and difference wave, which are ALSO IN THE RADIO FREQUENCY RANGE. These two waves will be 1001 KHz and 999KHz. These are known as SIDEBAND FREQUENCIES; the UPPER SIDEBAND and the LOWER SIDEBAND respectively. In Fig. 17-6, this modulation is shown as described.

The algebraic sum of the carrier wave and its sidebands results in the amplitude modulation wave. Note that the audio tone intelligence is present in both sidebands, since either sideband is the result of modulating a 1000 KHz signal with a 1000 Hz tone.

In Fig. 17-7, the waves are represented by their location on a frequency base. If a 2000 Hz tone was used for modulation, then sidebands

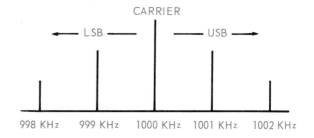

Fig. 17-7. Carrier and sideband locations for modulation tone of 1 KHz and 2 KHz.

would appear at 998 KHz and 1002 KHz. In order to transmit, using AM, a 5000 Hz tone of a picolo or a violin, sidebands at 995 KHz and 1005 KHz would be required. This would represent a frequency band width of 10 KHz necessary to transmit a 5000 Hz musical tone.

There is not that much space in the spectrum to permit all broadcasters to transmit and, if limited to musical tones of 5000 Hz, the output received and heard from your radio speaker would be VERY LOW FI. Indeed, it would be unpleasant to listen to. The BROADCAST BAND for AM radio extends between 535 KHz and 1605 KHz and is divided into 106 channels 10 KHz wide. A station is licensed to operate at a frequency in one of these channels.

The broadcast channels are carefully allotted to stations at sufficient distances from each

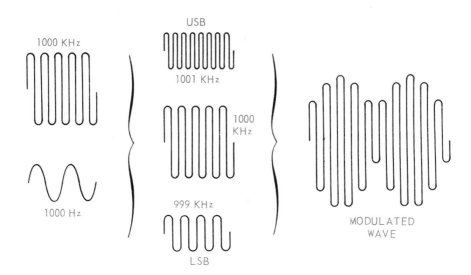

Fig. 17-6. Wave showing formation of sidebands and modulation envelope.

other to prevent interference. In order to improve the fidelity and quality of music under these limitations of 10 KHz band occupancy, a VESTIGIAL SIDEBAND FILTER is used to remove a large portion of one sideband. Remember that both sidebands contain the same information. By this means, frequencies higher than 5 KHz can be used for modulation. The fidelity is improved.

MODULATION PATTERNS

A radio transmitter is not permitted by law to exceed 100 percent modulation. This means that the modulation signal cannot cause the carrier signal to vary over 100 percent of its unmodulated value. Examine the patterns in Fig. 17-8. Note the amplitude of the modulated waves.

The 100 percent modulation wave variation is from zero to 2 times the peak value of the carrier wave. Overmodulation is caused when

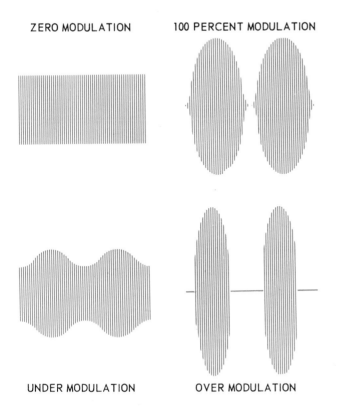

ZERO MODULATION   100 PERCENT MODULATION

UNDER MODULATION   OVER MODULATION

Fig. 17-8. Patterns of modulation are given for 0 and 100 percent, and for 50 percent (undermodulation) and overmodulation.

modulation increases the carrier wave to over two times its peak value and, at negative peaks, the waves cancel each other and leave a straight line of zero value. Overmodulation causes DISTORTION and INTERFERENCE called SPLATTER.

Percent of modulation may be computed by the formula:

$$\% \text{ Modulation} = \frac{e_{max} - e_{min}}{2e_c} \times 100$$

where:

e_{max} is the maximum amplitude of modulated wave.

e_{min} is the minimum amplitude of modulated wave.

e_c is the amplitude of unmodulated wave.

PROBLEM: A carrier wave has the peak value of 500 volts, and a modulating signal causes amplitude variation from 250 volts to 750 volts. What is the percent of modulation?

$$\% \text{ Mod.} = \frac{750 - 250}{2 \times 500} \times 100$$

$$= \frac{500}{1000} \times 100 = 50\%$$

INPUT POWER

The dc input power to the final amplifier of a transmitter is the product of voltage and current. To compute the power required by a modulator, use this formula:

$$P_{audio} = \frac{m^2 P_{dc}}{2}$$

where:

P_{audio} is the power of the modulator.
m is the percentage of modulation expressed as decimal.
P_{dc} is the input power to the final amplifier.

PROBLEM: What modulation power is required to modulate a transmitter with a dc power input of 500 watts to 100 percent?

$$P_{audio} = \frac{(1)^2 \; 500 \text{ watts}}{2} = 250 \text{ watts}$$

This represents a total input power of 750 watts. Note what happens under conditions of 50 percent modulation:

$$P_{audio} = \frac{(.5)^2 \; 500 \text{ watts}}{2} = 62.5 \text{ watts}$$

And the total input power is only 562.5 watts.

Where the modulation percentage is reduced to 50 percent, the power is reduced to 25 percent. This is a severe drop in power which decreases the broadcasting range of the transmitter. It is just good operation to maintain a transmitter as close to 100 percent modulation, without exceeding 100 percent. You may wish to know why the term INPUT POWER has been used. That is because any final amplifier is far from 100 percent efficient.

$$\%Eff = \frac{P_{out}}{P_{in}} \; x \; 100$$

If a power amplifier had a 60 percent efficiency and a P_{dc} input of 500 watts, its output power would approach:

$$P_{out} = \%Eff. \; x \; P_{in} = .6 \; x \; 500 = 300 \text{ watts}$$

Ham radio stations are limited by law to 1000 watts input power. Their output power is considerably less. As prescribed by law, one duty of a transmitter engineer is to monitor the input power of the transmitter at frequent intervals.

SIDEBAND POWER

Next, consider a transmitter with 100 percent modulation with a power of 750 watts. Our previous calculations show that 500 watts of this power is in the carrier wave and 250 watts added to produce the sidebands. Therefore, we can say that there are 125 watts, or one-sixth of the total power, in each sideband.

Now, each sideband contains the same information and each is a radio frequency wave which will radiate as well as the carrier wave. So why waste all this power? In single sideband transmission, this power is saved. The carrier and one sideband are suppressed and only one sideband is radiated. At the receiver end, the carrier is reinserted and the difference signal (the audio signal) is then detected and reproduced.

This text will not cover the methods of sideband transmission and reception, but you may wish to make an independent study of this very popular communication system.

TRANSISTORIZED TRANSMITTERS

The transistor holds the prominent position as a power amplifier for transmitters. However, only in recent years have transistors been developed with capabilities of handling large power requirements. As a result, our discussion of transistorized transmitters will be in the low power category.

Transistors will be found in audio circuits, oscillators and intermediate power amplifiers. See Fig. 17-9.

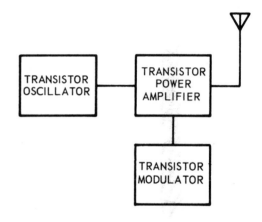

Fig. 17-9. Block diagram of a simple transistorized transmitter.

PHONO-OSCILLATOR PROJECT

The circuit of a transistorized transmitter project is illustrated in Fig. 17-10. This so-called PHONO-OSCILLATOR is a small transmitter which will broadcast to an AM broadcast receiver.

Basically, this project is designed to demonstrate the principles of radio waves and transmission and modulation. Stages Q_1 and Q_2 are audio voltage amplifiers. Q_3 is used for a radio frequency oscillator in the broadcast band. The audio signal is applied to the emitter of the oscillator which will modulate the amplitude of the oscillator.

Set up an AM radio a few feet away from your little transmitter. Tune the radio to around 600 KHz in the broadcast band (find a spot not occupied by a commercial broadcast station). Now, tune the phono-oscillator until a rushing sound is heard on the receiver. Attach the "mike" and broadcast to the radio. Slight retuning may be necessary.

If difficulty is experienced in getting both the oscillator and radio on the same frequency, connect your Code Oscillator (Chapter 16) as an audio input to the phono-oscillator. The tone will be easy to identify on the AM radio. Experiment with different lengths of antenna for the transmitter. Fig. 17-11 shows a printed circuit layout of the project.

DEMONSTRATION: Connect an oscilloscope from point A to ground. See Fig. 17-10.

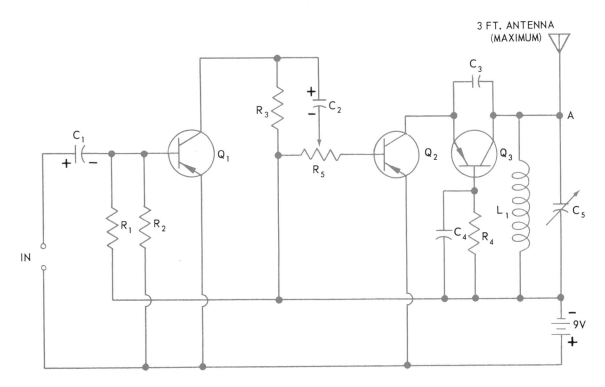

PARTS LIST FOR PHONO-OSCILLATOR

R_1 — 470 K Ω, 1/2W	C_3 — 100 pF, 50V dc
R_2, R_3 — 10 K Ω, 1/2W	C_4 — .02 μF, 50V dc
R_4 — 22 K Ω, 1/2W	C_5 — 0-365 pF, Variable
R_5 — 100 K Ω Potentiometer	L_1 — Loopstick (Miller 6300)
C_1, C_2 — 15 μF @ 15V dc	Q_1, Q_2, Q_3 — 2N408 or RCA SK 3003

Fig. 17-10. Schematic and parts list for phono-oscillator.

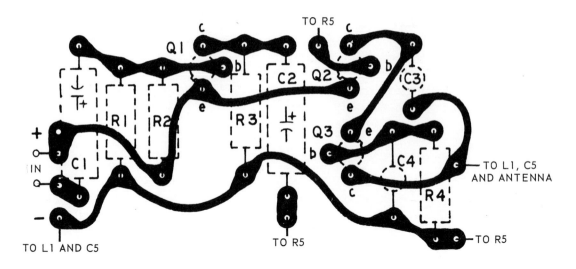

Fig. 17-11. Suggested circuit board layout for the phono-oscillator.

You will observe an rf oscillator output produced by tank circuit L, and C_5. Now connect a 3 ft. length of wire to point A. Place this wire antenna close to an AM broadcast receiver. Vary C_5 until a rushing sound is heard in the receiver.

Next, connect an audio generator to AF input and tune the generator to 400 Hz. An audible 400 Hz tone should be heard from the receiver. This transmitter will only radiate signals a few yards. It should not be connected to an outside antenna. Severe penalties are imposed for illegal broadcasting.

Fig. 17-12 pictures a 40 channel citizens band radio. Transistors are used both for receiver and transmitter. These will be discussed in more detail later in this chapter.

Fig. 17-12. A citizens band radio. (Motorola)

AM DETECTION

DETECTION or DEMODULATION is the process of removing the audio signal from a modulated radio wave to obtain half-wave pulses. It is a form of rectification.

AM WAVE   DETECTED AM WAVE   AVERAGE VALUE OF MODULATION SIGNAL

Fig. 17-13. Demodulation of an AM wave.

Study the wave forms in Fig. 17-13. The detected AM Wave will show half-wave pulses of rectified radio waves varying in amplitude. Between rectified pulses, the signal falls to zero. Detection is accomplished by applying the same principles as used in filtering the output of a power supply. The demodulated wave can be filtered to remove the pulses and raise the average value of the wave. This improvement is illustrated in Fig. 17-14.

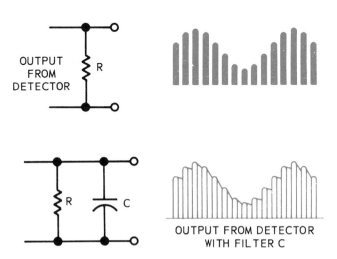

Fig. 17-14. Output of detector before and after filtering.

The semiconductor diode is used in the majority of detector applications in modern electronic circuits. A point contact diode is used to avoid shunting a radio frequency signal around a diode by its junction capacitance (the reactance of this capacitance would be at a low value in the radio frequency range).

A point contact diode, Fig. 17-15, effectively reduces the undesirable capacitance by means of a small contact area. An N type germanium wafer is mounted on a small contact plate. Impressed against this crystal is a fine beryllium-copper or phosphor-bronze wire called a catwhisker. During the manufacturing process, a relatively high current is passed through the catwhisker and the crystal. This current forms a small region of P type crystal around the contact point. The theory of this semiconductor action is exactly the same as described in Chapter 12. Review this again.

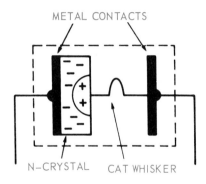

Fig. 17-15. The construction of a point contact diode.

The point contact diode can be made extremely small, and it adapts itself well to miniaturized and portable equipment. It requires no heating like the vacuum tube diodes. Many will be used in projects you will construct.

The symbols used for diodes are shown in Fig. 17-16. The circle around the one at left indicates that it is enclosed in a small glass envelope.

Fig. 17-16. Symbols used for diodes.

There is always considerable confusion as to the direction of electron flow through a diode. ELECTRON FLOW always is in the direction against the large arrow head. See Fig. 17-17.

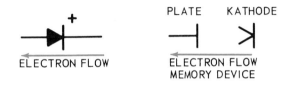

Fig. 17-17. Electron flow or conduction direction is indicated by the arrows.

If you have previously studied vacuum tubes, you will recall that electrons always flow from cathode to plate. By separating the diode symbol, as in Fig. 17-17, and taking some liberty with the spelling of CATHODE, you will remember that electron flow is from KATHODE TO PLATE. But diodes are always marked with a positive end (+). Nevertheless, electron flow within the diode will be from the positive to negative.

A SIMPLE RADIO RECEIVER

In order to take advantage of the crystal detector, it is necessary to select the radio station which is broadcasting at a specific frequency. This can be done by a tuned circuit. In Fig. 17-18, a circuit of a simple radio is illustrated.

Radio waves of many frequencies from as many broadcasting stations cut across and induce a very small voltage in the antenna. This causes currents to flow up and down from antenna to ground. The antenna lead-in wire is connected to L_1, Fig. 17-18, which is part of an antenna coil in the receiver. The other end of L_1 is connected to ground.

The oscillating currents of many frequencies produce varying magnetic fields around L_1. But L_1 is closely linked to L_2. Therefore, these induced voltages appear across L_2 by transformer action. Now, the tank circuit $L_2 C_1$ will be tuned to the desired frequency of the station to be heard. At this resonant frequency, the tank circuit will have maximum response, and its developed voltage will be MAGNIFIED many

times, depending on the Q of the circuit. Review Chapter 10. This signal is now detected by diode D and the audio output is filtered by C_2 and reproduces the "Sound of Music" in the earphones.

The coils of L_1 and L_2 usually are wound on a single core. In a parts catalog, it will be called an ANTENNA TRANSFORMER or just an ANTENNA COIL, Fig. 17-19. You must remember that the selection of L_2 and C_1 will depend upon the band of frequencies you wish to hear. Similar tuning circuits are used for TV, FM, police calls and all radio services. However, values of L_2 and C_1 must tune to the frequency desired.

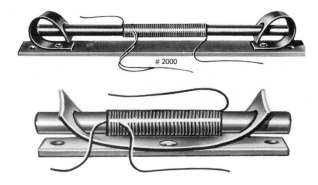

Fig. 17-19. Antenna coils used in transistor radios. (J.W. Miller Co.)

TRANSISTOR DETECTION

A transistor also maybe used as a detector, with the added advantage of signal amplification. A transistor used in this application would

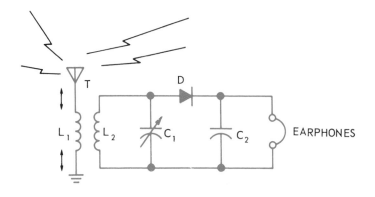

Fig. 17-18. The schematic of a simple crystal receiver.

necessarily have to be biased at cutoff or operated as Class B. You will recall that Class B amplification is the same as half-wave rectification. See Chapter 14.

The circuit of a transistor detector, Fig. 17-20, is a simple amplifier circuit. The emitter is at ground potential, and the voltage divider $R_F R_B$ will establish the proper bias. This bias holds the transistor just slightly above cutoff.

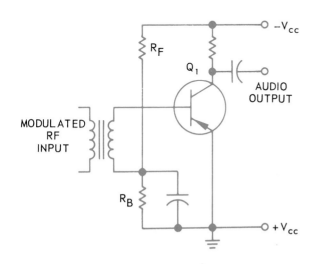

Fig. 17-20. Circuit of a transistor detector.

An incoming AM wave will cause the emitter-base voltage and current to vary. The transistor will conduct only when its base is made more negative. The positive half-cycles of the AM wave will be cut off. Capacitor C serves as a bypass for radio frequencies.

A TRANSISTORIZED RADIO

In Fig. 17-21, many of the previous lessons have been applied. This small radio receiver employs diode detection and two stages of transistor amplification with transformer coupling. You will recognize the tuning circuit $L_1 C_1$. The tuned signal is detected by diode D_1. It is necessary to have C_2 in the circuit to block the dc bias of Q_1 from grounding out through antenna coil L_1.

The remainder of the circuit shown in Fig. 17-21 is straightforward. Its components have been discussed in Chapter 14, except for potentiometer R_2. This is a gain control or volume control, which is a variable resistance shunted across the primary of T_1. Potentiometer R_2 is designed to determine the amount of signal voltage transferred to the next stage by transformer T_1.

A TRANSISTORIZED SUPERHETERODYNE RECEIVER

Many years ago, it was found that the gain and selectivity of a radio could be designed with acceptable performance for a single tuned frequency. When a station of another frequency

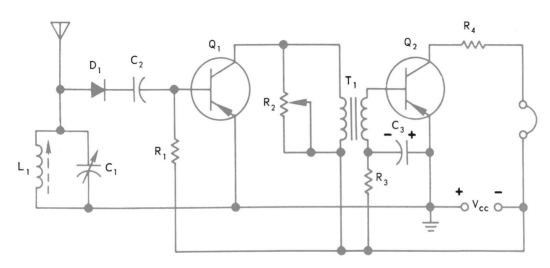

Fig. 17-21. A transistorized radio with two-stage amplification.

was desired, less than maximum performance resulted unless stage by stage tuning was performed. The old TUNED RADIO FREQUENCY (TRF) radio is now a museum piece or collector's item.

Then, it was discovered that if all tuned incoming signals were converted to a single frequency called the if frequency or INTERMEDIATE FREQUENCY, then stages and coupling circuits could be used to produce the best results. The added convenience of "one dial" tuning made it possible for anyone to tune a radio regardless of any technical knowledge.

This type of radio receiver employs the superheterodyne circuit. In this circuit, Fig. 17-22, the station is tuned by the usual methods. Then, this signal is MIXED with a LOCAL OSCILLATOR signal, and heterodyned signals produce a constant intermediate frequency.

When a station is "tuned-in", the local oscillator is varied in frequency, so the DIFFERENCE SIGNAL, or if, always is the same. In Fig. 17-22, the tuning variable capacitor and the oscillator tuning capacitor are ganged together. Both are turned by the same tuning knob. They TRACK together and always maintain the correct if. This frequency is 455 KHz for most home radios.

Assume that a radio is tuned to a 1000 KHz station. The local oscillator will be tuned to 1455 KHz. The difference frequency if is 455 KHz. If you tune to a station at 600 KHz, the oscillator will also tune to 1055 KHz. The if is still 455 KHz. If you tune a station at 1400 KHz, the oscillator will be 1855 KHz and the if is 455 KHz.

Beyond convenience, the advantages of the superheterodyne system are many. When the if amplifiers are designed for maximum gain at 455 KHz, all incoming signals tuned in by the radio will have maximum response. A block diagram, with signals, of the superheterodyne receiver is shown in Fig. 17-22.

AUTOMATIC VOLUME CONTROL

Unless the broadcasting stations are quite close to your home, the radiated signals have a tendency to vary in strength. This is due to fading and atmospheric conditions. When this occurs, frequent adjustment of the manual volume control is necessary to assure pleasant listening. It is better to have the volume of a receiver automatically adjusted by electronic circuitry. Then, any level of loudness will be maintained without further manual control. This is called AUTOMATIC VOLUME CONTROL or AVC.

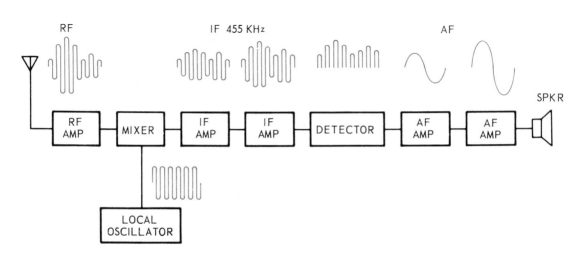

Fig. 17-22. Block diagram of a typical superheterodyne receiver.

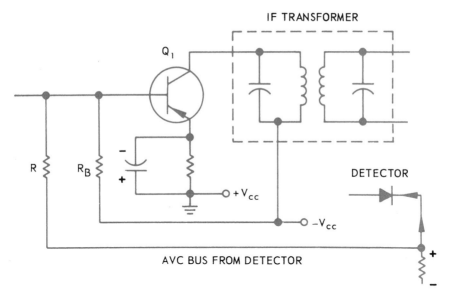

Fig. 17-23. Connection of AVC line to if amplifier.

In transistorized circuits, AVC is easy to set up since the gain of a transistor amplifier depends upon the emitter current. However, a transistor is a current controlled device and some power from some source must be used.

In most cases, the AVC voltage is secured from the detector stage of a transistorized circuit. The average value of the detected signal will vary according to the strength of the incoming signal. The current through the detector diode will produce a voltage across the detector load resistor which may be fed back to the previous if amplifiers to regulate their gain. A stronger signal will reduce the gain, and a weaker signal will increase the gain. A partial diagram, Fig. 17-23, shows the connection to an if amplifier to produce these results.

A strong positive voltage on the AVC line will make the base of Q_1 more positive. This reduces forward bias, reduces emitter and collector current and reduces amplifier gain. Resistor R is a limiting resistance to effectively control the amount of AVC feedback voltage.

Similar circuits are used in TV as an automatic gain control (AGC) to maintain electronically the video signal at a constant level. In this discussion on gain variation as the result of emitter-collector current, changes in input and output impedances have not been considered. More advanced texts will discuss these impedances which also affect amplifier gain.

FREQUENCY MODULATION

FM (frequency modulation) radio has become a very popular method of radio communication. It employs an entirely different system of superimposing intelligence such as sound and music on the radio frequency wave. The capabilities of frequency modulation allow relatively high audio sound to be transmitted and still remain within the legal spectrum space assigned to the broadcast station. Also, the general acceptance of stereophonic sound has been encouraged by FM transmission of dual channels of sound by multiplex systems.

Beginning with a constant amplitude continuous wave signal, the frequency is made to vary at an audio rate. This is shown graphically in Fig. 17-24.

Each broadcasting station is assigned a CENTER FREQUENCY in the FM band (92.1 to 107.9 MHz). This is the frequency to which you would tune your radio. Now study the wave forms in Fig. 17-25 to discover the exact meaning of FM.

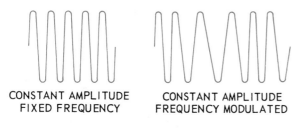

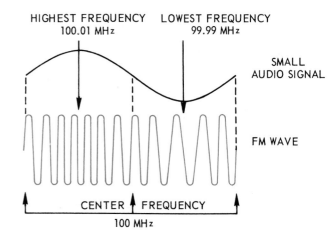

CONSTANT AMPLITUDE
FIXED FREQUENCY

CONSTANT AMPLITUDE
FREQUENCY MODULATED

Fig. 17-24. For FM, the frequency of the wave is varied at an audio rate.

The amount of variation of frequency from each side of the center frequency is called the FREQUENCY DEVIATION and is determined by the amplitude or strength of the audio modulating wave. In Fig. 17-25, a small audio signal causes the frequency of the carrier wave to vary between 100.01 MHz and 99.99 MHz, and the deviation is ± 10 KHz.

In the second example, Fig. 17-25, a stronger audio signal causes a frequency swing between 100.05 MHz and 99.95 MHz, or a deviation of ± 50 KHz. The stronger the modulation signal, the greater the FREQUENCY DEPARTURE and the greater the band occupancy.

The RATE at which the frequency varies from its highest to lowest frequency depends upon the FREQUENCY OF THE AUDIO MODULATING SIGNAL. Two graphic examples are found in Fig. 17-26.

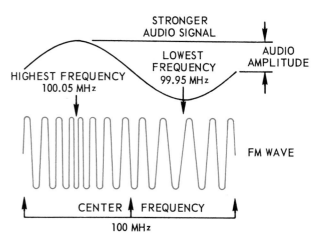

Fig. 17-25. The amplitude of the modulating signal determines the frequency swing from center frequency.

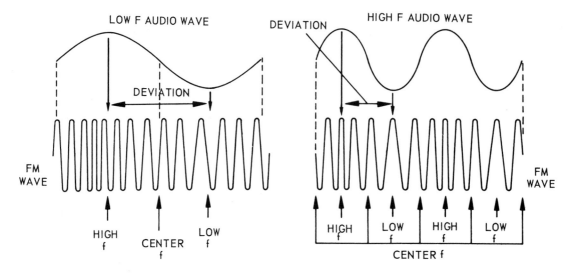

Fig. 17-26. The rate of frequency variation depends upon the frequency of the audio modulating signal.

If the audio signal is 1000 Hz, the carrier wave goes through its maximum deviation 1000 times per second. If the audio signal is 100 Hz, then the frequency changes at a rate of 100 times per second. Notice that the modulating frequency does not change the amplitude of the carrier wave.

When a signal is frequency modulated, there is also the formation of sidebands. However, the number of sidebands produced depends upon the frequency and amplitude of the modulating signal. Each sideband on either side of the center frequency is separated by the amount of the frequency of the modulating signal. This is illustrated in Fig. 17-27.

Note that the power of the carrier frequency is considerably reduced by the formation of sidebands which take power from the carrier. The amount of power taken depends upon the MAXIMUM DEVIATION and the modulating frequency.

A station may be assigned a center frequency and stay within its maximum permissable deviation. Yet, the formation of sidebands really is the determining factor on the bandwidth required for transmission. In FM, the bandwidth is specified by the frequency range between the upper and lower SIGNIFICANT SIDEBANDS. A significant sideband is one which has an amplitude of one percent or more of the unmodulated carrier.

NARROW BAND FM

If maximum deviation of a carrier wave is limited so that the FM wave occupies the same space as an AM wave carrying the same intelligence, it is called NARROW BAND FM. Some distortion is present in the received signal. It is quite satisfactory for voice communications, but not for high fidelity music.

MODULATION INDEX

The MODULATION INDEX IS THE RELATIONSHIP BETWEEN THE MAXIMUM CARRIER DEVIATION AND THE MAXIMUM MODULATING FREQUENCY. It is expressed as:

$$\text{Modulating index} = \frac{\text{maximum carrier deviation}}{\text{maximum modulating frequency}}$$

By the use of this index, the number of significant sidebands and the bandwidth of the FM signal may be calculated. The complete index may be found in more advanced texts. Examples of the use of the modulation index are given in the following table:

Mod. Index	No. of Sidebands	Bandwidth
.5	2	4 x F
1	3	6 x F
5	8	16 x F
10	14	28 x F

Where F is the modulating frequency.

EXAMPLE: The amplitude of a modulating signal causes a maximum deviation of 10 KHz.

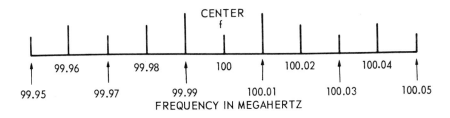

Fig. 17-27. Sidebands generated by a 10 KHz modulating signal on a 100 MHz carrier wave.

The frequency of the modulating signal is 1000 Hz. Then:

$$M \text{ Index} = \frac{10000}{1000} = 10$$

Based on the table, the FM signal would have 14 significant sidebands and occupy a bandwidth of 28 KHz.

PERCENT OF MODULATION

The percent of modulation has been arbitrarily stated as a maximum deviation of ± 75 KHz for commercial FM radio. For the FM sound transmission in television, it is limited to ± 25 KHz.

FM DETECTION

In the AM radio, the detector had to be sensitive to amplitude variations of wave. A FM detector must be sensitive to frequency variations and remove this intelligence from the FM wave. In other words, the FM detector must produce a varying amplitude and frequency audio signal from a FM wave.

To approach this method of detection, consider the diagram in Fig. 17-28. Assume a circuit which has a maximum response at its

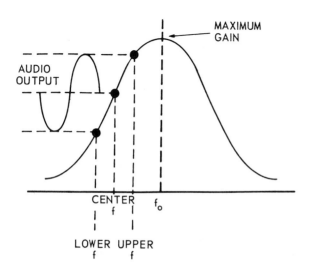

Fig. 17-28. These curves demonstrate slope detection.

resonant frequency. All frequencies other than resonance will have a lesser response. So, if the center frequency of a FM wave is on the SLOPE of the resonant response curve, a higher frequency will produce a higher response in voltage and a lower frequency will produce a lower voltage response.

Examine the curves in Fig. 17-28. You will see that the amplitude of the output wave is the result of the maximum deviation of the FM signal. Note, too, that the frequency of the audio output depends on the rate of change of frequency of the FM signal.

DISCRIMINATOR CIRCUITS

For ease in understanding, the discriminator in Fig. 17-29 uses three tuned circuits. In this circuit, $L_1 C_1$ is tuned to the center frequency. $L_2 C_2$ is tuned to above the center frequency. $L_3 C_3$ is tuned to below center frequency by an equal amount.

At center frequency, equal voltages are developed across the tuned circuits, and D_1 and D_2 conduct equally. The voltages across R_1 and R_2 are equal and opposite in polarity and the circuit output is zero.

If the input frequency increases above center, $L_2 C_2$ will develop a higher voltage. Then, D_1 will conduct more than D_2 and unequal voltages will develop across R_1 and R_2. The difference between these voltage drops will be the audio signal.

Therefore, the output is a voltage wave varying at the rate of frequency change at the input, and its amplitude depends upon the maximum deviation. The capacitors across the output of the discriminator filter out any remaining radio frequencies.

The discriminator in Fig. 17-30 is a typical circuit you will encounter in FM receivers. L_1 and C_1 are tuned to center frequency. At frequencies above resonance, the tuned circuit becomes more inductive. At frequencies below

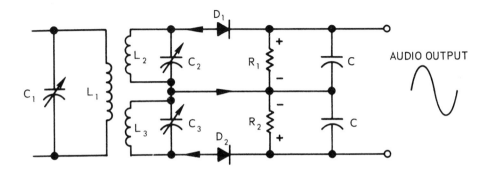

Fig. 17-29. A FM discriminator circuit using semiconductor diodes.

resonance, the circuit becomes more capacitive. The out-of-phase conditions produce resultant voltages that determine which diode will conduct. The output is an audio wave, as in the more elementary discussion.

This discussion is meant only to familiarize you with this type of detection. In advanced courses, you will study it in more detail.

It will interest you to know that each diode in the discriminator must have equal conduction capabilities. This means that semiconductor diodes used must be in MATCHED PAIRS.

THE RATIO DETECTOR

The FM detector illustrated in Fig. 17-31 is called RATIO DETECTOR. In this circuit, the diodes are connected in series with the tuned circuit. At center frequencies, both diodes conduct during half-cycles. The voltage across R_1 and R_2 charges C_1 to output voltage. C_1 remains charged because the time constant of

$C_1 R_1$ and R_2 is much longer than the period of the incoming waves. C_2 and C_3 also charge to the voltage of C_1.

When both D_1 and D_2 in Fig. 17-31 are conducting equally, the charge of C_2 equals C_3 and they form a voltage divider. At the center point between C_2 and C_3, the voltage is effectively zero.

A frequency shift either below or above center frequency will cause one diode to conduct more than the other. As a result, the voltages of C_2 and C_3 will become unequal. Yet, they will always total the voltage of C_1. This change of voltage at the junction of C_2 and C_3 is the result of the RATIO of the unequal division of charges between C_2 and C_3. This will vary at an audio rate the same as the rate of change of the FM signal.

Again in Fig. 17-31, note the charge on C_1. It is the result of the amplitude of the carrier wave or signal strength. It is charged by half-wave

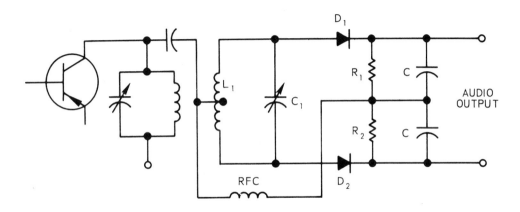

Fig. 17-30. A typical Foster-Seeley discriminator using a special transformer designed for this purpose.

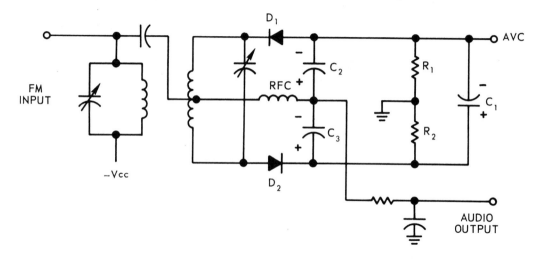

Fig. 17-31. A typical ratio detector circuit using semiconductor diodes.

rectification of the FM signal. Therefore, it is a fine point to pick off an automatic volume control voltage to feedback to previous stages to regulate stage gain.

NOISE LIMITING

Another particular advantage of FM radio is the fact that the receivers are sensitive to and can detect frequency variations and not amplitude variations. Most noise and interference to radio reception is in the form of NOISE SPIKES. These are amplitude variations that have little effect on the FM detector. Therefore, FM reception is relatively free of noise and disturbances.

To hold the FM signal at a CONSTANT AMPLITUDE before detection in a discrimina-

tor circuit, a LIMITER is used as a previous stage. A schematic of a transistorized LIMITER STAGE is shown in Fig. 17-32.

A limiter is nothing more than an OVER-DRIVEN amplifier stage. If the incoming signal reaches a certain amplitude in voltage, it drives the transistor to cutoff or to saturation when the voltage is opposite in polarity. At either of these points, gain cannot increase. Consequently, the output is confined within these limits. Any noise spikes in the form of amplitude modulation would be clipped off.

THE FM RECEIVER

A block diagram of a complete FM receiver is illustrated in Fig. 17-33. Each block is labeled

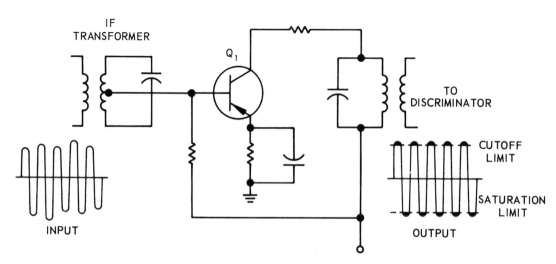

Fig. 17-32. A circuit from a limiter stage before a discriminator.

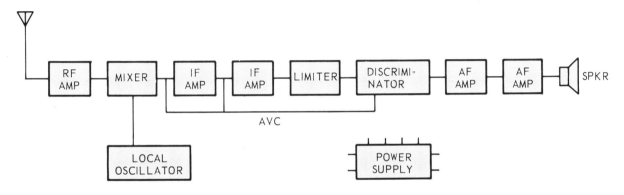

Fig. 17-33. The block diagram of a typical FM radio.

to designate its function in the system. These individual blocks or stages have all been described in previous chapters of this text.

The intermediate frequency used in FM radio is 10.7 MHz. The same heterodyne principles are involved as with the AM receiver.

CITIZENS BAND RADIO

"Breaker-breaker."

"What's your handle?"

"Red Bug. Is there an X-ray machine ahead?"

"10-4. County mountie at post 179."

"Thanks, 8s and other good numbers."

SMOKEY BEAR—a state policeman

MAMA BEAR—female police person

COUNTY MOUNTIE—a county policeman

LOCAL YOKEL—a city policeman

TIJUANA TAXI—marked police car

PLAIN WHITE WRAPPER—unmarked white police car, may be any color

EYE IN THE SKY—helicopter

BEAR'S DEN—police station

PICTURETAKER—policeman with radar

X-RAY MACHINE—radar

GREEN STAMPS—money

GREEN STAMP HIGHWAY—toll road

MONFORT LANE—inside lane on four-lane highway

8's AND OTHER GOOD NUMBERS—so long and signing off

HANDLE—code name of CB'er

POST—milepost on interstate highways

SHAKE THE TREES AND RAKE THE LEAVES—lead truck should watch for police and traffic ahead, and rear truck will watch mirrors from the rear

FRONT DOOR—lead truck of several running together.

BACK DOOR—rear truck of several running together

ROCKING CHAIR—the middle truck

18 WHEELER—tractor-trailer

4 WHEELER—automobile

PICKUMUP TRUCK—a pickup truck

HAMMER DOWN—going full speed

DROP THE HAMMER—accelerate to full speed

CHICKEN COOP—truck weigh station

FOOT WARMER—a CB radio with more than legal power

WALL TO WALL AND TREETOP TALL—very good reception

HAVE YOU GOT YOUR EARS ON?—is your CB on?

10-1	Receiving Poorly	**10-21**	Call By Telephone	**10-41**	Please Tune To Channel . . .
10-2	Receiving Well	**10-22**	Report in Person to . . .	**10-42**	Traffic Accident At . . .
10-3	Stop Transmitting	**10-23**	Stand By	**10-43**	Traffic Tieup At . . .
10-4	OK, Message Received	**10-24**	Completed Last Assignment	**10-44**	I Have A Message For You (or . . .
10-5	Relay Message	**10-25**	Can You Contact . . .	**10-46**	Assist Motorist
10-6	Busy, Stand By	**10-26**	Disregard Last Information	**10-50**	Break Channel . . .
10-7	Out Of Service, Leaving Air	**10-27**	I Am Moving To Channel . . .	**10-70**	Fire At . . .
10-8	In Service, Subject To Call	**10-28**	Identify Your Station	**10-73**	Speed Trap At . . .
10-9	Repeat Message	**10-29**	Time Is Up For Contact	**10-75**	You Are Causing Interference
10-10	Transmission Completed, Standing By	**10-30**	Does Not Conform To FCC Rules	**10-77**	Negative Contact
		10-33	EMERGENCY TRAFFIC AT THIS STATION	**10-82**	Reserve Room For . . .
10-11	Talking Too Rapidly			**10-84**	My Telephone Number Is . . .
10-12	Visitors Present	**10-34**	Trouble At This Station, Help Needed	**10-85**	My Address Is . . .
10-13	Advise Weather/Road Conditions			**10-89**	Radio Repairman Needed At . . .
10-16	Make Pickup At . . .	**10-35**	Confidential Information	**10-92**	Your Transmitter Is Out Of Adjustment
10-17	Urgent Business	**10-36**	Correct Time Is . . .		
10-18	Anything For Us?	**10-37**	Wrecker Needed At . . .	**10-93**	Check My Frequency On This Channel
10-19	Nothing For You, Return To Base	**10-38**	Ambulance Needed At . . .		
10-20	My Location Is . . .	**10-39**	Your Message Delivered	**10-200**	Police Needed At . . .

Fig. 17-34. Special CB terms and 10 signals. (KRACO)

Citizens band (CB) language is commonplace on our highways today. To understand some of the special terms used, refer to Fig. 17-34.

In 1947, the Federal Communications Commission (FCC) established the Citizens Radio Service to permit personal short range radio communications, signaling and remote control by radio signals. For the first time, CB provided a convenience to practically anyone for business or personal activities. The original Class D band of CB had 23 channels. During the early years of citizens band, there was limited usage. In the early 1970s, during the energy crisis, citizens band became very popular. In July, 1976, the

FCC increased the number of CB channels from 23 to 40. The output power of a 40 channel CB is 4 watts. See Fig. 17-35. A typical 40 channel CB radio is shown in Fig. 17-36.

Basically, citizens band radios come in three styles:

1. Automobile type (12V dc).
2. Base station type or type used at home (115V ac). See Fig. 17-37.
3. Walkie-talkie type. See Fig. 17-38.

Modern CB receivers are very complex instruments. Most 40 channel CBs use the most advanced circuitry possible for superior performance. Phase-locked loop circuits are used in

23 CHANNEL CB OPERATION

CHANNEL NO.	FREQ. IN MHz	CHANNEL NO.	FREQ. IN MHz	CHANNEL NO.	FREQ. IN MHz
1	26.965	9	27.065	17	27.165
2	26.975	10	27.075	18	27.175
3	26.985	11	27.085	19	27.185
4	27.005	12	27.105	20	27.205
5	27.015	13	27.115	21	27.215
6	27.025	14	27.125	22	27.225
7	27.035	15	27.135	23	27.255
8	27.055	16	27.155		

40 CHANNEL CB OPERATION (TRANSMIT)

CHANNEL NO.	VCO OUTPUT MHz	TX OSCILLATOR MHz	PLL MIXER OUTPUT MHz	CHANNEL NO.	VCO OUTPUT MHz	TX OSCILLATOR MHz	PLL MIXER OUTPUT MHz
1	26.965	29.515	2.55	21	27.215	29.515	2.30
2	26.975	29.515	2.54	22	27.225	29.515	2.29
3	26.985	29.515	2.53	23	27.255	29.515	2.26
4	27.005	29.515	2.51	24	27.235	29.515	2.28
5	27.015	29.515	2.50	25	27.245	29.515	2.27
6	27.025	29.515	2.49	26	27.265	29.515	2.25
7	27.035	29.515	2.48	27	27.275	29.515	2.24
8	27.055	29.515	2.46	28	27.285	29.515	2.23
9	27.065	29.515	2.45	29	27.295	29.515	2.22
10	27.075	29.515	2.44	30	27.305	29.515	2.21
11	27.085	29.515	2.43	31	27.315	29.515	2.20
12	27.105	29.515	2.41	32	27.325	29.515	2.19
13	27.115	29.515	2.40	33	27.335	29.515	2.18
14	27.125	29.515	2.39	34	27.345	29.515	2.17
15	27.135	29.515	2.38	35	27.355	29.515	2.16
16	27.155	29.515	2.36	36	27.365	29.515	2.15
17	27.165	29.515	2.35	37	27.375	29.515	2.14
18	27.175	29.515	2.34	38	27.385	29.515	2.13
19	27.185	29.515	2.33	39	27.395	29.515	2.12
20	27.205	29.515	2.31	40	27.405	29.515	2.11

Fig. 17-35. Table lists first 23 channels and current 40 channels for CB radios.

some CBs to make them extremely stable. Single sideband models have the capability of converting 40 channels to 120 channels. Large, light emitting diodes (LED) digital channel readouts are found on some models.

There are many different types of antennas designed for CB mobile use. Make your selection on the basis of the type of installation or car mount desired and the antenna specifications. A vertical whip normally has a 360 deg.

radiation pattern. It can be mounted on the rear bumper, rear fender or rear deck lid. It is very important that you use the correct type of transmission line. It should be of the coaxial type and should have an impedance equal to the antenna impedance. See Fig. 17-39 for a typical CB automobile antenna.

Often, it is better to mount the antenna on the left side of the car, rather than on the right side, to minimize contact with trees and other

Fig. 17-36. A modern, 40-channel citizens band radio. (E.F. Johnson Co.)

Fig. 17-37. A base station citizens band radio. (Cobra Div. of Dynascan Corp.)

Fig. 17-38. Walkie-talkie citizens band radio. (E.F. Johnson Co.)

Fig. 17-39. A mobile CB antenna. (Panasonic)

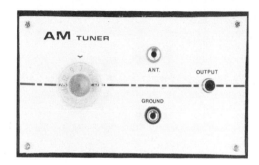

Fig. 17-40. Front view of AM tuner.

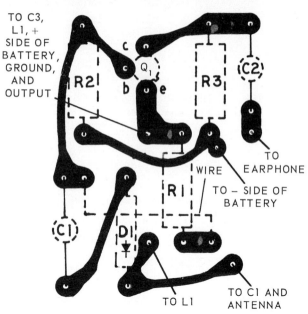

Fig. 17-42. A suggested circuit board layout for the AM tuner.

low-clearance obstructions. Generally, the better the antenna, the better the communications over greater distances.

A full 1/4 wavelength antenna is usually more efficient than the shorter versions equipped with a loading coil to electrically make up for the shorter length. However, the shorter antennas can provide adequate service and be less prone to damage from contact with external obstructions. Some short antennas can be more centrally located on the car. The car body acts as a ground plane and tends to shift the radiation pattern to favor a diagonal line,

running from right to left for an antenna mounted on the right front or left rear portion of the car. For an antenna mounted on the left front or right rear of the car, the pattern will follow the diagonal line from left to right. For a more circular pattern, the antenna would have to be centrally mounted on the car.

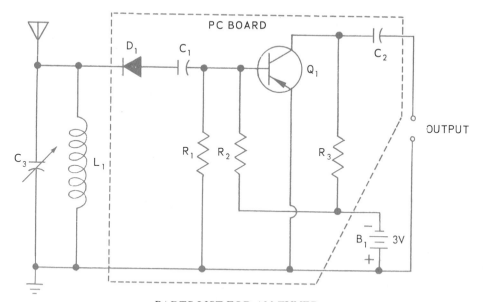

PARTS LIST FOR AM TUNER

R_1 — 33 K Ω, 1/2W resistor
R_2 — 220 K Ω, 1/2W resistor
R_3 — 47 K Ω, 1/2W resistor
C_1, C_2 — .01 μ F @ 50V dc capacitor
C_3 — 0-365 pF variable capacitor

D_1 — IN34 diode or RCA SK 3087
L_1 — loopstick, Miller 6300 or equivalent
Q_1 — 2N408 transistor or RCA SK 3003
B_1 — 3V power supply or two "A" cells

Fig. 17-41. Schematic and parts list for AM tuner.

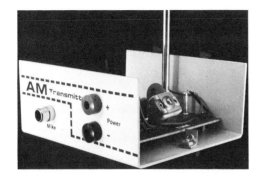

Fig. 17-43. AM Transmitter.

Effective antenna height, clearance and directional characteristics are affected by terrain, buildings, tunnels, bridges, etc. Generally, the fewer obstructions and the higher the ground level, the greater the range of communications.

AM TUNER

You can build a transistorized radio receiver. This project includes an AM tuner, detector and a one-stage amplifier. You may use it with an earphone. Stations will be much louder when the tuner is fed directly into the amplifier, such as the basic transistor amplifier in Chapter 14.

A rather good antenna will improve the performance of this radio, but it works well with four or five feet of wire as an antenna. The tuner is illustrated in Fig. 17-40. The parts list and diagram are given in Fig. 17-41. All these projects may be made on etched circuit boards, Fig. 17-42.

AM TRANSMITTER

Here is an AM transmitter to go with the AM tuner you just built. This is a simple design that requires a minimum of electronic parts. The transmitting distance is approximately 50 ft. See Figs. 17-43 and 17-44.

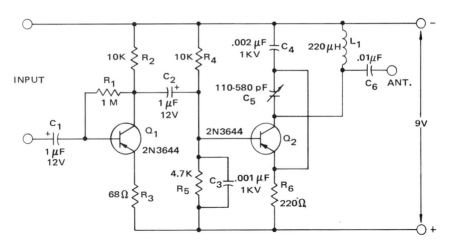

PARTS LIST FOR AM TRANSMITTER

R_1 — 1 M Ω, 1/2 resistor (10 percent tol.)
R_2, R_4 — 10 K Ω, 1/2W resistor (10 percent tol.)
R_3 — 68 Ω, 1/2W resistor (10 percent tol.)
R_5 — 4.7 K Ω, 1/2W resistor (10 percent tol.)
R_6 — 220 Ω, 1/2W resistor (10 percent tol.)
C_1, C_2 — 1 μF, 12V electrolytic capacitor or Sprague TE1120
C_3 — .001 μF, 1KV disc capacitor or Centralab DD102
C_4 — .002 μF, 1 KV disc capacitor or Centralab DD202
C_5 — 110-580 pF trimmer capacitor or Arco 467

C_6 — .01 μF, 1 KV disc capacitor or Centralab DD103
Q_1, Q_2 — 2N3644 transistor
L_1 — 220 μH rf choke or Miller 9230-76 18 AWG Buss wire for Soldering L_1 on printed circuit
Misc. — antenna, printed circuit material, crystal microphone (A basic set of parts for this AM transmitter is available from Hickok Teaching Systems, Wheeling Ave., Woburn, MA 01801. This does not include an antenna or crystal microphone.)

Fig. 17-44. Schematic and parts list for AM Transmitter.

A printed circuit layout appears in Fig. 17-45. Parts placement for the printed circuit is shown in Fig. 17-46.

Once you have completed the printed circuit, connect a crystal microphone to the input.

Fig. 17-45. Printed circuit layout for AM Transmitter. (Hickok)

Connect a battery or power supply to the transmitter and a 36 in. piece of wire or a small telescoping antenna to it.

To operate the tranmitter, speak into the microphone and tune your AM radio across the band. When you find the spot on the dial where you receive the signal, adjust C_5 to a point where you receive the best signal.

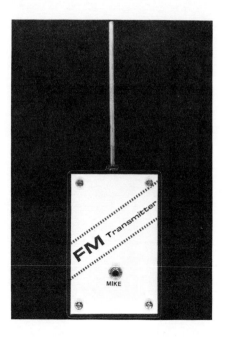

Fig. 17-47. FM Transmitter.

FM TRANSMITTER

An FM transmitter is an excellent project for the experimenter. This little project uses many of the same component parts that were in the AM transmitter. See Fig. 17-47. The schematic for the FM transmitter and the parts list are given in Fig. 17-48. The component parts placement is shown in 17-49.

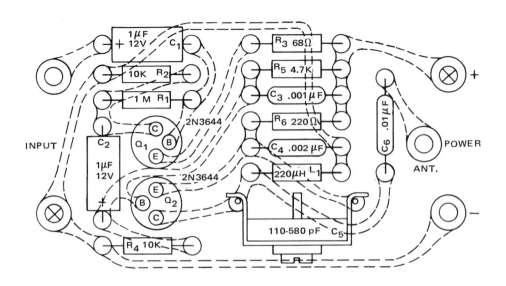

Fig. 17-46. Parts placement for AM Transmitter on PC board (top view).

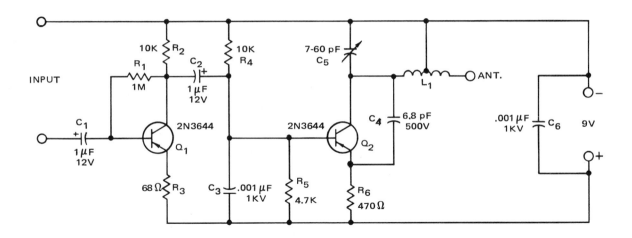

PARTS LIST FOR FM TRANSMITTER

R_1 — 1 M Ω, 1/2W resistor (10 percent tol.)
R_2, R_4 — 10 K Ω, 1/2W resistor (10 percent tol.)
R_3 — 68 Ω, 1/2W resistor (10 percent tol.)
R_5 — 4.7 K Ω, 1/2W resistor (10 percent tol.)
R_6 — 470 Ω, 1/2W resistor (10 percent tol.)
C_1, C_2 — 1 μF, 12V electrolytic capacitor or Sprague TE1120
C_3, C_6 — .001 μF, 1 KV disc capacitor or Centralab DD102
C_4 — 6.8 pF, 500V mica capacitor or Arco CM15

C_5 — 7-60 pF trimmer capacitor or Arco 404
L_1 — coil tuning
18 AWG Buss wire for L_1. See Fig. 17-44.
Q_1, Q_2 — 2N3644 transistors
Misc. — antenna, printed circuit, crystal microphone (A basic set of parts for this FM Transmitter are available from Hickok Teaching Systems, Wheeling Ave., Woburn, MA 01801. This does not include an antenna or crystal microphone.)

Fig. 17-48. Schematic and parts list for FM Transmitter.

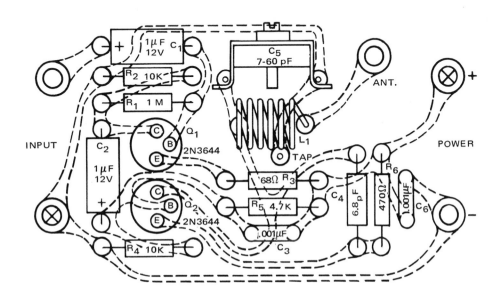

Fig. 17-49. Parts placement for FM Transmitter (top view).

A printed circuit layout is shown in Fig. 17-50. See Fig. 17-51 for details of how to make coil L_1. See Fig. 17-52 for a top view of printed circuit.

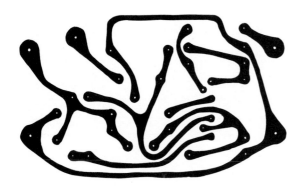

Fig. 17-50. Printed circuit layout for FM Transmitter. (Hickok)

Fig. 17-52. Top view of FM Transmitter PC module. (Hickok)

The operation of this project is similar to the AM transmitter. Place an AM receiver nearby and turn it on. Adjust the tuner to a spot in the middle of the FM broadcast range. Connect a power supply or battery to the power terminals and a 24 to 36 in. antenna to the unit. Hook up a crystal microphone to the input, then adjust the C_5 unit. The signal is heard in the FM receiver.

TEST YOUR KNOWLEDGE

1. What is the wavelength in metres of a 21 MHz radio wave?
   a. 1.428 metres.
   b. 14.28 metres.
   c. 142.80 metres.
2. The frequency of an 11 metre wave is _____ MHz. (Note: This band of frequencies is assigned to the Citizens Band.)
3. A radio frequency wave of 455 KHz must

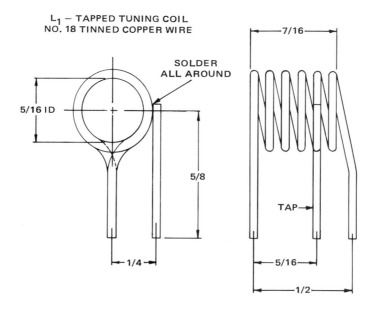

Fig. 17-51. Construction details for coil L_1 in FM Transmitter. (Hickok)

be heterodyned with what frequency to produce a 500 Hz audio tone signal?

4. What is meant by the term "zero beating?"

5. _____ watts modulation power is required for 100 percent modulation of a 1000 watt amplifier.

6. _____ watts modulation power is required for 75 percent modulation of a 1000 watt amplifier.

7. _____ watts modulation power is required for 50 percent modulation of a 1000 watt amplifier.

8. Draw wave forms showing 100 percent modulation, overmodulation and undermodulation.

9. Why is demodulation similar to half-wave rectification?

10. What is the principle of the transistorized superheterodyne receiver?

11. What are some of the advantages of FM over AM?

12. The discriminator is a typical circuit you will encounter in _____ receivers.

13. The if used in FM radio is _____ MHz.
    a. 10.7 MHz.
    b. 17.7 MHz.
    c. 21.7 MHz.

14. _____ or demodulation is the process of removing the audio signal from a modulated radio wave.

15. Name the three styles of citizens band radios.

16. Why has citizens band radio become so popular in recent years?

Chapter 18

SWITCHING AND LOGIC CIRCUITS

The modern miracle of the "computer" is made possible by the rapid switching action of the transistor. Computers may be required to switch circuits at the rate of several million times per second.

In this chapter, you will become familiar with:

1. The terminology used in switching circuits.
2. The characteristics of a switching transistor.
3. The applications of diodes and transistors in logic circuits.
4. Some of the basic switching circuits used in electronics.

SWITCHING

SWITCHING is not a mysterious term. The typical example is the ON or OFF switch found most anywhere to turn the lights ON or OFF. Switching may also be expressed as "opening" or "closing" a circuit. With a switch "open," it offers INFINITY RESISTANCE with very little if any current flow. A "closed" switch offers near zero resistance and maximum current flow.

A mechanical light switch does require some manual effort to make it "change its state" from ON to OFF or vice versa. In the case of the electromagnetic switch, such as the RELAY, the change of state could be accomplished by a current through its operating coil. Refer to Chapter 5. This could be a steady current or a PULSE of current.

In either of these applications, the switching rate must be comparatively slow due to the inertia of the mechanical parts. It would be impossible to turn a light ON or OFF manually more than two or three times per second. Circuits may be switched at unbelievable speeds by means of transistors.

THE TRANSISTOR AS A SWITCH

An examination of the curves of a typical switching transistor, Fig. 18-1, will disclose two states at which the transistor could operate as a switch:

ON state — with the transistor at saturation current with minimum resistance. I_B over 300 microamps.

OFF state — with the transistor cut off and minimum current. Any current would be leakage current $I_B = -I_{CO}$.

In our earlier discussions of transistor action, we discovered that these conditions may be obtained by changing the bias of the transistor. Actually, in the cutoff state, the emitter-base junction is in reverse bias state. In the saturation state, the emitter-base junction is forward biased. Operated in this manner, the circuit would be called a SATURATED SWITCHING CIRCUIT. This method has a disadvantage since there is a slight delay to change a transistor from conducting in saturation to its cutoff state.

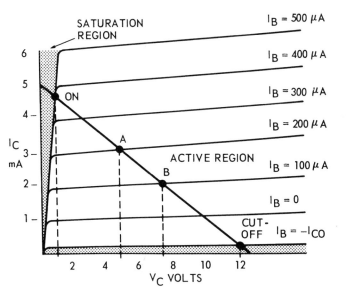

Fig. 18-1. The characteristic curves of a switching transistor showing saturation region, active region and cutoff.

For extremely rapid switching applications, it is better not to drive the transistor to saturation. Rather, it should approach some point in the "active region of the transistor." See points A and B in Fig. 18-1. Either of these points of bias may produce a voltage change (V_c) or current change (I_c) of sufficient magnitude for a useful output.

EXAMPLE: Consider point A. Switching from cutoff to 3 milliamps would produce a change in collector voltage of approximately 8 volts with this particular load resistance and load line. In order to turn the transistor ON, the emitter-base bias current would be 200 microamps.

A change in the dc level of voltage applied as a bias between the emitter-base of a transistor would change its state from ON to OFF or vice versa. In Fig. 18-2, the common emitter circuit is drawn. Input pulses are indicated with the corresponding output pulses. Follow this action:

1. When the input is at zero, the emitter-base voltage is also zero, and the transistor is cut off. With no collector current, there is no voltage drop across R_L and the collector voltage approaches the value of $-V_{cc}$.

2. A negative 6 volt pulse applied to the input makes the base 6 volts negative in respect to the emitter. This is severe forward bias, and the transistor is driven to saturation. The collector current I_c causes a voltage drop across R_L similar to the value of $-V_{cc}$, and the collector voltage approaches zero. Moving from $-V_{cc}$ to zero is in a positive direction, and the output is inverted when compared to the input pulse.

The resistor in the base circuit in Fig. 18-2 is a current limiting resistor. To compute the value of this resistor, assume a value of $-V_{cc}$ as -6 volts. At saturation, when $V_{R_L} = -V_{cc}$, the saturation can be found since $R_L = 1000$ ohms.

$$I_c = \frac{-V_{cc}}{R_L} = \frac{-6 \text{ volts}}{1000} = 6 \text{ milliamps}$$

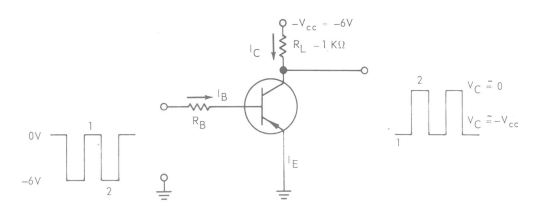

Fig. 18-2. The common emitter circuit is used to demonstrate switching action. R_B is a base current limiting resistor.

What base current is required to produce a collector current of 6 milliamps? A review of information found in Chapter 13 will show that the current gain of the common emitter circuit is:

$$\beta_{ac} = \frac{\Delta_{i_c}}{\Delta_{i_b}} \quad \bigg| \quad V_{CE} \text{ constant}$$

and β_{ac} is referred to as h_{fe}.

Rewriting the equation, we find that:

$$h_{fe} = \frac{\Delta_{i_c}}{\Delta_{i_b}} \quad \text{or} \quad \Delta_{i_b} = \frac{\Delta_{i_c}}{h_{fe}}$$

Assuming the current gain of this selected transistor is 50 (this value of h_{fe} will be found in the specifications of the transistor), then the base current will be:

$$I_B = \frac{6 \text{ mA}}{50} = 120 \ \mu A$$

The voltage drop across the emitter-base junction is very small and can be disregarded for practical purposes. The value of the current limiting resistor will become:

$$R_B = \frac{-6V}{120 \ \mu A} = \frac{-6V}{1.2 \times 10^{-4}} = 50 \text{ kilohms}$$

When the transistor is in the OFF condition and zero volts are applied between E and B, the current through R_B will be zero.

PULSE TERMINOLOGY

We are now working with changes in dc levels of voltage and seemingly square waves or pulses. First, look at the ideal pulse wave in Fig. 18-3. Then, familiarize yourself with the following terms.

LEADING EDGE. Moving from left to right, this is the edge where the pulse rises from zero to maximum.

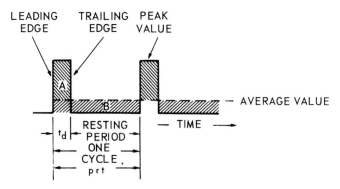

Fig. 18-3. Terminology of an ideal pulse wave.

TRAILING EDGE. This is the edge where the pulse drops from maximum to zero.

PEAK VALUE. The maximum peak value of pulse.

AVERAGE VALUE. This is the level indicated by the dashed line. It encloses an area equal to the pulse area: area A = area B.

PULSE DURATION. The time that the pulse remains above a certain amplitude of its maximum value (usually 90 percent).

RESTING PERIOD. The time interval between pulses.

CYCLE. Sequences of events or changes between the beginning of one pulse to the beginning of the next pulse.

PULSE REPETITION RATE (prr). The number of cycles per second. The term "frequency" is not used.

PULSE REPETITION TIME (prt). The period of time interval for one cycle.

DUTY CYCLE. This is the ratio between the average value of the pulse and the peak value.

$$\text{duty cycle} = \frac{\text{average value}}{\text{peak value}}$$

Expressed in another way, duty cycle refers to the ratio between ON time to OFF time.

$$\text{duty cycle} = \frac{t_d \text{ (pulse duration)}}{\text{prt (pulse repetition time)}}$$

or

$$\text{duty cycle} = t_d \times \text{prr}$$

RISE TIME (t_r). The time required for buildup or rise of wave. It is the time period between voltage change from 10 percent to 90 percent of peak value. See Fig. 18-4.

FALL TIME (t_f). The time required for a pulse to drop or decay from 90 percent of its peak value to 10 percent.

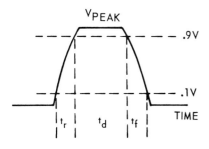

Fig. 18-4. Terminology of the rise and decay of a pulse.

OTHER TYPES OF WAVE FORMS

The ideal square wave is difficult to approach. The actual wave may resemble a wave similar to Fig. 18-5. You will understand this as you realize that the leading edge and trailing edge of a wave cannot be perfectly vertical. A minute fraction of time must elapse as the wave builds up or decays. If the rise time is extremely rapid, the leading edge would approach a straight vertical line.

At this point, we will illustrate and describe other common types of wave forms that find wide application in electronic circuitry. It is not our purpose to explore deeply into the generation of these waves, but rather to give you a speaking knowledge of them. Some have already been discussed in earlier chapters of this text. Study Figs. 18-5 to 18-11.

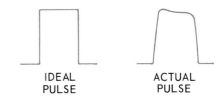

Fig. 18-5. The comparison of an ideal pulse to an actual pulse.

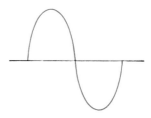

Fig. 18-6. A SINE WAVE. Curve is a function of the angle theta. See Chapter 6.

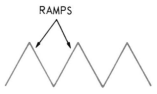

Fig. 18-7. TRIANGULAR WAVE FORM. The positive and negative ramps have the same slant and time duration.

Fig. 18-8. SAWTOOTH WAVE. It is similar to the triangular wave except the positive ramps and negative ramps have unequal slopes and the rise time is greater than the fall time.

Fig. 18-9. TRAPEZOIDAL WAVE FORM. This wave consists of a positive step and a positive ramp followed by a negative step.

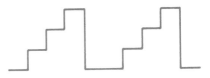

Fig. 18-10. STAIRCASE WAVE FORM. This wave is a series of positive steps followed by a negative step.

Fig. 18-11. EXPONENTIAL WAVE FORM. The rise and fall of this wave is dictated by a mathematical equation involving a variable exponent. See Universal Time Constant Chart, Chapter 9, Fig. 9-15.

GATING CIRCUITS

Two value circuits or BINARY LOGIC circuits are the foundation of most modern digital computers. The following discussions will introduce you to the fascinating field of computer technology and provide an insight into its circuitry. The computer may seem extremely complicated and involved. Nevertheless, it is the combination of many quite simple circuits working together.

A switch may be either ON or OFF. In binary logic, these conditions are represented by 1 and 0 respectively. Gating circuits produce an output for a given input or inputs. The purpose of the gate may be to permit the passage of a signal or prevent the passage of a signal. The word GATE applies in electronics just as a gate in a fence around your garden. The gate must be opened to allow you to pass.

AND GATE

Starting with the more simple types, the AND gate may be understood by the switching circuit of Fig. 18-12. In this case, both switch A and switch B must be closed in order to light the lamp. Considering the binary system as ON equals 1 and OFF equals 0, the TRUTH TABLE

shown in Fig. 18-13 can be constructed. A and B columns indicate the switch positions. C is the output or the lamp. ZERO means the lamp is OUT. ONE means the light is ON.

A	B	C
0	0	0
1	0	0
0	1	0
1	1	1

Fig. 18-13. Truth Table for AND gate.

This AND gate is written as an equation in Boolean Algebra as: $AB = C$. This means that A and B equal output C. Rather than draw the circuitry for the gate, a symbol is used. See Fig. 18-14. Note that the AND gate may have more than two inputs. Inputs must be present at A and B, or at A, B and C, to get output D.

Fig. 18-14. Logic symbols for AND gate.

In Fig. 18-15, the TRUTH TABLE is expanded to include another term (D).

Fig. 18-16 presents a circuit which will perform the AND function by using diodes. Follow this action:

1. In its present state, both diodes D_1 and D_2 are conducting. The voltage across R is equal to 5 volts and the voltage at C is zero.
2. A 5 volt pulse applied to A will cut off D_1, but D_2 still conducts and voltage at C is zero.
3. A 5 volt pulse applied to B will cut off D_2, but D_1 still conducts and voltage at C is zero.

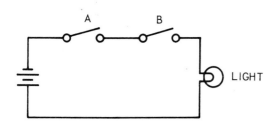

Fig. 18-12. This switch circuit describes an AND gate.

A	B	C	D
0	0	0	0
0	0	1	0
0	1	0	0
0	1	1	0
1	0	0	0
1	0	1	0
1	1	0	0
1	1	1	1

Fig. 18-15. Expanded Truth Table for AND gate.

4. When 5 volt pulses are applied to both A and B, then D_1 and D_2 are cut off. No current flows, there is no voltage drop across R and the voltage at C is 5 volts.

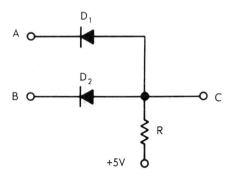

Fig. 18-16. A diode AND gate circuit.

Fig. 18-17 illustrates the circuit of the AND gate, using transistors. The transistor has the advantage of signal amplification with each gate. Follow this action:

1. A positive pulse would forward bias the EB junction of Q_1 and permit it to conduct. However, Q_2 is cut off, so no current will flow in the series circuit. No voltage will appear at C.

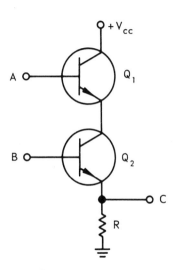

Fig. 18-17. Circuit of a direct coupled transistor logic (DCTL) AND gate.

2. A positive pulse at the base of Q_2 would cause conduction, but Q_1 is cut off.
3. When positive pulses appear at both the bases of Q_1 and Q_2, they both conduct and an output voltage appears at C equal to the voltage drop across R.

Another AND gate circuit drawn in Fig. 18-18 is called the DTL or diode-transistor logic circuit. It combines both the diode logic gate and the transistor amplifier. The explanation of this circuit will be left to you. It is not difficult.

OR GATES

A second popular gate circuit is the OR gate. The theory of the OR gate is illustrated in Fig.

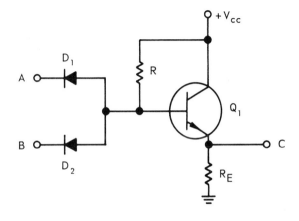

Fig. 18-18. Circuit of a diode-transistor-logic (DTL) AND gate.

299

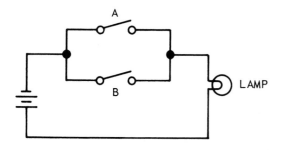

Fig. 18-19. This switch circuit describes an OR gate.

18-19. When either A or B switch is closed, the lamp will turn on.

The TRUTH TABLE for the OR gate circuit is shown in Fig. 18-20. Again, the number 1 indicates ON, and 0 indicates OFF. This circuit is symbolized in Fig. 18-21.

A	B	C
1	0	1
0	1	1
1	1	1
0	0	0

Fig. 18-20. Truth Table for OR gate.

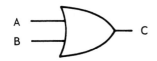

Fig. 18-21. Logic symbol for OR gate.

In Boolean Algebra, the equation is A + B = C. The plus sign in the equation is read as OR. With an input at either A or B or both, there is an output at C.

The diode OR gate is represented in Fig. 18-22. Without an input, there is no current flow, and the voltage across R is zero or no output. If a positive pulse is applied to either A

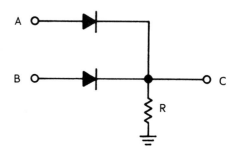

Fig. 18-22. A diode OR gate circuit.

or B, the respective diode will conduct and a voltage output equal to I x R will appear at C.

In Fig. 18-23, the same OR gate is drawn schematically using transistors. Since these are NPN transistors, a positive pulse applied to the base of either transistor will drive it to conduction, and an output will appear across the emitter resistor.

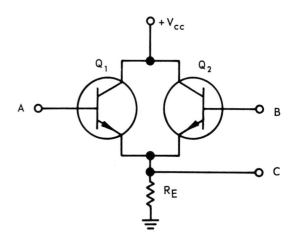

Fig. 18-23. An OR gate circuit, using transistors.

The combination diode-transistor OR gate is illustrated in Fig. 18-24. It combines both the diode and transistor circuits. The transistor amplifies the signal and compensates for the loss due to circuit components.

INVERTERS OR NOT GATES

Many times in computer logic circuits, it is necessary to invert a signal. In other words, you may want to change a 0 to 1 or a 1 to a 0. A

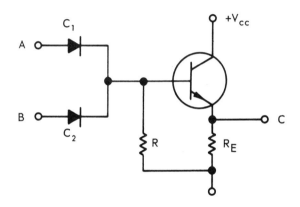

Fig. 18-24. A diode transistor OR gate circuit (DTL).

circuit which performs this function is an INVERTER or NOT GATE. Fig. 18-25 shows the logic symbol for the NOT gate.

Fig. 18-25. Symbol for inverter or NOT gate.

The circuit to perform this simple function has already been studied. We know that a common emitter transistor circuit will invert its input signal. Consider Fig. 18-26.

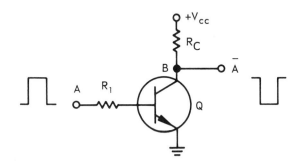

Fig. 18-26. A common emitter inverter circuit.

With no signal applied to the base of Q, the transistor is cut off, and the voltage at B is similar to $+V_{cc}$. A positive signal to the base of Q will turn on the transistor. It is an NPN transistor. The voltage at B will drop and approach zero. Resistor R_1 is only a current limiting resistor for the base circuit.

NAND GATES

A variation of the AND gate is the NOT AND or NAND gate. This circuit combines the INVERTER and the AND gate as illustrated in Fig. 18-27.

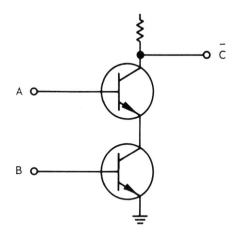

Fig. 18-27. A transistorized NAND gate circuit.

With only the AND gate, a signal at A and B would produce an output C. With the inverter added, a signal at A and B produces no output since the one will be inverted to zero. The truth table for the NAND gate is shown in Fig. 18-28. The symbol appears in Fig. 18-29.

A	B	C
0	0	1
0	1	1
1	0	1
1	1	0

Fig. 18-28. Truth Table for NAND gate.

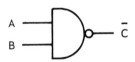

Fig. 18-29. Logic symbol for NAND gate.

NOR GATES

The NOR gate is a variation of the OR gate, and it is symbolized in Fig. 18-30.

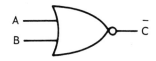

Fig. 18-30. Logic symbol for NOR gate.

With only the OR gate, a signal at A or B would produce a signal at C or 1. With the inverter added, a signal at 1 or at A or B will produce a 0 at C. Truth table for a NOR gate is shown in Fig. 18-31. The transistor circuit to perform this logic is drawn in Fig. 18-32. A positive signal applied to the base of either transistor will cause it to conduct. The common emitter circuit inverts the signal. As a consequence, a 1 signal at A or a 1 signal at B will produce a 0 signal at C.

A	B	C
0	0	1
0	1	0
1	0	0
1	1	0

Fig. 18-31. Truth Table for NOR gate.

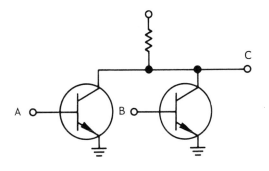

Fig. 18-32. A transistor NOR gate.

The previous discussion on switching transistors and simple logic circuits may whet your interest in this fascinating field. You are encouraged to read advanced electronics books for a more complete study.

FLIP-FLOPS

In Chapter 16, you studied multivibrator circuits, which are readily adaptable to counting circuits. These FLIP-FLOPS can be combined with gate circuits to perform an almost unlimited number of logic functions. These circuits are the foundation of the modern computer. See Fig. 18-33.

TRANSISTORIZED IGNITION SYSTEMS

A widely used development in the field of transistorized switching is found in automotive ignition systems. In older automobiles, the primary circuit of the ignition coil was opened and closed by the cam-operated breaker points in the distributor. This system, although universally used, had the disadvantages of points, arcing, points bouncing and other problems inherent in any mechanical type of switch.

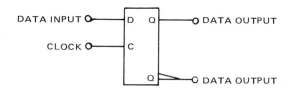

Fig. 18-33. Symbol for one type of FLIP-FLOP.

Fig. 18-34 illustrates the basic ignition system using a transistor to switch the primary of the coil. When the points are OPEN, the voltage at the base of Q_1 is similar to supply voltage. There is no current in the circuit and no drop across either R_4 or R_3. However, current does flow through R_2 and rectifier CR_1 which holds the emitter of Q_1 slightly negative. The transistor is reverse biased and cut off.

When the breaker points close, the voltage at the base of Q_1 becomes more negative than the emitter due to voltage drops across the divider

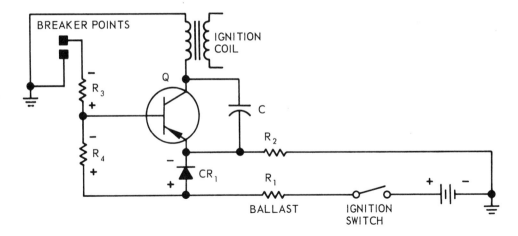

Fig. 18-34. The basic transistorized automotive ignition system. The primary circuit is switched by the transistor.

R_3 and R_4. The transistor is turned on and primary current flows in the ignition coil.

R_1 is a ballast resistor to limit the primary current. The capacitor C from emitter-to-collector takes care of any transient voltage spikes and protects the transistor.

Another type of transistorized ignition system eliminates the breaker points. The transistor is turned on and off by a voltage developed by a pulse from a magnet and coil as a toothed wheel interrupts the magnetic field.

Fig. 18-35 shows the circuit for a four cylinder engine. The interrupter wheel is on the distributor shaft and rotates with it. Each time one of the sections of the wheel passes the magnet and coil, it interrupts the magnetic field and induces a voltage in the coil. This voltage makes the base of Q_1 more negative and turns on the transistor. CR_1 prevents currents from flowing in the reverse direction in the base circuit. C absorbs any transient spikes of voltage developed across CR.

SCR IGNITION SYSTEM

A more sophisticated ignition system employing both transistors and an SCR is illustrated in Fig. 18-36. You will recognize, at first, the power oscillator or dc-ac converter which

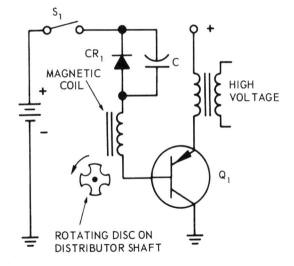

Fig. 18-35. A magnetic pulse switches the transistor on and off in this ignition system.

uses transistors Q_1 and Q_2. A circuit similar to this was studied in Chapter 16.

The output of the oscillator is increased by step-up transformer T_1 to approximately 350 volts, then rectified to dc by the bridge rectifier circuit. The storage capacitor C_1 charges to the full voltage of the rectifier.

The trigger circuit consists of Q_3 and associated components. Between ignition firing points, Q_3 is held at cutoff by bias network R_1 and R_2. When the breaker points open, the

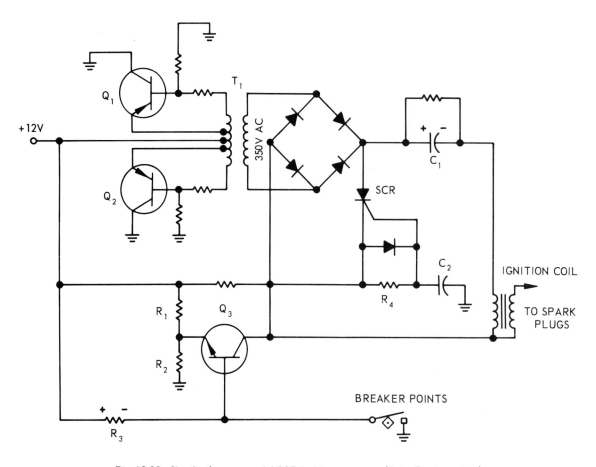

Fig. 18-36. Circuit of a commercial SCR ignition system. (Delta Products, Inc.)

voltage across R_3 drops to zero and forward biases Q_3 into conduction.

When Q_3 conducts, a gating pulse at the junction of R_4 and C_2 "fires" the SCR into conduction. C_1 is rapidly discharged through

the primary of the ignition coil. A high voltage appears across the ignition coil secondary due to the turns ratio.

The resistor across C_1 is a bleeder to permit C_1 to discharge when the ignition system is off.

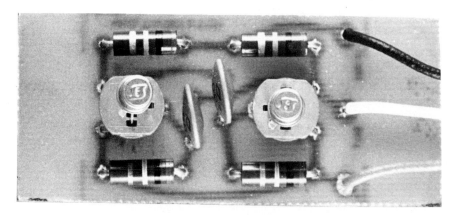

Fig. 18-37. Completed FET multivibrator constructed on an etched circuit board.

This circuit is a classical example of electronics in the automotive field. Each day, the need for understanding electronics increases, and all trade and industry technicians must include these studies in their basic training.

MULTIVIBRATOR PROJECT

Field effect transistors are again used in the free running multivibrator circuit and project illustrated in Fig. 18-37. On the oscilloscope, the output of this circuit will be a series of square wave pulses. The multivibrator, in fact, can be called a square wave generator.

The FET multivibrator circuit schematic and list of component parts are shown in Fig. 18-38. The printed circuit layout and parts placement are illustrated in Fig. 18-39. Consult the Appendix for the circuit etching process.

LED BLINKER PROJECT

Build a LED Blinker, a simple and interesting project that uses some of the latest electronic

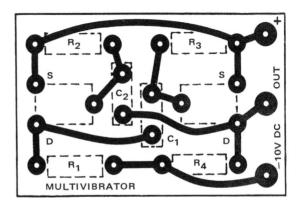

Fig. 18-39. Multivibrator circuit and parts placement.

components. Heart of the project is a LM-3909 integrated circuit that causes the light emitting diode (LED) to blink for long periods of time on a 1.5 volt battery. The current drain for the blinker should be so low that this device should blink for as long as the typical shelf life of the AAA cell.

The circuit for the LED blinker requires a minimum number of components. The schematic and parts list are shown in Fig. 18-40. A printed circuit layout appears in Fig. 18-41.

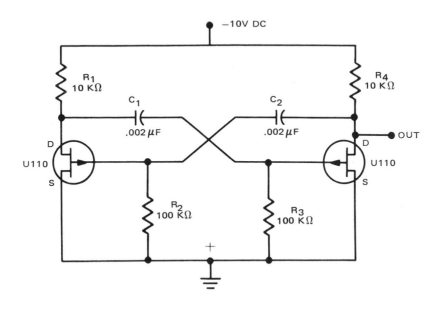

PARTS LIST FOR FET MULTIVIBRATOR

R_1, R_4 — 10 K Ω resistors   FET — U 110 Field effect transistor
R_2, R_3 — 100 K Ω resistors E_1 — −10V dc power source
C_1, C_2 — .002 μ F capacitors

Fig. 18-38. Schematic and parts list for FET Multivibrator.

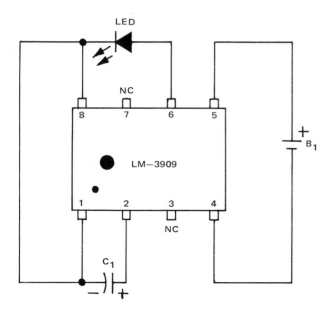

PARTS LIST FOR LED BLINKER

IC_1 — LM-3909 integrated circuit
(National Semiconductor)
LED — 1 1/2V NSL5027 light emitting diode
C_1 — 300 μF, 3V dc PC electrolytic capacitor
PC — 2 1/2 x 2 in. printed circuit board
B_1 — 1 1/2V AAA cell

Fig. 18-40. Schematic and parts list for LED Blinker.

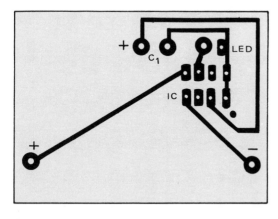

Fig. 18-41. Printed circuit layout for LED Blinker.
(Walter Deal, Old Dominion University, Norfolk, VA)

TEST YOUR KNOWLEDGE

1. There are two "states" at which a switching transistor will operate. Name them.
2. Switching may be expressed as _____ or _____ a circuit.

3. In pulse terminology, _____ is the sequence of events or changes between the beginning of one pulse to the beginning of the next pulse.
4. _____ is the time interval between pulses.
   a. Pulse pause.
   b. Resting time.
   c. Fall time.
5. _____ is the time required for buildup of the wave.
   a. Rise time.
   b. Gain.
   c. Amplitude.
6. Binary logic circuits are also called "two value" circuits. True or False?
7. Draw a truth table for this circuit.

8. Write the Boolean equation for:

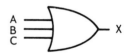

9. Using logic symbols, represent this equation:

$$(A + B) \quad (C + D) = X$$

10. Using logic symbols, represent this equation:

$$AB + CD = X$$

11. Using logic symbols, draw this diagram and equation: "If it does not rain and I get out of school at three o'clock, we will play basketball if I can find four other boys."

Hint: Yes = 1   No = 0

12. Draw the symbol for this equation:

$$AB + C + D = X$$

13. Write the equation for this symbol:

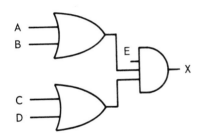

14. Write the equation for this symbol:

15. Draw the symbol for this equation:

$$(A + B)\,CD = X$$

16. Write the binary number for numbers from 1 to 16.

1. _____   9. _____
2. _____   10. _____
3. _____   11. _____
4. _____   12. _____
5. _____   13. _____
6. _____   14. _____
7. _____   15. _____
8. _____   16. _____

Chapter 19

INTEGRATED CIRCUITS

Probably one of the most amazing and unbelievable accomplishments in the field of electronics is the miniaturization of circuits. They are so small, in fact, that their actual construction must be done by technicians using microscopes. These tiny circuits that are revolutionizing the electronics field are called INTEGRATED CIRCUITS (ICs).

In this chapter, you will become familiar with:

1. The advantages of ICs over discrete electronic components.
2. How integrated circuits are made.
3. The various types of integrated circuits.
4. Some applications for ICs.
5. Whether or not integrated circuits require special handling or mounting precautions.

Integrated circuits were invented in 1958 by Texas Instruments, Inc.

ACTIVE VS PASSIVE ELECTRONICS

It used to be the common practice of electronics manufacturers to produce all components separately, then assemble the complete device by wiring the various components together with metallic conductors. With the development of microelectric circuits, it is possible that all of the various components and their interconnections are now fabricated on a single substrate. These microelectronic devices (ICs) are made up of transistors, resistors, diodes, capacitors and similar components. The advent of these microelectronic circuits have not, for the most part, changed the purpose of the basic units. Instead, they provide an integrated circuit that is smaller, more reliable and has a higher number of components in a given area. All the components of the integrated circuit are fabricated in a crystal of silicon (an excellent semiconductor element).

An ACTIVE device is an electronic component which can change its response to an external signal. Some common active devices are vacuum tubes, transistors and integrated circuits. All of these devices are capable of "gain." PASSIVE devices do not have the ability to change their output response based on the external input signal. Some common passive components are resistors, capacitors and inductors.

An integrated circuit usually is made up of both active and passive devices, all fabricated into a complete circuit.

ADVANTAGES AND DISADVANTAGES OF INTEGRATED CIRCUITS

An integrated circuit (IC) is a packaged electronic circuit. It usually contains transistors and possibly diodes, resistors or capacitors. These components are connected to form a complete device, such as an amplifier or detector and limiter. Usually, integrated circuits are made into a very small piece of silicon that may be referred to as a "chip." Fig. 19-1 shows a schematic diagram of an integrated circuit housed in a package about .3 in. (7.6 mm) by .18 in. (4.6 mm).

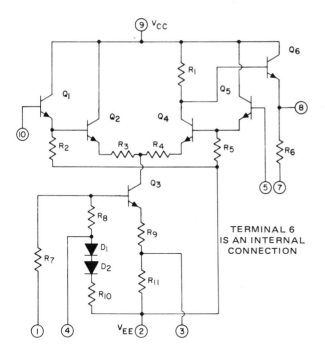

Fig. 19-1. Schematic diagram of CA3002 integrated circuit if amplifier. (RCA)

The size of the IC is an advantage over the equivalent number of individual components. See Fig. 19-2. Another advantage is WEIGHT. Manufacturers are able to pack much more circuitry into an integrated circuit microminature package. This cuts down on the weight of the circuit.

COST is another major advantage of integrated circuits over discrete components. By mass production, these devices are now comparable in cost to individual transistors. RELIABILITY, too, is a major advantage of ICs over individual components. This means that ICs usually will work for longer periods of time without giving trouble. An IC logic gate is 100,000 times more reliable than a vacuum tube logic gate, and 100 times more reliable than a transistor logic gate.*

Integrated circuits do have some disadvantages: Inductors (coils) usually are not suitable for integrated circuit packaging. Another limitation is that ICs function at fairly low operating voltages. Also, ICs currently are limited by the amount of power that they can handle. ICs, like diodes and transistors, are quite delicate and cannot withstand rough handling or excessive heat. A socket or heat sink (mass of metal used to carry away heat from a component) must be used when soldering ICs to a circuit board.

In summary, the advantages of the integrated circuit far outweigh the disadvantages or limitations. Integrated circuits are used in many modern applications of electronics, such as electronic watches or calculators, Fig. 19-3.

TYPES OF INTEGRATED CIRCUITS

Basically, integrated circuits may be classified according to two types:

1. DIGITAL.
2. LINEAR.

A digital integrated circuit is one which performs repetitive (one after the other) operations. ICs usually operate from either ON or OFF conditions, Fig. 19-4. These digital circuits sometimes are referred to as switching circuits. See Fig. 19-5. Digital signals generally are at two levels, each signal being discrete intervals

Fig. 19-2. A typical integrated circuit compared to the size of a golf tee. (Texas Instruments)

---

* John S. Mayo, "The Role of Microelectronics in Communication," MICROELECTRONICS (New York: Scientific American, 1978), p.100.

Fig. 19-3. A modern calculator, which uses integrated circuits. (Texas Instruments)

Fig. 19-5. Digital (switching) integrated circuits. (Texas Instruments)

Fig. 19-6. Operational amplifier linear ICs. (Texas Instruments)

apart. Usually, digital signals are binary (two state). See Chapter 20.

Linear integrated circuits are amplifier types. They control varying voltages, rather than operate as switches. Linear ICs are similar in operation to a transistor type amplifier, where the main function is to increase the output power, current or voltage in a circuit. Some typical linear ICs may be audio amplifiers, if amplifiers or operational amplifiers (OP AMP). See Fig. 19-6. Sometimes, linear ICs are called analog ICs.

HOW INTEGRATED CIRCUITS ARE MADE

Integrated circuits are formed by masking, etching and diffusion on a monolithic substrate of silicon. It is far beyond the scope of this text to study the design and detailed manufacturing processes in making integrated circuits. However, a simplified explanation will permit you to appreciate the impact these circuits now have in the science of electronics. See Fig. 19-7.

Integrated circuits are made on a thin slice of silicon from one to two inches in diameter. A normal slice may contain from 100 to 600 circuits side by side, all processed at the same time. After processing, the slice is divided into separate circuits called "dies." The number of completed ICs will depend upon the "die" size.

Fig. 19-4. Typical digital signals.

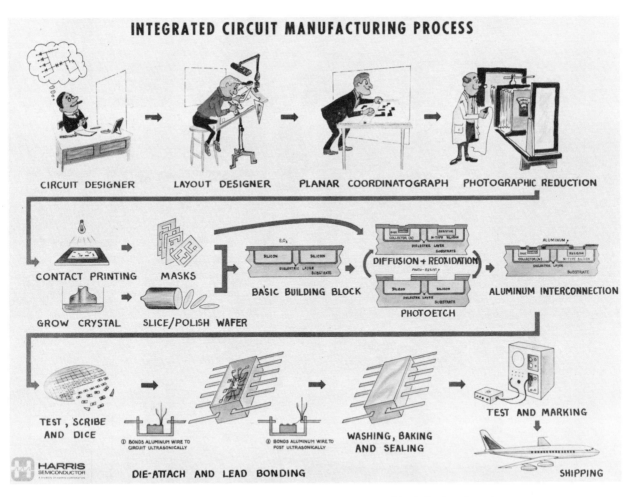

INTEGRATED CIRCUIT MANUFACTURING PROCESS

CIRCUIT DESIGNER   LAYOUT DESIGNER   PLANAR COORDINATOGRAPH   PHOTOGRAPHIC REDUCTION

CONTACT PRINTING   MASKS   BASIC BUILDING BLOCK   DIFFUSION + REOXIDATION   ALUMINUM INTERCONNECTION

GROW CRYSTAL   SLICE/POLISH WAFER   PHOTOETCH

TEST, SCRIBE AND DICE

① BONDS ALUMINUM WIRE TO CIRCUIT ULTRASONICALLY

② BONDS ALUMINUM WIRE TO POST ULTRASONICALLY

WASHING, BAKING AND SEALING   TEST AND MARKING

HARRIS SEMICONDUCTOR

DIE-ATTACH AND LEAD BONDING   SHIPPING

Fig. 19-7. Integrated circuit manufacturing process. (Harris Semiconductor)

These could be in squares of 40 to 60 mils. A photo of a silicon slice with completed circuits ready for scribing and separation is shown in Fig. 19-8.

Study the following steps in making a simple integrated circuit, starting with a wafer of P-type silicon as a substrate.

1. First, an epitaxial layer ot N type silicon about .25 microns thick is grown in the wafer. NOTE: Epitaxy is the physical placement of materials on a surface. This layer ultimately becomes the collector for transistors or an element of a diode or capacitor.
2. Next, a thin coat of silicon dioxide is grown over the N type material by exposing the wafer to an oxygen atmosphere at about 1000 deg. C. Steps 1 and 2 are illustrated in Fig. 19-9.

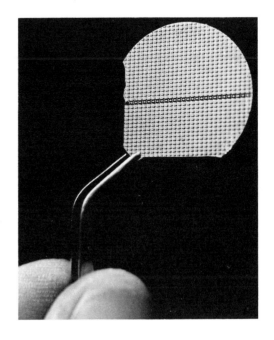

Fig. 19-8. Over one thousand complete integrated circuits are manufactured on a chip of silicon less than one inch in diameter. Each individual circuit can hardly be seen by the naked eye. (Pacific Telephone Co.)

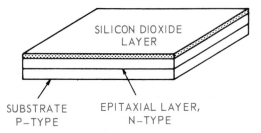

Fig. 19-9. Steps 1 and 2 in the process of making an IC.

3. To prepare the wafer for isolation between various components, the wafer is covered with photo-resist and exposed under ultraviolet light through a specific photographic mask. The nature of the mask depends upon the circuits to be made.

4. Then, the wafer is etched with hydrofluoric acid and unexposed areas of silicon dioxide are etched away. Steps 3 and 4 are shown in Fig. 19-10.

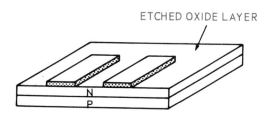

Fig. 19-10. The first masking and etching is for isolation of components.

5. Next, the wafer is subjected to a diffusion process using boron. The boron diffuses into and forms a P type material on all areas not protected by the silicon dioxide. Sufficient time is allowed for diffusion completely through the epitaxial layer to the P type substrate. The wafer now appears as in Fig. 19-11, with isolated

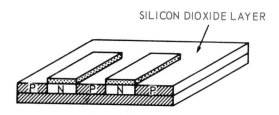

Fig. 19-11. N type material islands remain after P diffusion.

islands of N type material. Isolation is realized by the formation of the NP junctions around each island, and there are back-to-back diodes between each N type island.

6. During diffusion, a new layer of silicon dioxide forms over the diffused P type areas as well as on the top of the islands.

7. Using the photo-resist coating again and exposure under a specified mask, areas in the N type islands are etched away. Once again, the wafer is subjected to a P type diffusant and areas are formed for transistor base regions, resistors or elements of diodes or capacitors.

8. Again the wafer is given an oxide coating. Refer to Fig. 19-12.

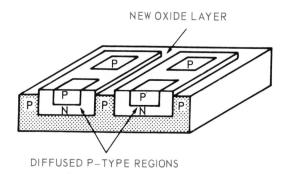

Fig. 19-12. P type regions are diffused in the N type islands.

9. The wafer is again masked, exposed and etched to open windows in the P type regions. Now a phosphorus diffusant is used to produce N type regions for transistor emitters, and regions for diodes and capacitors. Small windows are also etched through to the N-layer for electrical connections. The total wafer is again given the oxide coating. See Fig. 19-13.

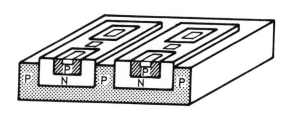

Fig. 19-13. Emitters are diffused into the P type regions.

10. For this particular IC, the monolithic circuit is complete except for the interconnections. A thin coating of aluminum is vacuum deposited over the entire circuit. The aluminum coating is then sensitized and exposed through another special mask. After etching, only the interconnecting aluminum forms a pattern between transistors, diodes and resistors and pads for wires to connect the wafer to an external circuit.
11. The wafers are now scribed with a diamond-tipped tool and separated into individual circuits.

After separation, the individual circuits are mounted on a ceramic wafer and leads (.001 in. or 0.025 mm diameter) are attached. The circuit may be mounted in a small can or flat package.

There are other methods of forming integrated circuits. The described method is called "epitaxial-diffused process." Second and third methods are "diffused collector process" and the "triple diffusion process."

NUMBER OF COMPONENTS PER CIRCUIT

The ability of electronic manufacturers to put more and more components into a small space is astounding. Since they were invented, the number of components that could fit into a standard size integrated circuit has dramatically increased each year. A method of classifying the amount of "component density" in integrated circuits is as follows:

SMALL SCALE INTEGRATION (SSI) — 0 to 30 gates per IC
MEDIUM SCALE INTEGRATION (MSI) — 31 to 100 gates per IC
LARGE SCALE INTEGRATION (LSI) — more than 100 gates per IC

The development of these semiconductor technologies in the electronic field is shown in Fig. 19-14.

Very large scale integration (VLSI) is now a reality. With this technique, it is possible to have thousands of components per square centimetre.

INTEGRATED CIRCUIT SYMBOLS AND PACKAGES

Integrated circuits come in various sizes and shapes. As the technology becomes more developed, the packages become larger to house

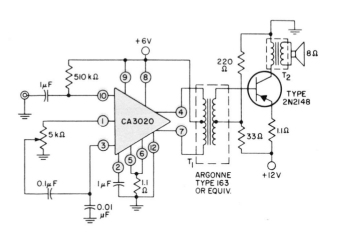

Fig. 19-15. A 4 watt audio amplifier. (RCA)

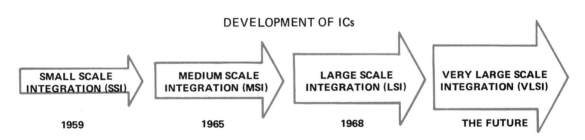

Fig. 19-14. Development of integrated circuits.

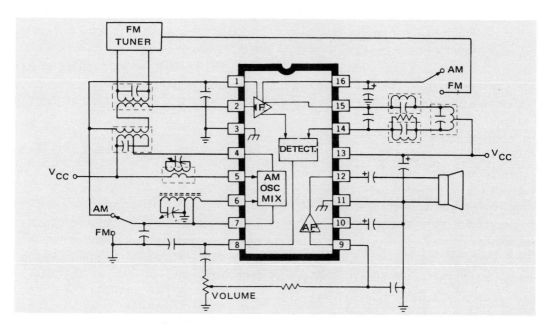

Fig. 19-16. AM/FM radio system.   (Sprague)

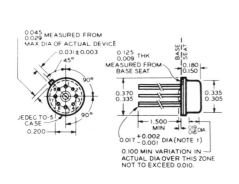

(TO-99)
D PACKAGE: 8-PIN METAL CAN

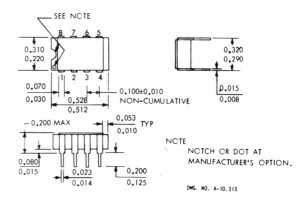

'H' PACKAGE: 8-PIN HERMETIC DUAL IN-LINE

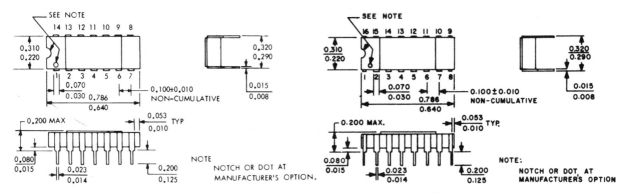

'H' PACKAGE: 14-PIN HERMETIC DUAL IN-LINE

'H' PACKAGE: 16-PIN HERMETIC DUAL IN-LINE

Fig. 19-17. Some integrated circuit pin numbering systems.   (Sprague)

medium scale integration (MSI) and large scale integration (LSI) devices.

Symbols for integrated circuits are very simple. Usually, they are either a triangle or a rectangle. Some basic IC symbols are shown in Figs. 19-15 and 19-16.

The numbering systems for integrated circuits' pins vary, depending on the manufactur-

er. There is an effort to unify the symbols so that they are the same throughout the industry. Fig. 19-17 shows the numbering system for a "D" package and three sizes of "H" package integrated circuit pins. Note the starting dot or space for the numbering of the IC pins.

Package designs for integrated circuits are usually round, square or rectangular. Fig. 19-18 shows some typical IC package designs with the

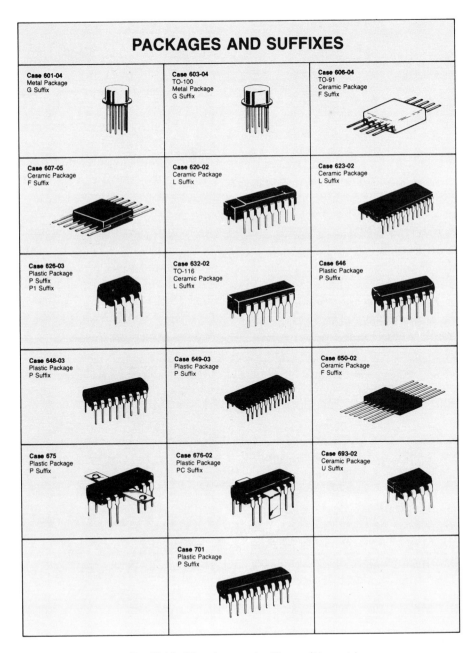

Fig. 19-18. IC packages and suffixes. (Motorola)

type of suffix given to the family of the package design. Fig. 19-19 shows an experimenter board that accepts various types of IC packages, as well as other types of components.

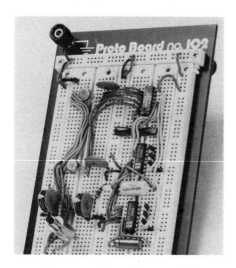

Fig. 19-19. An experimenter board with ICs and other components. (Continental Specialties Corp.)

TYPICAL APPLICATIONS FOR INTEGRATED CIRCUITS

Microelectronic technology has developed very rapidly in the past decade. Smaller and smaller electronic components perform ever increasingly complex electronic functions. We can see many benefits to our lives. These applications are wide-ranging, and there are many electronic consumer products on the market today that did not exist in the recent past.

Probably, one of the widely accepted applications of microelectronic technology is the digital watch. Every digital watch has an integrated circuit as its "brain." It is common for these watches to tell the hour, minute, second, day and month. Fig. 19-20 shows a digital wristwatch/stopwatch. Digital watch readouts are either light emitting diode (LED) or liquid crystal (LC).

Another popular application of microelectronic technology is the calculator. These con-

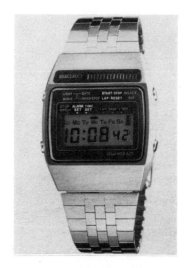

Fig. 19-20. Digital wristwatch. (Seiko Time Corp.)

sumer products are very popular. They range in complexity from a basic model that performs addition, subtraction, multiplication and division to the scientific pocket caluculator that is programmable. See Fig. 19-21.

Modern electronic test instruments use electronic circuits. Digital meters, frequency counters and logic probes are some examples of test equipment that utilize integrated circuits. Fig. 19-22 shows a miniature digital volt-ohmmeter being used to test the voltage in a medical electronic device.

Fig. 19-21. A scientific pocket calculator. (Hewlett-Packard)

Fig. 19-22. Miniature volt-ohmmeter. (Hewlett-Packard)

Electronic games for the home have evolved from microelectronic and computer know-how. Video games that have to be connected to home television sets have become very popular. These games vary from tennis to ping-pong to electronic pinball. New games are being designed each year for the consumer market. Fig. 19-23 shows a popular video game.

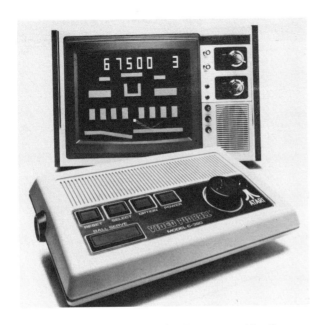

Fig. 19-23. Electronic television games. (Atari)

Hand held, calculator type games are also becoming popular. These self-contained games require no TV hookup. Each game operates off of a battery or an ac adaptor, and they feature LED display, precision lens, a variety of sound effects and an electronic printed circuit.

In the electronic football game shown in Fig. 19-24, the computer is on defense. You control the ball carrier's direction and speed as you cut

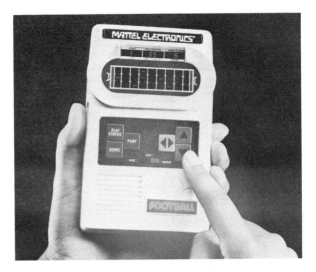

Fig. 19-24. Pocket size electronic football game. (Mattel Toys)

back and forth, avoid tacklers, break for daylight and score a touchdown. You can even hear a musical "Charge" tune. At the press of a button, you can find out the time remaining to play, the score, down, field position and yards to go for a first down. Video games will serve as entry level products for future personal computer uses.

An AM-FM radio integrated circuit, developed by General Electric, is an excellent example of how microelectronic circuits can be used to reduce size and cost, and provide improved performance over a conventional tube or transistor radio. Fig. 19-25 shows the schematic of the IC. Fig. 19-26 pictures the IC and a pin identification diagram.

The AM-FM IC contains all the radio circuit's active components, excluding two transis-

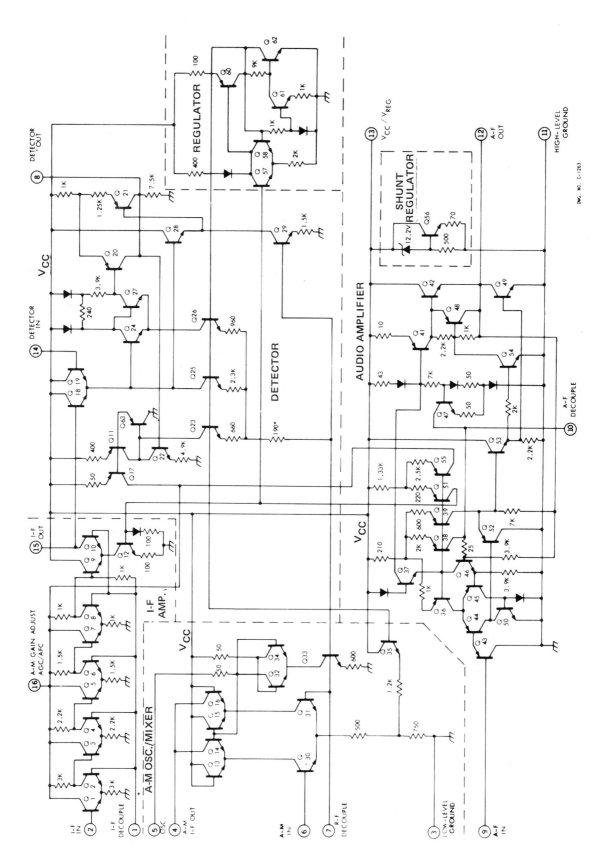

Fig. 19-25. Schematic of AM/FM integrated circuit. (Sprague)

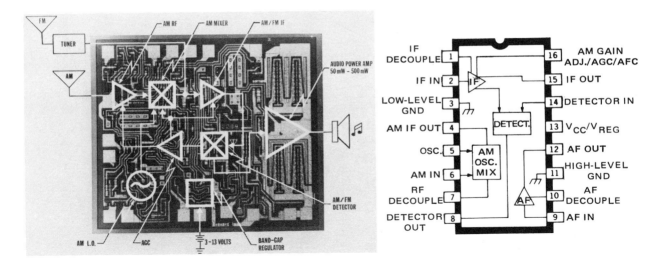

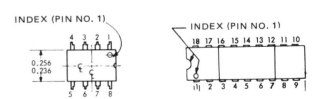

Fig. 19-26. Photo and pin diagram of the AM/FM integrated circuit. (Sprague)

tors in the FM mixer/oscillator section. One of the interesting characteristics of this integrated circuit is its ability to operate over a wide supply voltage range. Typically, it can operate from voltages up to 13V dc and down to as little as 2V dc.

PRECAUTIONS IN USING INTEGRATED CIRCUITS

When building experimenter circuits that have ICs, be very careful when you are soldering them into a printed circuit. It is always a good idea to use a small soldering pencil of 30 watts or less to solder ICs, transistors, diodes and SCRs. If possible, use a socket or heat sink (mass of metal used to carry heat away) for ICs to avoid damage because of excessive heat.

Some ICs can be damaged by static electric charges. Read the manufacturer's specification sheet and observe any special handling suggestions.

Be sure to observe correct polarity of the battery or power supply when making these connections to an integrated circuit. Also, take care to insert the IC into the circuit properly. Fig. 19-27 shows some typical IC pin locations or index.

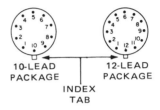

Fig. 19-27. Pin locators (index) on IC packages.

DIGITAL CLOCK PROJECT

Electronic digital clocks are used in many applications. This particular clock, Fig. 19-28, is easy to build since a minimum amount of wiring and component parts are needed. The project features a MOS-LSI integrated circuit, which uses a four digit, .5 in. light emitting diode (LED) display. The IC and other components are mounted on a single printed circuit board. You need to build a case and install the transformer and push button setting switches.

Fig. 19-29 shows a wiring diagram and parts list for the digital clock. Front and rear views of

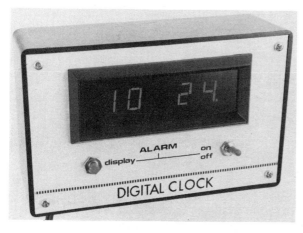

Fig. 19-28. Completed Digital Clock.

PARTS LIST FOR DIGITAL CLOCK

*CM — MA1002A clock module
 (National Semiconductor)
*T_1 — 102P3 transformer (120V primary,
 16V @ 23 mA and 5.2V @ 250 mA
 secondaries)
*S_1-S_6 — SPST push button switches
 (normally open)
*S_7 — SPT toggle switch
Misc. — line cord, cabinet, wire,
 digital display bezel

OPTIONAL PARTS FOR ALARM CIRCUIT

Q_1 — 2N3904 transistor
R_1 — 680 Ω, 1/2W resistor
B_1 — 6V buzzer with 6V battery or four
 "D" cells

*These items may be ordered from Digi-Key Corp., Box 677, Thief River, MN 57601 or Quest Electronics, P.O. Box 4430C, Santa Clara, CA 95054.

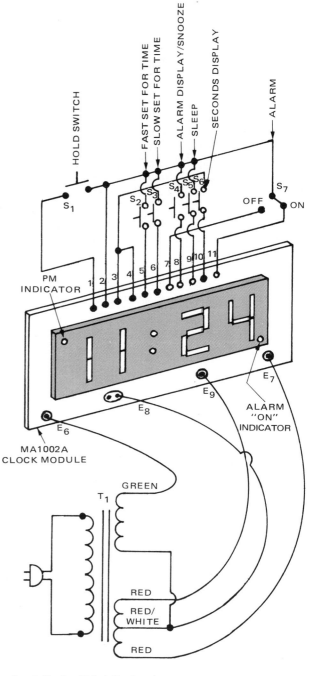

Fig. 19-29. Wiring diagram and parts list for Digital Clock.

the MA1002A printed circuit module are pictured in Fig. 19-30.

An optional alarm circuit for the digital clock is shown in Fig. 19-31. The MA1002A module provides an alarm (E_3) and radio control (E_1) output. These outputs are current sources limited to a maximum current of 5 mA dc. Depending upon your application, it may be necessary to interface these lines with the devices you wish to control.

Fig. 19-31 shows interface circuits which can switch up to 150 mA at 26 volts dc maximum.

Fig. 19-30. Front and rear views of MA1002A printed circuit. (National Semiconductor)

The circuit at left could be used to control a transistorized radio. V_A would be the radio's power supply, with the transistor 2N3904 acting as a switch to control the radio. The circuit will turn on the radio when either the alarm or radio outputs are on.

The circuit at right in Fig. 19-31 provides two outputs. The alarm output is independent of the radio output and can be used to drive an audio oscillator. The radio output actually is the composite of the radio and alarm outputs. It is on when either the alarm or radio outputs are on.

FUNCTIONAL DESCRIPTION

The MA1002A module provides four basic selectable display modes: These are summarized by National Semiconductor in Fig. 19-32.

Colon: This 12 hour display model is furnished with a colon display which flashes at a 1 Hz rate. (Fixed colon units are available on special order.)

Alarm "ON" Indicator: Setting the alarm switch to ON lights a dot in the lower right hand corner of the display.

AM/PM Indicator: PM time indication is given by a dot in the upper left hand corner of the display. Indication applies for both time and alarm display modes.

Power Failure Indication: Power failure is indicated by the entire display flashing at a 1 Hz rate. Contact to either the FAST or SLOW time set control cancels this indication.

Zero Blanking: Zeroes appearing in the first digit are blanked.

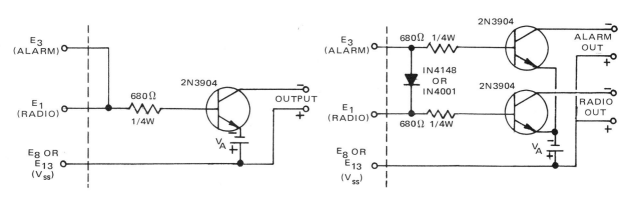

Fig. 19-31. Optional alarm circuit. (Digi-Key Corp.)

MA1002A DISPLAY MODELS

SELECTED DISPLAY MODE	DIGIT NO. 1	DIGIT NO. 2	DIGIT NO. 3	DIGIT NO. 4
TIME	10s OF HOURS AND AM/PM	HOURS	10s of MINUTES	MINUTES
SECONDS	BLANKED	MINUTES	10s OF SECONDS	SECONDS
ALARM	10s OF HOURS AND AM/PM	HOURS	10s OF MINUTES	MINUTES
SLEEP	BLANKED	BLANKED	10s OF MINUTES	MINUTES

Fig. 19-32. MA1002A display modes. If more than one display mode input is applied, the display priorities are: Sleep (overrides all others), Alarm, Seconds, Time (no other mode selected).

CONTROL FUNCTIONS

Setting of Time, Alarm Time, Seconds and Sleep Timer registers is accomplished by selecting the appropriate display mode and simultaneously contacting one or both of the FAST and SLOW time setting switches. This is summarized in Fig. 19-33.

Alarm ON/OFF Switch: The Alarm ON/OFF switch is an SPDT switch. The ON position lights the alarm set indicator; the OFF position disables the alarm output latch and silences the alarm. The alarm output will continue for 59 minutes unless cancelled by the Alarm ON/OFF switch or inhibited by the Alarm Display/Snooze button.

Alarm Display/Snooze Button: This momentary switch has four functions: displays the alarm time; enables setting of alarm time (in conjunction with fast or slow set switches); cancels the Sleep (Radio) output; and inhibits the alarm output for a period of between 8 and 9 minutes (Snooze function). The Snooze alarm feature may be used repeatedly during the 59 minute alarm enable period.

Sleep Display/Timer Button: A momentary contact displays the time remaining in the sleep register and enables programming the desired sleep time by simultaneously using the Fast or Slow buttons. The Sleep (Radio) output is latched on for the interval programmed, which may be up to 59 minutes. The Sleep output

MA1002A CONTROL FUNCTIONS

SELECTED DISPLAY MODE	CONTROL INPUT	CONTROL FUNCTION
TIME	SLOW	MINUTES ADVANCE AT 2 Hz RATE
	FAST	MINUTES ADVANCE AT 60 Hz RATE
	BOTH	MINUTES ADVANCE AT 60 Hz RATE
ALARM/ SNOOZE	SLOW	ALARM MINUTES ADVANCE AT 2 Hz RATE
	FAST	ALARM MINUTES ADVANCE AT 60 Hz RATE
	BOTH	ALARM RESETS TO 12:00 AM (12 HOUR FORMAT)
	BOTH	ALARM RESETS TO (0)0:00 (24-HOUR FORMAT)
SECONDS	SLOW	INPUT TO ENTIRE TIME COUNTER IS INHIBITED (HOLD)
	FAST	SECONDS AND 10s OF SECONDS RESET TO ZERO WITHOUT A CARRY TO MINUTES
	BOTH	TIME RESETS TO 12:00:00 AM (12-HOUR FORMAT)
SLEEP	SLOW	SUBTRACTS COUNT AT 2 Hz
	FAST	SUBTRACTS COUNT AT 60 Hz
	BOTH	SUBTRACTS COUNT AT 60 Hz

Fig. 19-33. MA1002A control functions. When setting time, sleep minutes will decrease at rate of time counter, until the sleep counter reaches 00 minutes (sleep counter will not recycle).

may be cancelled by contacting the Alarm Display/Snooze button, Resetting the time-of-day will decrease sleep minutes on the Sleep Timer, which will not recycle past 00.

Control Priorities: In the absence of Display Control switch inputs, the display shows the time-of-day information. If more than one mode is simultaneously selected, the priorities are as shown in Fig. 19-32.

OUTPUTS

Sleep (Radio): A positive current source output controlled by the sleep timer. This output can be used to switch on an NPN power transistor for controlling a radio or other appliance.

Alarm: A positive current source output controlled by the alarm comparator and enable circuit. This output may be used to control an alarm oscillator, wake-to-radio function, or start an appliance at a predetermined time.

TEST YOUR KNOWLEDGE

1. An _____ electronic device is one that can change its output response to an external input signal.
2. A _____ electronic device is one that does not have the ability to change its output response to an external input signal.
3. What are four advantages of integrated circuits over individual components?
4. Name the two basic types of integrated circuits and explain the function of each type.
5. What basic type of material is used to make integrated circuit substrates?
   a. Carbon.
   b. Silicon.
   c. Boron.
6. What is the primary difference between the various types of IC integration (SSI, MSI or LSI)?
7. Draw a symbol for an 8-pin integrated circuit in a rectangular package showing the locator index.
8. Name some IC applications that are discussed in this chapter. Also name some applications that are not listed in this chapter.
9. A soldering pencil of less than _____ should be used to solder integrated circuits on a printed circuit.
10. Why does heat damage an IC or a transistor?
11. Define the term "heat sink."
12. How long have ICs been in the electronics market?
    a. Since 1950.
    b. Since 1954.
    c. Since 1958.

Chapter 20

COMPUTERS AND MICROCOMPUTERS

Since the beginning of time, humans have always been fascinated with making complicated tasks easier. In the past few decades, the computer has helped humans extend their knowledge to a great degree. It is not an exaggeration to say that most of the technological achievements in the past ten years have depended on the computer.

This chapter will discuss electronic computers and microcomputers. Specifically, you will:

1. Study the history and development of computers.
2. Learn about the basic functional blocks in a computer.
3. Explore the basic differences between computers and microcomputers.
4. Discuss the uses of modern microcomputers.

HISTORY OF COMPUTERS

One of the first attempts by humans to count by use of an instrument was around 400 BC. This device, called an ABACUS, was made of a board with movable beads. See Fig. 20-1.

The founder of modern calculating machines was Blaise Pascal. In 1642, he invented a device that used gears and wheels to add and subtract. Pascal's calculating machine used the decimal (base 10) numbering system. In 1671, Gottfried Leibniz invented a machine that would mechanically multiply and divide.

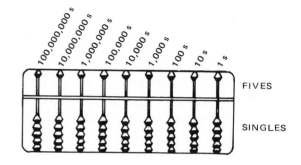

Fig. 20-1. The Japanese Abacus was an early means of keeping count by moving beads.

Another genius in the evolution of the computer was Charles Babbage. In 1833, he attempted to build the Difference Engine, an automatic adding machine, fifty years before a practical adding machine could be produced. The major problem encountered in constructing this device was in the accurate machining of the gears and wheels. Because his device was never successful, it became known as "Babbage's Folly." The concept of data STORAGE was introduced for the first time in this machine. Another innovation in the Babbage machine was the MILL that worked on the data and the CONTROL used to automatically control the operation of data.

The first successful commercial adding machine was produced in 1885 by William S. Burroughs. In 1890, Herman Hollerith made the first data-processing machine to use punched cards. This concept of using cards to store data was the forerunner of today's computer punched cards.

In 1938, Claude Shannon of the Massachusetts Institute of Technology (MIT) stated that switching circuits (like relays) could be used to evaluate logic statements. As you know, a relay is an electromechanical switch. The on and off conditions of these devices can be used to stand for yes/no or the 0 and 1 conditions in the binary numbering system. Mechanical relays have many built-in problems such as speed of operation and relatively short life.

The logic system used in most modern computers is binary. The binary numbering system was explained in Chapter 18.

The MARK I built in 1944 at Harvard University was a modern version of the Babbage Difference Engine. MARK I was a room-sized electromechanical calculating machine that used relays to compute by binary arithmetic.

ENIAC was the first true electronic computer. Developed at the University of Pennsylvania in 1945, it weighed 30 tons and contained about 18,000 vacuum tubes. The various tubes acted somewhat like relays in that they could be conducting (1) or not conducting (0).

One of the major technological breakthroughs in the development of modern computers was the invention of the transistor in 1948 at Bell Telephone Laboratories. Another major step was the concept of the stored program, which began with Babbage and was greatly refined in this century by Jon von Newmann.

The invention of the integrated circuit in 1958 was still another important development. It made modern "computers on a chip" possible. The microelectronics revolution (discussed in Chapter 19) has made it possible for us to have personal computers that have much more computing capability than the first ENIAC computer. Today's microcomputer is 20 times as fast as ENIAC. It occupies 1/30,000 the

volume, consumes about 100 watts of power, and costs about 1/12,000 as much as the first ENIAC computer.

COMPUTER MAKEUP

Today's computer is made up of two basic ingredients: HARDWARE and SOFTWARE. The actual computer machinery is the hardware, while the instructions that tell the computer what to do are software. This will be discussed in more detail later in this chapter.

A very basic function block diagram for a computer is shown in Fig. 20-2.

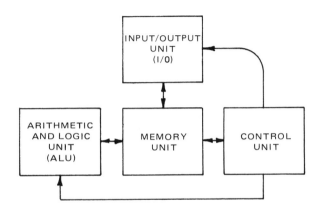

Fig. 20-2. Basic block diagram for a computer.

INPUT/OUTPUT UNITS

The input/output (I/O) unit feeds data and instructions into and out of the computer memory. A common form of input/output is the punched card, Fig. 20-3. In the punched card, no hole registers 0 and a punched hole represents 1 in the binary system.

Another typical input/output device is the keyboard printer. Sometimes the keyboard may be in the same unit as the output printer terminal. See Fig. 20-4. Also, the keyboard may be connected to a cathode ray tube video terminal, as shown in Fig. 20-5.

In summary, the input/output provides a means of communicating with the computer, or

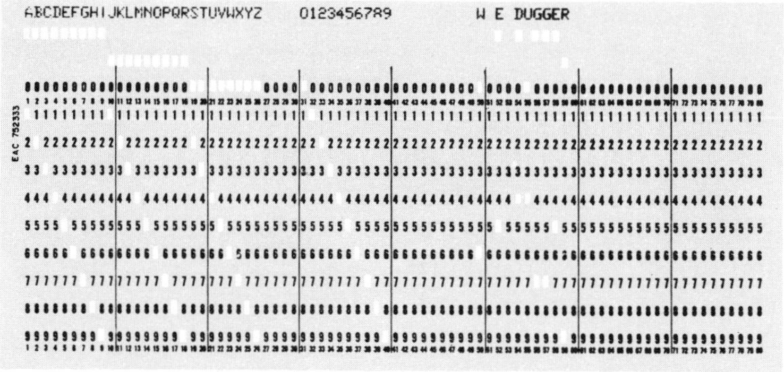

ABCDEFGHIJKLMNOPQRSTUVWXYZ   0123456789   W E DUGGER

Fig. 20-3. Typical punched card for use in computer input/output.

Fig. 20-4. Keyboard printer terminal.   (Digital)

it may be the method where the computer communicates with the outside world, another computer, or peripheral units. PERIPHERAL units work as partners with a large computer but are not part of it. A microcomputer designed for use in business applications is shown in Fig. 20-6.

MEMORY

The computer's memory is a storage for the data. This data, or the information on how to use it, usually is in binary form. Some general types of information stored in a computer's

Fig. 20-5. Computer for personal use features a central processor and keyboard video terminal.   (Hewlett-Packard)

Fig. 20-6. Compact memory computer is especially designed for business use.   (Digital Equipment Corp.)

memory would be fixed or variable instructions, fixed or variable messages, or data.

The binary data is stored in the memory in such a manner that each bit (0 or 1) or group of bits can be easily retrieved or obtained. A BIT is

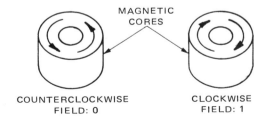

COUNTERCLOCKWISE FIELD: 0   CLOCKWISE FIELD: 1

Fig. 20-7. Saturation of magnetic core determines binary information.

a name for a Binary digIT, and it is the information equal to one binary decision. This means a bit could be a 0 or 1 or a YES or NO.

Until the early 1970s, the main type of computer memory was the MAGNETIC CORE, Fig. 20-7. If the magnetic core is magnetized (or saturated) in one direction, this represents one binary digit (0). Likewise, if the core is saturated in the opposite direction, the opposite binary digit (1) is represented. A single plane of magnetic cores of a core memory system is shown in Fig. 20-8.

Another type of memory is provided by magnetic tape or discs. This can be in reel-to-reel type, tape cartridge or disc memory. These

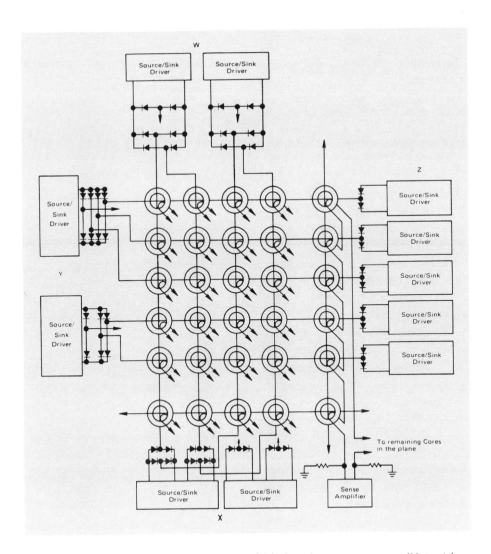

Fig. 20-8. Simplified diagram of a single plane (bit) of a core memory system. (Motorola)

types of storage devices store information in bulk, and they have unlimited capacity. One problem with them is the slow speed in information retrieval. Fig. 20-9 shows a typical desk type computer that uses a magnetic cassette cartridge as a memory.

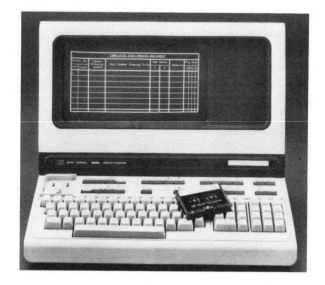

Fig. 20-9. Desk type computer with magnetic cassette memory. (Hewlett-Packard)

The semiconductor memory is becoming more and more popular, especially in microcomputers. These have extremely high speed but are somewhat limited in memory capacity. However, for small business, office or home computers, semiconductor memories are excellent. In the semiconductor memory, binary bits of logic are stored in the form of electronic signals. These signals are stored in a semiconductor device such as a flip-flop integrated circuit. Semiconductor memories lose their data whenever the power is turned off in the circuit. Fig. 20-10 shows a semiconductor memory that will store approximately 16,384 bits of information.

The term READ in computer terminology means to transmit data out from the memory to some other peripheral computer device. Likewise, WRITE means to transmit data or record it into a memory from some outside computer device.

Fig. 20-10. A Semiconductor Memory. (Hewlett-Packard)

One method of classifying memories is according to how the data is stored and retrieved. Some examples of different types of computer memories are:

1. RANDOM ACCESS MEMORY (RAM). These are read/write memories where data (0 or 1) can be stored (written in), then retrieved again (read out).
2. READ-ONLY MEMORY (ROM). This is a memory where the data can be continuously read out (retrieved). The ROM contains permanently stored (or rarely altered) programs. Inexpensive pocket calculators usually have read-only memories that are programmed by the manufacturer.
3. PROGRAMMABLE READ-ONLY MEMORY (PROM). This is an alterable or programmable read-only memory.

A computer memory is organized into a number of WORDS which are a string of characters. Computer words may be eight or sixteen bits in length. The word is the smallest number of bits that may be written into or read out of a memory in one cycle. A BYTE is a set of common binary digits which is acted upon as a unit but being shorter than a computer word.

The exact location of each word in the computer memory is called the ADDRESS.

CONTROL UNIT

The control unit in the computer takes the instructions and generates the necessary signals to make the Arithmetic and Logic Unit (ALU) or the whole computer accomplish a function or task. See Fig. 20-2. When a computer or microcomputer is under the control of a program, it shifts between two basic cycles:

1. FETCH CYCLE. The memory sends a word of data (instruction) to the control unit.
2. EXECUTE CYCLE. This tells the computer to execute or carry out the instruction.

The instruction FETCH and EXECUTE cycles usually alternate one after the other, such as:

1. FETCH Data of program counter
2. EXECUTE Load data into general register

1. FETCH Data of program counter
2. EXECUTE Load into accumulator

The timing of these cycles usually is done in the control unit.

ARITHMETIC AND LOGIC UNIT

The arithmetic and logic unit (ALU), Fig. 20-2, contains devices such as adders for arithmetic and logic operations on data received. ALU contains REGISTERS, which are temporary or short term storage circuits for digital data that have the capacity of storing one computer word. A basic type of register is the SHIFT REGISTER, in which the stored data can be moved (or shifted) to the left or to the right. Shift registers can handle data in serial or parallel form. In serial data, the data is handled one bit as a time. In parallel, all bits are handled all at once or all together.

In digital computers, the Arithmetic and Logic Unit give the basic instructions to the other parts of the computer. The instructions may be add, subtract, load, store, etc. The ALU is most important to the overall operation of the computer.

MINICOMPUTERS VS MICROPROCESSORS

Small computers are becoming very important in today's electronic age. A MICROCOMPUTER is a small computer. A MICROPROCESSOR is a computer central arithmetic and logic unit (ALU) and control unit on a single integrated circuit "chip." This single IC chip may hold tens of thousands of transistors, resistors, diodes and connecting circuitry. Fig. 20-11 shows a 16-bit microprocessor IC with full microcomputer capability.

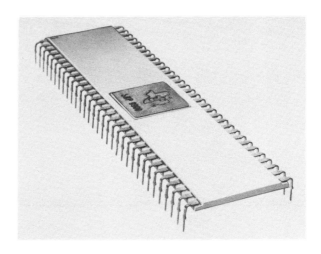

Fig. 20-11. Microprocessor integrated circuit.
(Texas Instruments)

The job of the microprocessor is to:

1. Receive data in the form of binary digits (0s and 1s).
2. Store the data for processing later.
3. Deliver the results to the user on an output device such as a cathode ray tube, on a typewriter terminal or through a two dimensional plotter.

Fig. 20-12 shows a block diagram of a microprocessor as part of the complete microcomputer. Modules and breadboards may be used to

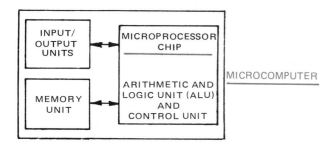

Fig. 20-12. Microcomputer block diagram.

make up the most basic actual computer. These may include microprocessor integrated circuits on a printed circuit board, Fig. 20-13.

A true microcomputer can be made up of a small desktop computer, which includes packaged modules, a power supply and expansion of

Fig. 20-13. Microcomputer module.   (Heath Co.)

memory and interface capability. See Fig. 20-14. They can be connected to a total system, using a keyboard display console, other memories and line printers, Fig. 20-15. Some of these components are shown in Fig. 20-16, where technicians are working on a space technology problem.

CALCULATORS

The electronic calculator first appeared on the scene in 1964 as a hefty, high-priced office

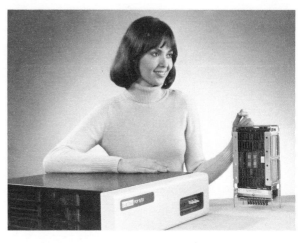

Fig. 20-14. Microcomputer for business applications. (Digital Equipment Corp.)

machine. Today, calculators not much bigger than your hand are low enough in price for most budgets.

This development of the calculator was brought about by technological improvements which allow one tiny integrated circuit to carry more than 10,000 transistors and to perform a variety of mathematical functions.

Fig. 20-15. Packaged computer system features floppy disk storage and video terminal.   (Digital Equipment Corp.)

Fig. 20-16. A microcomputer system being used in a space industry application. (NASA)

Some pocket-sized calculators, Fig. 20-17, can be programmed like larger computers. They can be used in medicine, science, engineering, surveying, statistics and mathematics. Because users can write their own programs easily, these calculators also can be used in areas such as business, education and navigation.

COMPUTER HARDWARE-SOFTWARE

Modern computers can be described in terms of hardware or software. "Hardware" is the actual physical equipment (material components) associated with the computer. "Software" refers to the stored program of step-by-step sequence of instructions in the memory of the computer. In the actual situation, the program includes the 0s and 1s (8 or 16 bits long) that tell the computer the sequence of operations.

The actual process of writing the program so that the computer knows what the instructions are is called "programming." The computer LANGUAGE is the actual form of the set of representations or instructions the computer is fed in order to make it do certain things. Some common computer languages are BASIC, FORTRAN, ALGOL and COBOL. Programs can be entered into the computer by normal input devices such as floppy discs, Fig. 20-18, cassettes or keyboard.

Fig. 20-17. A hand-held scientific calculator that can be programmed. (Hewlett-Packard)

Fig. 20-18. A computer using floppy discs. (Hewlett-Packard)

The process of writing a program is very difficult and sometimes mistakes are made. DEBUGGING is the process of correcting the mistakes made in the computer program.

USES OF MICROCOMPUTERS

There are so many potential uses of micro-computers today, and so many possible new applications for them in the future, that one is overwhelmed by the possibilities. Home computers are becoming very popular, which can add to the variety of applications of modern computers.

Computers may be used in the home in the future to help keep your checkbook and bank account records accurate. Also, computer stored TV games may be enjoyed by members of the household. The computer can assist with the family meal plan, or help figure special diets for any member of the family, or assist in your education where special programmed courses can be taken at home. The household environment may be controlled in the future by the home computer so that the maximum energy savings can take place.

Your automobile may have a computer controlled engine in which the ignition timing and the amount of fuel being consumed are adjusted to fit the type of driving you are doing. This type of computer controlled engine is already available in some automobiles.

Many types of telephone equipment are already computer controlled. In the future, you may have a telephone that memorizes the most frequently called numbers. Also, these phones can keep a record of the last called number.

The evolution of electronic computer technology over the last few years has been very rapid. With the development of the transistor and integrated circuits, we have seen some startling developments. Looking to the future, the field of electronics can be very exciting. The computer could help in making the future better for all humankind.

TEST YOUR KNOWLEDGE

1. Draw a basic block diagram for a computer.
2. With reference to computers, what is "hardware?"
3. What is "software?"
4. The computer's _____ is a storage for the data.
5. A _____ is a name for a binary digit.
6. What are the two basic cycles of a control unit?
   Explain each.
7. Name two types of computer memories and explain what makes them different.
8. The exact location of each word in the computer memory is called the _____.
9. The term _____ in computer terminology means to transmit data out from the memory to some other peripheral computer device.
   a. Read.
   b. Write.
10. The term _____ means to transmit data or record it into a memory for some outside computer device.
    a. Read.
    b. Write.
11. Semiconductor memories lose their data whenever the power is turned off in the circuit. True or False?
12. What is the difference between a microcomputer and a microprocessor?
13. What is computer "language?"
14. Explain the term "debugging."
15. List three input devices for computers.
16. List three output devices for computers.
17. Two methods of storing data in a computer memory are _____ and _____.
18. Name some future applications of microcomputers not described in this chapter.

Chapter 21

CAREER OPPORTUNITIES IN ELECTRONICS

Career opportunities in the electronics industry are almost unlimited. The field of electronics offers many advantages to a person who is interested in an excellent career.

Today, the electronics industry is divided into four main market areas: government products, industrial products, consumer products and components. The government purchases widely different products, such as missile and space guidance systems, communications systems and other electronic goods used in medicine, education, crime detection and traffic control.

Electronic products also have become an important part of daily business operations. Industrial purchases include computers, radio and television broadcasting equipment and production control equipment.

Consumer products probably are the most familiar types of electronic products. Every day, thousands of people buy television sets, radios, microwave ovens, stereos and calculators.

Components are needed to manufacture and repair electronic products. Some of the most well-known components are capacitors, switches, transistors, relays, television picture tubes and amplifiers.

CAREERS AVAILABLE IN THE FUTURE

There are many opportunities for semiskilled, skilled and professional people. See Fig. 21-1. Today, there is very little opportunity for the unskilled worker. Employment in electronics manufacturing is expected to increase faster than the average for all industries. In addition to the jobs resulting from employment growth, large numbers of openings will arise as experienced workers retire, or take jobs in other industries.

A young person aiming at a career in the electronics field, especially in the areas of research and development, should possess a number of personal qualities:

1. An intense curiosity about the nature of things and how they work.
2. An analytical mind, and a desire to solve problems.
3. A genuine interest in people and an urge to serve them.
4. A capacity for hard work, self-discipline and the ability to overcome failure.

Most of the Nation's workers are in industries that produce services. The production of goods requires only about one-third of the country's work force. In general, job growth is expected to continue to be faster in the service-producing industries than in the goods-producing industries. However, among industry divisions within both the goods-producing and service-producing sectors, the growth pattern will continue to vary.

EDUCATION AND YOUR CAREER

Education is very important to anyone entering a technical field such as Electronics. A high

Fig. 21-1. A trained computer operator researching flight of aircraft. (NASA)

school education has become a minimum standard for American workers. Thus, a high school graduate is in a better competitive position in the job market than a non-graduate.

Although training beyond high school has been the standard for some time for many professional occupations, other areas of work also require more than a high school diploma. As new, automated equipment is introduced on a wider scale, skill requirements are rising for clerical and other jobs. Employers increasingly are demanding better trained workers to operate complicated machinery.

In many areas of sales work, developments in machine design, use of new materials and the complexity of equipment are making greater technical knowledge a requirement. Because many occupations are becoming increasingly complex and technical, Figs. 21-2 and 21-3, occupational training such as that obtained through apprenticeship, junior and community colleges and post-high school vocational education courses is becoming more and more important to young people preparing for successful careers.

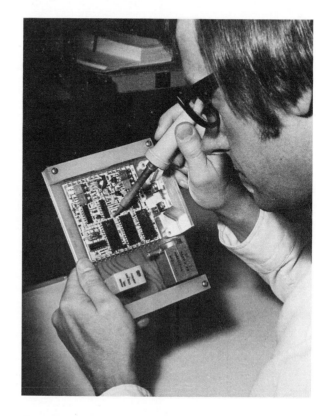

Fig. 21-2. An electronics engineer works on a mock-up of an electronics advanced automotive digital control unit. (Bendix Research Laboratories)

Young persons who do not get good preparation for work will find the going more difficult in the years ahead. Employers will be more likely to hire workers who have at least a high school diploma. Furthermore, present experience shows that the less education and training workers have, the less chance they have for obtaining steady jobs.

In addition to its importance in competing for jobs, education makes a difference in lifetime income. According to the most recently available data, those who have college degrees can expect to earn: about two and three quarters (2 3/4) times the amount likely to be earned by workers who had less than 8 years of schooling; nearly twice the amount earned by workers who had 1 to 3 years of high school; and more than 1 1/2 times as much as high school graduates.

Clearly, the completion of high school pays a dividend. A worker who had only 1 to 3 years of high school could expect to earn 13 percent more than workers who had an elementary school education. However, a high school graduate could look forward to a 39 percent lifetime income advantage over an individual completing elementary school.

In summary, young people who have acquired skills and a good basic education will have a better chance for interesting work, good wages and steady employment. Therefore, getting as much education and training as one's abilities and circumstances permit should be a top priority for today's youth.

HOW TO PLAN A CAREER IN ELECTRONICS

The United States Department of Education (USDE) has identified 15 career clusters. These clusters are categories for classifying jobs. They include: Agri-Business and Natural Resources; Business and Office; Communications and Media; Construction; Consumer and Homemaking; Environment; Fine Arts and Humanities; Health; Hospitality and Recreation; Manufacturing; Marine Science; Marketing and Distribution; Personal Services; Public Services; Transportation. Careers in electronics cut across all 15 career clusters.

You should start as early as possible to plan your career. What you want to become has a definite relationship on what you need to do to prepare for that career. If electronics is your choice, make plans carefully for your career. Talk with your parents and friends about your hopes and plans. Try to meet people who work in the electronics field. Make use of the professional experience of industrial education and science teachers and counselors. Read as much as you can about the subject*. Your school library and public libraries contain many helpful references. Also see Fig. 21-4.

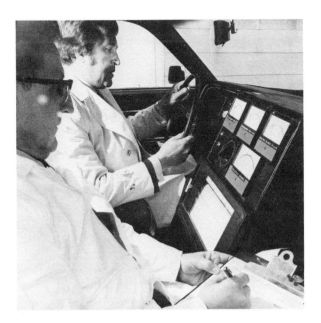

Fig. 21-3. Technicians in test car monitor the effects of high temperatures on engine components. Car is equipped with 24 channels of permanent readback data printed on strip recorder.

* Much of the information in Chapter 21 is from the Electronics Industries Association (EIA), 2001 Eye Street, N.W., Washington, DC 20006, and from the OCCUPATIONAL OUTLOOK HANDBOOK, U.S. Department of Labor, Bureau of Labor Statistics. The authors recommend that you order a free career guidance brochure from EIA titled "Futures Unlimited," which describes the electronics service technician careers. Also, it is recommended that you consult the OCCUPATIONAL OUTLOOK HANDBOOK for valuable career planning information.

EDUCATION/EXPERIENCE LEVEL

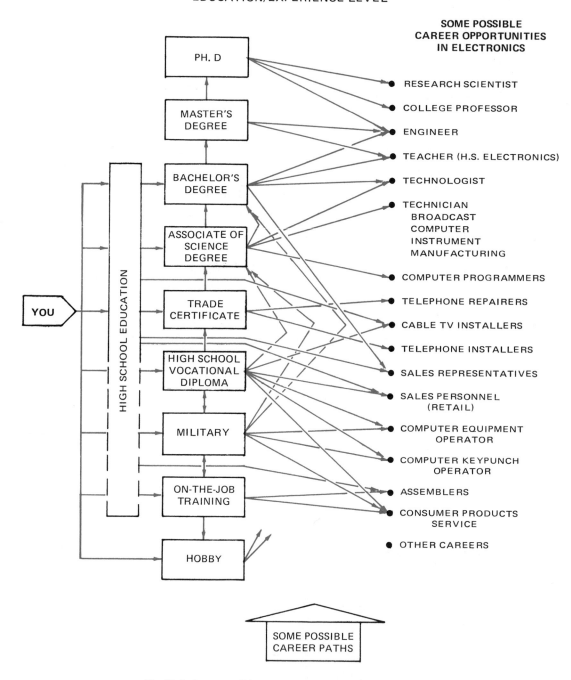

Fig. 21-4. Some possible career opportunities in electronics.

Contact your guidance counselor and instructors who will help you obtain further information about employment in the electronic service technician area. Additional information may be obtained from local dealers who sell and service consumer electronic products. Also contact local consumer electronic service associations, and manufacturers who operate their own service centers. Technical and vocational schools that offer courses in television and radio repair, or electronics, can also provide helpful information about training.

Colleges or universities usually have a part in "career nights" at local high schools each year. Talk to their representatives and, later, visit the college or university of your choice (if you plan to enter a career that requires this level of education).

Career Education is an important element in many schools today. Take advantage of special mini courses, career days or whatever is available in your school that may help you to better plan your career in electronics.

There are many effective things you can do while still in high school to prepare yourself for this career. You can learn to develop: orderly study habits; an inquiring, logical mind; and the ability to express yourself effectively. You should enroll in electricity and electronics courses, join your school science and radio club. Get interested in electronic kit-building (a vast variety of low cost project kits are available). Consider becoming an amateur radio operator. Perhaps there is no more rewarding hobby or better way to prepare for a career in electronics.

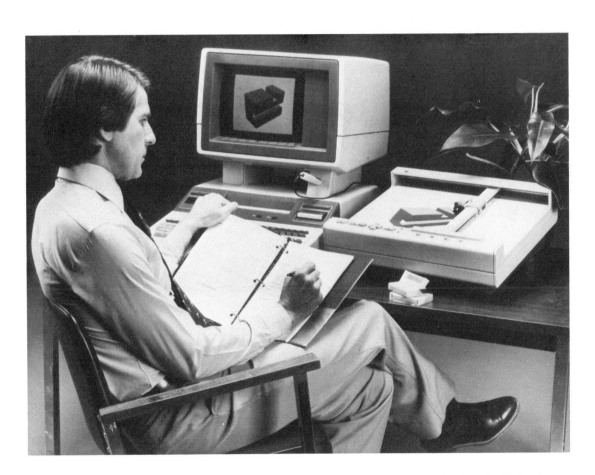

An engineer works with a microcomputer using a graphic plotter to display his research. (Hewlett-Packard)

REFERENCE SECTION

Appendix 1
SCIENTIFIC NOTATION, PREFIXES,
AND CONVERSIONS

In working with electronic equations, typical methods of arithmetic lead to many errors and the incorrect placement of decimal points. It is advisable to master the "power of ten" or scientific notation. In its more simple form, a large number is converted to a number between one and ten and multiplied by 10 at a certain power. Examples:

$$365 = 3.65 \times 10^2$$
$$3650 = 3.65 \times 10^3$$
$$36500 = 3.65 \times 10^4$$

Study these tables:

$$1 = 1 \times 10^0$$
$$10 = 1 \times 10^1 \qquad\qquad .1 = 1 \times 10^{-1}$$
$$100 = 1 \times 10^2 \qquad\qquad .01 = 1 \times 10^{-2}$$
$$1000 = 1 \times 10^3 \qquad\qquad .001 = 1 \times 10^{-3}$$
$$10000 = 1 \times 10^4 \qquad\qquad .0001 = 1 \times 10^{-4}$$
$$100000 = 1 \times 10^5 \qquad\qquad .00001 = 1 \times 10^{-5}$$
$$1000000 = 1 \therefore 10^6 \qquad\qquad .000001 = 1 \times 10^{-6}$$

When multiplying two or more numbers, it is only necessary to multiply the numbers and add the exponents of ten. Examples:

$$4 \times 10^3 \ \times 6 \times 10^5 = 24 \times 10^8 \qquad\qquad 3 \times 10^6 \ \times 3 \times 10^{-2} = 9 \times 10^4$$
$$8 \times 10^{-2} \times 6 \times 10^{-4} = 48 \times 10^{-6} \qquad\qquad 3 \times 10^{-6} \times 3 \times 10^2 = 9 \times 10^{-4}$$

When dividing numbers, the powers of ten in the denominator of a fraction can be moved to the numerator by changing the sign of the exponent. Also, the powers of ten can be moved from numerator to denominator by changing the sign of the exponent. Examples:

$$\frac{4 \times 10^2}{2 \times 10^{-2}} = \frac{4 \times 10^2 \times 10^2}{2} = 2 \times 10^4$$

$$\frac{4 \times 10^2 \times 2 \times 10^{-6}}{2 \times 10^{-3} \times 2 \times 10^3} = \frac{8 \times 10^2 \times 10^{-6} \times 10^3 \times 10^{-3}}{4} = 2 \times 10^{-4}$$

Most equations in electronics are written to use quantities in their fundamental units: current in amperes, volts in volts and ohms in ohms. If a term is given in other than its fundamental form, a conversion must be made. If you have milliamperes, it must be changed to amperes. If you have microfarads, it must be changed to farads.

Look at some common prefixes and find their meaning.

Pico	— one millionth of one millionth of basic unit	Pico $= \times 10^{-12}$
Micro	— one millionth of basic unit	Micro $= \times 10^{-6}$
Milli	— one thousandth of basic unit	Milli $= \times 10^{-3}$
Kilo	— one thousand times basic unit	Kilo $= \times 10^{3}$
Mega	— one million times basic unit	Mega $= \times 10^{6}$

This may be explained in another way by prefix chart in Fig. A1-1.

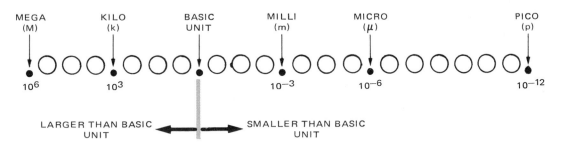

Fig. A1-1. Chart shows how prefixes of basic units are carried out by zeros in both directions.

To apply powers of ten to conversions, follow these examples.

$$
\begin{aligned}
1 \text{ Amp} &= 1 \times 10^{3} \text{ milliamps} = 1 \times 10^{6} \text{ microamps} \\
1 \text{ Milliamp} &= 1 \times 10^{-3} \text{ amps} = 1 \times 10^{3} \text{ microamps} \\
1 \text{ Microamp} &= 1 \times 10^{-3} \text{ milliamps} = 1 \times 10^{-6} \text{ amps}
\end{aligned}
$$

and

$$
\begin{aligned}
1 \text{ Ohm} &= 1 \times 10^{-3} \text{ kilohms} = 1 \times 10^{-6} \text{ megohms} \\
1 \text{ Kilohm} &= 1 \times 10^{3} \text{ ohms} = 1 \times 10^{-3} \text{ megohms} \\
1 \text{ Megohm} &= 1 \times 10^{6} \text{ ohms} = 1 \times 10^{3} \text{ kilohms}
\end{aligned}
$$

Summarizing these examples: To change

Unit to milli unit, $\times 10^{3}$	milli unit to unit, $\times 10^{-3}$
Unit to micro unit, $\times 10^{6}$	micro unit to unit, $\times 10^{-6}$
Unit to pico unit, $\times 10^{12}$	pico unit to unit, $\times 10^{-12}$
Milli unit to micro unit, $\times 10^{3}$	micro unit to milli unit, $\times 10^{-3}$
Milli unit to pico unit, $\times 10^{9}$	pico unit to milli unit, $\times 10^{-9}$
Unit to kilo unit, $\times 10^{-3}$	kilo unit to unit, $\times 10^{3}$
Unit to mega unit, $\times 10^{-6}$	mega unit to unit, $\times 10^{6}$
Kilo unit to mega unit, $\times 10^{-3}$	mega unit to kilo unit, $\times 10^{3}$

It is better to "think through" your conversion problems and solve by understanding. Do not try to memorize this table. To avoid errors when changing a number to "powers of ten," it is wise to change the number first, then multiply by the correct power of ten.

For example: Change 319 pF to farads:

Step 1 3.19×10^{2} pF
Step 2 $3.19 \times 10^{2} \times 10^{-12}$
Step 3 3.19×10^{-10} farads

Change .001 μF to farads:

Step 1 $.001 = 1 \times 10^{-3} \mu$F
Step 2 $1 \times 10^{-3} \times 10^{-6}$ farads
Step 3 1×10^{-9} farads

Appendix 2
SIGNS AND SYMBOLS

One of the more difficult tasks the beginner in electronics must overcome is the memory and understanding of the many signs, symbols and subscripts used to identify components and parameters of circuits. There has been an attempt to standardize these symbols, but there still exists much confusion among manufacturers and users of components. The following symbols are IEEE (Institute of Electrical and Electronic Engineers) approved.

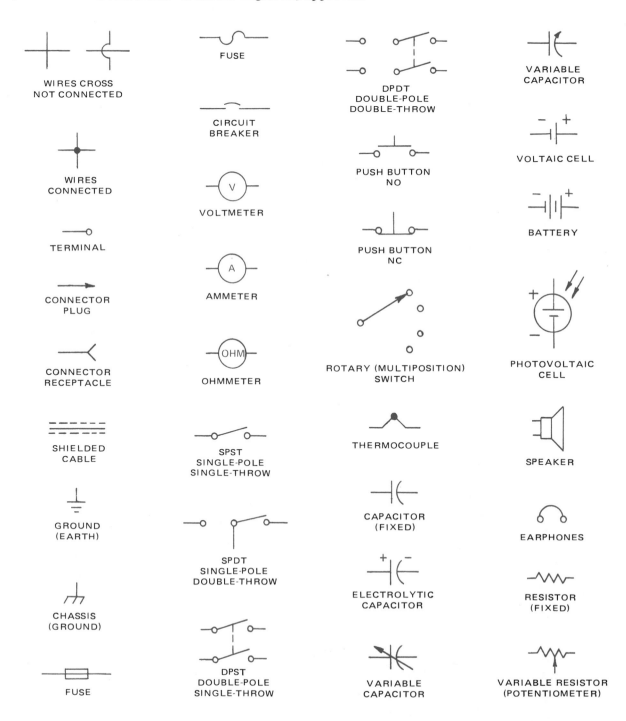

VARIABLE RESISTOR
(RHEOSTAT)

TAPPED
RESISTOR

PIEZOELECTRIC
CRYSTAL

MOTOR,
DC

MOTOR,
AC

GENERATOR
(DC)

GENERATOR
(AC)

TRANSFORMER
AIR CORE

TRANSFORMER
IRON CORE

COIL OR
RF CHOKE
(INDUCTOR)

COIL (INDUCTOR)
IRON CORE

COIL (INDUCTOR)
(VARIABLE)

INCANDESCENT
LAMP

SIGNAL
LAMP

NEON
LAMP

TRANSISTOR
NPN

TRANSISTOR
PNP

FIELD EFFECT
TRANSISTOR
(FET)
(N - CHANNEL)

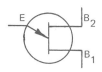

FIELD EFFECT
TRANSISTOR
(FET)
(P - CHANNEL)

UNIJUNCTION
TRANSISTOR
(UJT)
WITH N BASE

UNIJUNCTION
TRANSISTOR
(UJT)
WITH P BASE

SINGLE
ENDED
AMPLIFIER

PHOTOTRANSISTOR
(NPN)

AND GATE

NAND GATE

OR GATE

NOR GATE

INVERTER (NOT)
GATE

FLIP FLOP
(D TYPE)

DIODE

ZENER DIODE

SILICON
CONTROLLED
RECTIFIER
SCR

TRIAC

LIGHT EMITTING
DIODE
(LED)

STANDARD ABBREVIATIONS
AND SYMBOLS

TRANSISTOR SYMBOLS

I_C — collector current
I_E — emitter current
I_B — base current
V_{CC} — supply voltage
V_C — collector voltage
V_E — emitter voltage
V_B — base voltage
V_{CE} — collector-emitter voltage
V_{CB} — collector-base voltage
V_{EB} — emitter-base voltage
I_{CBO} — saturation current, reverse leakage current
h_{fe} — common emitter current gain
h_{fb} — common base current gain
P_C — collector power dissipation
A_p — power gain
A_i — current gain
A_V — voltage gain
Z_i — input impedance
Z_o — output impedance

LETTER SYMBOLS

C — capacitance
L — inductance
R — resistance
E or V — voltage, potential difference, EMF
I — current
X — reactance
X_C — capacitive reactance
X_L — inductive reactance
G — conductivity in mhos (siemens)
M — mutual inductance
Z — impedance
Q — operating point, quality factor, transistor
F — magnetic force
H — magnetizing force
B — flux density
CEMF — counter electromotive force
MMF — magnetomotive force
A — gain
S — sensitivity, stabilization factor
K — relay
rms — root-mean-square, effective value
T — transformer

GREEK SYMBOLS

μ — mu amplification factor, permeability, micro
Φ — Phi magnetic flux
$\propto$ — alpha current gain, CB transistor
β — beta current gain, CE transistor
λ — lambda wave length
ω — omega angular velocity
θ — theta phase angle
π — Pi 3.1416
Δ — Delta a change of
p — rho resistivity
Ω — Omega ohms
η — eta efficiency

MATHEMATICAL SYMBOLS

x multiplied by
÷ divided by
+ positive, add
— negative, subtract
= equal to
≠ does not equal
≅ approximately equal to
∠ angle
± plus or minus
$\sqrt{}$ square root
^{2} square
∞ infinity
Δ increment or decrement

Appendix 3
COLOR CODES

COLOR	NUMERICAL FIGURE	DECIMAL MULTIPLIER		VALUE TOLERANCE
		Power of 10	Multiplier Value	
Blk	0	10^0	1	
Brn	1	10^1	10	
Red	2	10^2	100	
Orn	3	10^3	1,000	
Yel	4	10^4	10,000	
Grn	5	10^5	100,000	
Blu	6	10^6	1 Million	
Vio	7	10^7	10 Million	
Gra	8	10^{-2} (alternate)	0.01	
Wht	9	10^{-1} (alternate)	0.1	
Sil	–	10^{-2} (preferred)	0.01	±10%
Gld	–	10^{-1} (preferred)	0.1	±5%
None	–	—	—	±20%

Fig. A3-1. Color code for resistors and capacitors.

The tendency for manufacturers to mark the value on the component is becoming increasingly popular. However, the color code for the determination of values of resistors and capacitors is still widely used. This color code is displayed in Fig. A3-1, and is supplied through the courtesy of the Electronics Industry Association (EIA).

HOW TO READ THE CODE

Secure several resistors from stock and study them. You will find that they have four and sometimes only three colored bands around them near one end of the resistor. Hold a resistor in your hand so that the colored bands are on the left, Fig. A3-2.

Assume that the bands from left to right are colored BROWN, BLACK, GREEN, SILVER. Consult the table and you will find that Brown means ONE. This is the first number of the value. The second band, Black, is ZERO. This is the second number of the value. The third band, Green, is the multiplier and means you should multiply the first two significant figures by ten to the fifth power. The fourth band of SILVER means that its tolerance is within 10 percent of its indicated value. This is summarized as shown in Fig. A3-2.

If the fourth band were gold, it would be a 5 percent resistor. If the fourth band is missing, it is a 20 percent resistor.

Try this one. You have a three band resistor which is Red, Violet, Gold. Its value is:

$$27 \times 10^{-1} \text{ or } 2.7 \text{ ohms}$$

Practice with several assorted resistors. You should MEMORIZE the color code.

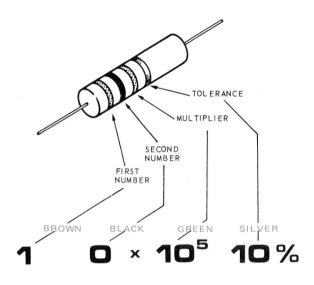

Fig. A3-2. Reading the color code from left to right.

CAPACITORS

Most capacitors made today have numerical markings, so codes are unnecessary. There is one variety, the MICA capacitor, which you will find with colored dots to indicate its value. The sketch in Fig. A3-3 shows the six-dot system.

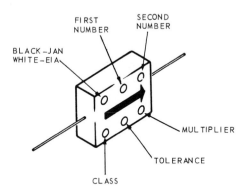

Fig. A3-3. Colored dots make up color code of MICA capacitors. To read code, arrow must point to right.

On the face of the capacitor will be an arrow or the name of the manufacturer. Hold the capacitor so that the arrow points to the right, or so that you can read the manufacturer's name.

    FIRST DOT: Code to use
        White - EIA
        Black - Joint-Army-Navy- JAN
    SECOND DOT: First figure of value
    THIRD DOT: Second figure of value
    LOWER RIGHT DOT: Multiplier

The other dots will give the tolerance or classification of the capacitor. The value of the capacitor will be in picofarads (pF).

DIODES AND RECTIFIERS

Color codes are also used to determine type of a semiconductor diode. In all cases, the first number and the letter N will not be coded. Only the number of the diode is coded.

The CATHODE end of the diode will be identified by a double width first band of color. Or, the color bands will be equal in width, but grouped at the cathode end of the diode. The code shall read from cathode to anode. See Fig. A3-4.

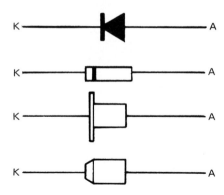

Fig. A3-4. Color code of semiconductors involves use of color bands to be read from K to A.

Appendix 4
TRIGONOMETRY

Trigonometry will make your electronic computation easier, more accurate and enjoyable. It is not difficult to learn.

Trigonometry is based on the ratios between the sides of a right triangle with reference to either of the two acute angles. We are only concerned with three of these ratios, and they are expressed as:

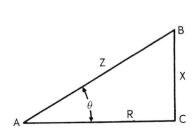

$$\text{Sine} \angle \theta = \frac{BC}{AB} \text{ or } \frac{X}{Z}$$

$$\text{Cos} \angle \theta = \frac{AC}{AB} \text{ or } \frac{R}{Z}$$

$$\text{Tan} \angle \theta = \frac{BC}{AC} \text{ or } \frac{X}{R}$$

Note that X will represent reactance in ohms, R equals resistance in ohms and Z equals impedance in ohms.

If any two values in any of the above equations are known, then the unknown may be found.

EXAMPLE 1. X = 500 Ω and R = 500 Ω. Find angle θ . Use the equation which contains the two known values. This is the tangent function.

$$\text{Tan} \angle \theta = \frac{X}{R} = \frac{500}{500} = 1$$

Consult the "trig" tables under the tangent column and find the degrees of the angle whose tangent is 1. θ = 45 deg.

EXAMPLE 2. Find the value of Z in problem 1. We could use either the sine or cosine function. Using the cosine,

$$\text{Cos} \angle \theta = \frac{R}{Z} \text{ and } Z = \frac{R}{\cos \theta}$$

Find the cosine of angle 45 deg. in the table. It is .707.

$$Z = \frac{500}{.707} = 707 \text{ ohms}$$

EXAMPLE 3. The impedance Z of a circuit is 1000 Ω and the angle θ = 30 deg. Find values of R and X.

$$\text{Sin} \angle \theta = \frac{X}{Z} \qquad X = \text{Sin} \angle \theta \times Z$$

$$\text{Sin} \angle 30 \text{ deg.} = .5 \qquad X = .5 \times 1000 \,\Omega = 500 \,\Omega \,, \ \text{Cos} \angle 30 \text{ deg.} = .8660$$

$$\text{Cos} \angle \theta = \frac{R}{Z} \qquad R = .8660 \times 1000 \,\Omega = 866 \text{ ohms}$$

Proof: Tan $\angle$ 30 deg. $= \dfrac{X}{R} = \dfrac{500}{866} = .5774$

Consult the "trig" tables. Tan $\angle$ 30deg. = .5774

EXAMPLE 4. The peak voltage of a sine wave is 10 volts. The instantaneous voltage at any point during the cycle is: e = Sin $\angle \theta$ x E_{peak} . What is the voltage at 70 deg.?

Sin$\angle$70 deg. = .9397

e = .9397 x 10 volts = 9.397 volts

NATURAL TRIGONOMETRIC FUNCTIONS

Angle	Sine	Cosine	Tangent	Angle	Sine	Cosine	Tangent
1°	.0175	.9998	.0175	46°	.7193	.6947	1.0355
2°	.0349	.9994	.0349	47°	.7314	.6820	1.0724
3°	.0523	.9986	.0524	48°	.7431	.6691	1.1106
4°	.0698	.9976	.0699	49°	.7547	.6561	1.1504
5°	.0872	.9962	.0875	50°	.7660	.6428	1.1918
6°	.1045	.9945	.1051	51°	.7771	.6293	1.2349
7°	.1219	.9925	.1228	52°	.7880	.6157	1.2799
8°	.1392	.9903	.1405	53°	.7986	.6018	1.3270
9°	.1564	.9877	.1584	54°	.8090	.5878	1.3764
10°	.1736	.9848	.1763	55°	.8192	.5736	1.4281
11°	.1908	.9816	.1944	56°	.8290	.5592	1.4826
12°	.2079	.9781	.2126	57°	.8387	.5446	1.5399
13°	.2250	.9744	.2309	58°	.8480	.5299	1.6003
14°	.2419	.9703	.2493	59°	.8572	.5150	1.6643
15°	.2588	.9659	.2679	60°	.8660	.5000	1.7321
16°	.2756	.9613	.2867	61°	.8746	.4848	1.8040
17°	.2924	.9563	.3057	62°	.8829	.4695	1.8807
18°	.3090	.9511	.3249	63°	.8910	.4540	1.9626
19°	.3256	.9455	.3443	64°	.8988	.4384	2.0503
20°	.3420	.9397	.3640	65°	.9063	.4226	2.1445
21°	.3584	.9336	.3839	66°	.9135	.4067	2.2460
22°	.3746	.9272	.4040	67°	.9205	.3907	2.3559
23°	.3907	.9205	.4245	68°	.9272	.3746	2.4751
24°	.4067	.9135	.4452	69°	.9336	.3584	2.6051
25°	.4226	.9063	.4663	70°	.9397	.3420	2.7475
26°	.4384	.8988	.4877	71°	.9455	.3256	2.9042
27°	.4540	.8910	.5095	72°	.9511	.3090	3.0777
28°	.4695	.8829	.5317	73°	.9563	.2924	3.2709
29°	.4848	.8746	.5543	74°	.9613	.2756	3.4874
30°	.5000	.8660	.5774	75°	.9659	.2588	3.7321
31°	.5150	.8572	.6009	76°	.9703	.2419	4.0108
32°	.5299	.8480	.6249	77°	.9744	.2250	4.3315
33°	.5446	.8387	.6494	78°	.9781	.2079	4.7046
34°	.5592	.8290	.6745	79°	.9816	.1908	5.1446
35°	.5736	.8192	.7002	80°	.9848	.1736	5.6713
36°	.5878	.8090	.7265	81°	.9877	.1564	6.3138
37°	.6018	.7986	.7536	82°	.9903	.1392	7.1154
38°	.6157	.7880	.7813	83°	.9925	.1219	8.1443
39°	.6293	.7771	.8098	84°	.9945	.1045	9.5144
40°	.6428	.7660	.8391	85°	.9962	.0872	11.4301
41°	.6561	.7547	.8693	86°	.9976	.0698	14.3006
42°	.6691	.7431	.9004	87°	.9986	.0523	19.0811
43°	.6820	.7314	.9325	88°	.9994	.0349	28.6363
44°	.6947	.7193	.9657	89°	.9998	.0175	57.2900
45°	.7071	.7071	1.0000	90°	1.0000	.0000	

Appendix 5
ETCHED CIRCUITS

Many projects in the text suggest the use of etched circuits. You should know how to make them. The older method of connecting components together by means of wire is replaced by copper conductive patterns etched on thin copper clad plastic board.

There are three general methods of making etched circuits:

1. The copper clad board is covered with a thin coating of light-sensitive photo chemical, similar to Kodak Photoresist. Then, the board is exposed to ultraviolet light, through a photographic mask. The board is developed in an etching bath that eats away the undesired parts of the thin copper, leaving only the conductive paths.
2. The art or circuit design may be actually printed on the copper clad board with acid resisting ink. A silk screen process is used. After the ink is dry, the board is etched in the usual way.
3. The conductive circuits may be painted on the copper clad board. After drying, the board is etched to remove the unwanted copper and leave only the painted circuit.

MAKING AN ETCHED CIRCUIT BOARD

1. Study the schematic diagram of the circuit, and place the components on a sheet of tracing paper. Use a colored pencil to carefully outline each resistor, capacitor, transistor, etc., in its proper place. Remove parts and label each outlined part with its component number such as R_3, C_1, etc. Using a pencil of another color, draw lines representing the wire connections between components. Check and recheck with the schematic drawing.
2. Carefully bend all leads of components downward, and determine the exact spacing between leads. At the point where each lead will be mounted, place a small circle or

pad about 1/8 to 1/4 in. in diameter. Connect the pads together according to your outline with a heavy black line about 1/8 in. wide. Make a neat drawing, using a ruler and appropriate curves.

NOTE: The size of the pads and width of conductive paths will depend on the circuit to be made. Study the photo in Fig. A5-1. Avoid crossover conductive paths. A rearrangement of components may eliminate crossovers.

3. Cut a piece of copper clad board the size of your circuit design. Using a 4H pencil and carbon paper, transfer the outline of the circuit to the copper clad. With a pointed pencil, mark all centers of holes to be drilled.
4. Drill all holes for component leads with a No. 40 drill. Enlarge with slightly larger drills if necessary. Remove all burrs.
5. Paint the circuit path, very neatly, with an

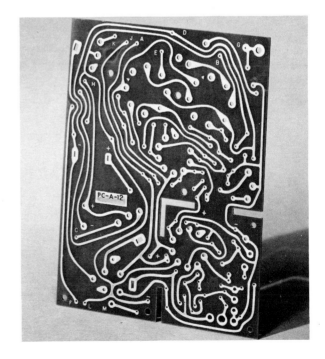

Fig. A5-1. A completed circuit board. Note the pads and conductive paths.

acid-resist paint. Use a No. 2 artist brush or special pen with acid-resist ink. Allow time to dry.

6. Etch the board in a solution of 1.41 specific gravity Ferric-Chloride ($FeCl_3$) for one to two hours. Do not leave in solution too long. Stir and turn the board frequently with an old pair of tweezers or tongs.

NOTE: Some shops warm the solution to hasten the etching.

7. Remove board, inspect and wash with clear water.
8. Clean off resist paint with rag and lacquer thinner or kerosene.
9. Replace pattern and mark all wire holes.

Drill with a No. 54 wire drill. Do not use center punch to mark holes.
10. Mount parts, solder and clip off excess leads. Do not use excessive heat, but avoid cold solder joints.

Throughout this text, suggested circuit board layouts are supplied. You are encouraged to improve upon them. When mounting components (such as a tuning capacitor or a potentiometer) on panels, attach wires to board at a pad and run directly to the component.

LESSON IN SAFETY: Keep ETCHANT from your clothes and skin. Etch in a shallow glass dish, with proper ventilation.

Appendix 6
ELECTRONIC PROJECT LAYOUT

In this book, there are many excellent projects for the student or experimenter. The appealing design of the project depends on careful planning and the wise use of many materials. The following suggestions will help you in the fabrication and design of your project.

1. Many different chassis designs are available already built through electronic hobby stores and catalogs. However, you may wish to design your own chassis and fabricate it with sheet metal equipment.
2. Chassis may be painted or covered with self-adhesive vinyl contact paper.
3. Rub-on decals can be purchased from local hobby stores or office supply stores. In the case of a mistake, use transparent tape to "lift" the decal from the surface.
4. Certain areas of the chassis can be highlighted by the use of thin chart tape. These come in solid colors and black, as well as striped designs. See Fig. A6-1.
5. Attractive knobs may be purchased to com-

Fig. A6-1. Some chart and graphic art tape styles.

pliment your chassis design.
6. Perforated metal or phenolic board may be used to cover a speaker where sound must exit.
7. Adhesive felt or rubber feet help keep the chassis from scratching surfaces. Also screw-on feet are available for larger chassis.
8. Always use grommets where wires come through the chassis.
9. Good tools are important in the construction and fabrication of electronic projects. Always use the correct tool for the job at hand.

ACKNOWLEDGMENTS

The preparation of a text of this nature is necessarily the result of many people in the field of electronics sharing their knowledge and experience. To those who have assisted us, we express sincere appreciation and hope that together we have made some significant contribution toward better education in schools.

1. 1969 Edition:
   We personally wish to thank Mr. George Parker, Mr. Roy Brixen, Mr. Wallace Britten and Mr. Douglas Jackson, graduate technicians and prospective industrial educational teachers at San Jose State College for their able assistance in the development and construction of the highly motivating projects contained in this text, and Virtue B. Gerrish (wife of Howard Gerrish) for typing the original manuscript. We also wish to recognize the advice and editorial assistance of the Buck Engineering Co., Lab-Volt Div., Farmingdale, New Jersey. The illustrations in the 1969 editions were done by Gary A. Spear, Mountain View, California.

2. 1979 and 1981 Editions:
   Specifically, we would like to thank Drs. Alan F. Clifford and John S. Roberts of the Virginia Polytechnic Institute and State University Chemistry Department for their help in semiconductor and theory of energy topics. Also special appreciation goes to Bruce Boal and Jack Moore of Hickok Teaching Systems for their assistance with selected projects and photographs.

Mr. Nicholar J. Carcich and Lab-Volt Systems, Division of Buck Engineering gave excellent support in supplying some photographs for this edition.

Thanks also goes to Carl Giegold of Blacksburg, Virginia, for certain project photographs, and to Tom Klare of the LRC Photo Lab at VPI and SU.

Typing was done by Carol Keeton and Teresa Greene.

Much sacrifice and understanding by the family of the author takes place during the writing period. For this, we would like to express appreciation to Carrie Dugger, wife of the co-author, and to his children, Edward, Cammie and Toy. Edward helped his father in the construction of many of the new projects in the 1979 edition.

Technical books are the result of many contributors, and we wish to thank these industries or associations for generously supplying illustrative material and technical information.

Allen-Bradley Co.
Atari
B.I.C. (British Industries Co.)
B & K Div. of Dynascan Corp.
Bally Mfg. Co.
Beckman Instruments Co.
Bell Telephone Co.
Bendix Research Laboratories
Bose Corp.
Buck Engineering, Lab-Volt Div.
Burgess Battery Co.
Centralab, Globe Union, Inc.
Cobra Div. of Dynascan Corp.
Continental Specialities Corp.
Delco Products, General Motors Corp.
Delta Products, Inc.
Digi-Key Corp.
Digital Equipment Corp.
Dynascan Corp.
Dynatech Corp.
The Electric Storage Battery Co.
Electronic Industries Association
Electronics Instruments Co., Inc.
Fenwall Electronics, Inc.
General Electric Co.
W.W. Grainger, Inc.
Graymark International, Inc.
Harris Semiconductor
Heathkit Corp.
Hewlett-Packard
Hickok Teaching Systems
International Rectifier Corp.
IRC, Inc.
E.F. Johnson, Inc.
Koss Corp.
KRACO
Lab-Volt
Lockheed Electronics Co.

Lockheed Missiles & Space Co.
Mattel Electronics
Microma
James Millen Co.
J.W. Miller Co.
Motorola
Motorola Semiconductor Products, Inc.
National Aeronautics & Space Administration
National Semiconductor Corp.
Ohmite Mfg. Co.
Pacific Telephone Co.
Panasonic
Philco-Ford Corp.
Philco Technical Institute
Potter and Brumfield
Radio Corp. of America
Radio Shack
Raytheon Co.
RCA
Sargent-Welch Scientific Co.
Seiko Time Corp.
Shure Bros., Inc.
Siliconix, Inc.
Sprague Products Co.
Stanford Linear Accelerator
L.S. Starrett Co.
Texas Crystals
Texas Instruments
The Triad Transformer Co.
Triplett
The Union Carbide Consumer Products Co.
The United Transformer Co.
U.S. Dept. of Energy
U.S. Pioneer Electronics Corp.
Westinghouse Electric Co.
Westinghouse Semiconductors
Weston Instruments, Inc.

DICTIONARY OF TERMS

AC: Alternating current.

ACCEPTOR CIRCUIT: Series tuned circuit at resonance. Accepts signals at resonant frequency.

ACCEPTOR IMPURITY: Impurity added to semiconductor material which creates holes for current carriers.

AC GENERATOR (alternator): Generator using slip rings and brushes to connect armature to external circuit. Output is alternating current.

AC PLATE RESISTANCE (symbol r_p): Variational characteristic of vacuum tube representing ratio of change of plate voltage to change in plate current, while grid voltage is constant.

ACTIVE DEVICE: An electronic component that can change its response to an external signal. Some typical active devices are transistors, diodes, ICs and tubes.

ADDRESS: The exact location of a word in the computer memory.

AGC: Abbreviation for AUTOMATIC GAIN CONTROL.

AIR-CORE INDUCTOR: Inductor wound on insulated form without metallic core. Self-supporting coil without core.

ALIGNMENT: The adjustment of tuned circuits in amplifier and/ or oscillator circuits so that they will produce a specified response at a given frequency.

ALNICO: Special alloy used to make small permanent magnets.

ALPHA: Greek letter ∝, represents current gain of a transistor. It is equal to change in collector current caused by change in emitter current for constant collector voltage.

ALPHA CUT-OFF FREQUENCY: Frequency at which current gain drops to .707 of its maximum gain.

ALTERNATING CURRENT (ac): Current of electrons that moves first in one direction and then in the other.

ALTERNATOR: An ac generator.

AMMETER: Meter used to measure current.

AMPERE (symbol I): Electron or current flow representing the flow of one coulomb per second past a given point in a circuit.

AMPERE-HOUR: Capacity rating measurement of batteries. A 100 ampere-hour battery will produce, theoretically, 100 amperes for one hour.

AMPERE TURN (IN): Unit of measurement of magnetomotive force. Represents product of amperes times number of turns in coil of electromagnet. F = 1.257 IN.

AMPLIFICATION: Ability to control a relatively large force by a small force. In a vacuum tube, relatively small variation in grid input signal is accompanied by relatively large variation in output signal.

AMPLIFICATION FACTOR: Expressed as μ (mu). Characteristic of vacuum tube to amplify a voltage. Mu is equal to change in plate voltage as result of change in grid voltage while plate current is constant.

AMPLIFIERS:

POWER. Electron tube used to increase power output. Sometimes called a current amplifier.

VOLTAGE. An electron tube used to amplify a voltage.

CLASS A. An amplifier biased so that plate current flows during entire cycle of input signal.

CLASS B. An amplifier biased so that plate current flows for approximately one half the cycle of input signal.

CLASS C. An amplifier biased so that plate current flows for appreciably less than half of each cycle of applied signal.

CLASS AB. A compromise between class A and class B.

DC AMPLIFIER. Directly coupled amplifiers. Amplifies without loss of dc component.

AF AMPLIFIER. Used to amplify audio frequencies.

IF AMPLIFIER. Used to amplify intermediate frequencies.

RF AMPLIFIER. Used to amplify radio frequencies.

AMPLITUDE: Extreme range of varying quantity. Size, height of.

AMPLITUDE MODULATION (AM): Modulating a transmitter by varying strength of rf carrier at audio rate.

AND CIRCUIT: A circuit with two or more inputs and all inputs must be present to produce an output signal.

ANGLE OF LEAD OR LAG: The angle between voltage and current vectors describing an ac circuit expressed in degrees. Example: In a capacitive circuit, the current leads the voltage by a certain angle.

ANGULAR PHASE: Position of rotating vector in respect to reference line.

ANGULAR VELOCITY (ω): Speed of rotating vector in radians per second. ω (omega) = $2\pi f$.

ANODE: Positive terminal, such as plate in electron tube.

ANTENNA: Device for radiating or receiving radio waves.

APPARENT POWER: Power apparently used in circuit as product of current times voltage.

ARITHMETIC AND LOGIC UNIT: A computer unit that contains such devices as adders and registers.

ARMATURE: Revolving part in generator or motor. Vibrating or moving part of relay or buzzer.

ASYMMETRICAL: A multivibrator producing unequal waves from each tube due to unequal conduction periods.

AT-CUT CRYSTAL: Crystal cut at approximately a 35 deg. angle with Z axis.

ATOM: Smallest particle that makes up a type of material called an element.

ATOMIC NUMBER: Number of protons in nucleus of a given atom.

ATOMIC WEIGHT: Mass of nucleus of atom in reference to oxygen, which has a weight of 16.

ATTENUATION: Decrease in amplitude or intensity.

AUTOMATIC GAIN CONTROL (AGC): Circuit employed to vary gain of amplifier in proportion to input signal strength so output remains at a constant level.

AUTOMATION: An industrial technique of employing automatic self-controlled machinery to replace human labor and control.

AVC: Automatic volume control.

AVERAGE VALUE: Value of alternating current or voltage of sine wave form that is found by dividing area under one alternation by distance along X axis between 0 and 180 deg. E_{avg} = .637 E_{max}.

AWG: American Wire Gauge – used in sizing wire by numbers.

BAND: Group of adjacent frequencies in frequency spectrum.

BAND PASS FILTER: Filter circuit designed to pass currents of frequencies within continuous band and reject or attenuate frequencies above or below the band.

BAND REJECT FILTER: Filter circuit designed to reject currents in continuous band of frequencies, but pass frequencies above or below the band.

BAND SWITCHING: Receiver employing switch to change frequency range of reception.

BANDWIDTH: Band of frequencies allowed for transmitting modulated signal.

BARRIER REGION: Potential difference across a PN junction due to diffusion of electrons and holes across junction.

BASE: The thin section between the emitter and collector of a transistor.

BATTERY: Several voltaic cells connected in series or parallel. Usually contained in one case.

BEAT FREQUENCY: The resultant frequency obtained by combining two frequencies.

BEAT FREQUENCY OSCILLATOR: Oscillator whose output is beat with continuous wave to produce beat frequency in audio range. Used in CW reception.

BEL: Unit of measurement of gain equivalent to 10 to 1 ratio of power gain.

BETA: Greek letter β, represents current gain of common-emitter connected transistor. It is equal to ratio of change in collector current to change in base current, while collector voltage is constant.

BIAS:

CATHODE SELF-BIAS. Bias created by voltage drop across cathode resistor.

FIXED. Voltage supplied by fixed source.

GRID LEAK. Bias created by charging capacitor in grid circuit. Bias level is maintained by leak resistor.

BIAS, FORWARD: Connection of potential to produce current across PN junction. Source potential connected so it opposes potential hill and reduces it.

BIAS, REVERSE: Connection of potential so little or no current will flow across PN junction. Source potential has same polarity as potential hill and adds to it.

BINARY: Number system having base of 2, using only the symbols 0 and 1.

BIT: Binary digit.

BLACK BOX: Box containing unknown and possibly complicated circuit.

BLEEDER: Resistor connected across power supply to discharge filter capacitors.

BRIDGE CIRCUIT: Circuit with series-parallel groups of components that are connected by common bridge. Bridge is frequently a meter in measuring devices.

BRIDGE RECTIFIER: Full-wave rectifier circuit employing four rectifiers in bridge configuration.

BRIDGE, WHEATSTONE: A bridge circuit for determining the value of an unknown component by comparison to one of known value.

BRUSH: Sliding contact, usually carbon, between commutator and external circuit in dc generator.

B-SUPPLY: Voltages supplied for plate circuits of electron tubes.

BYPASS CAPACITOR: Fixed capacitor which bypasses unwanted ac to ground.

CABLE: May be stranded conductor or group of single conductors insulated from each other.

CAPACITANCE: Inherent property of electric circuit that opposes change in voltage. Property of circuit whereby energy may be stored in electrostatic field.

CAPACITANCE, DISTRIBUTED: The capacitance in a circuit resulting from adjacent turns on coils, parallel leads and connections.

CAPACITIVE COUPLING: Coupling resulting from capacitive effect between components or elements in electron tube.

CAPACITIVE REACTANCE (X_c): Opposition to ac as a result of capacitance.

CAPACITOR: Device which possesses capacitance. Simple capacitor consists of two metal plates separated by insulator.

CAPACITOR INPUT FILTER: Filter employing capacitor as its input.

CAPACITY: Ability of battery to produce current over given length of time. Capacity is measured in ampere-hours.

CARRIER: Usually radio frequency continuous wave to which modulation is applied. Frequency of transmitting station.

CARRIER (in a semiconductor); Conducting hole or electron.

CASCADE: Arrangement of amplifiers where output of one stage becomes input of next, throughout series of stages.

CATHODE: Emitter in electron tube.

CATHODE RAY TUBE: Vacuum tube in which electrons emitted from cathode are shaped into narrow beam and accelerated to high velocity before striking phosphor-coated viewing screen.

CENTER FREQUENCY: Frequency of transmitted carrier wave in FM when no modulation is applied.

CENTER TAP: Connection made to center of coil.

CHARACTERISTIC CURVE: Graphical representation of characteristics of component, circuit or device.

CHIP: A very small piece of silicon containing transistors and possibly diodes, resistors or capacitors.

CHOKE COIL: A high inductance coil used to prevent the passage of pulsating currents, but allows dc to pass.

CHOKE, RF: A choke coil with a high impedance at radio frequencies.

CIRCUIT BREAKER: Safety device which automatically opens circuit if overloaded.

CIRCUIT, ETCHED: The method of circuit board production in which the actual conduction paths on a copper-clad insulation board are coated with an acid resist. The board is then placed in an acid bath and unprotected parts of the copper clad are eaten away, leaving the circuit conductors. Components are mounted and soldered between the conductors to form the completed circuit.

CIRCUIT INPUT FILTER: Filter employing choke as its input.

CIRCUIT, PLUG-IN: A total or part of a circuit, usually a printed circuit board, which can be plugged into a piece of equipment. It may be rapidly removed or replaced during service.

CIRCUIT, PRINTED: The method of printing circuit conductors on an insulated base. Component parts may also be printed or actual components soldered in place.

CIRCULAR MIL: Cross-sectional area of conductor one mil in diameter.

CIRCULAR MIL FOOT: Unit conductor one foot long with cross-sectional area of one circular mil.

CIRCULATING CURRENT: Inductive and capacitive currents flowing around parallel circuit.

CITIZENS' BAND (CB): A band of frequencies allotted to two-way radio communications by private citizens. Operators are not required to pass technical examinations.

COEFFICIENT OF COUPLING (K): Percentage of coupling between coils, expressed as a decimal.

COLLECTOR: In a transistor, the semiconductor section which collects the majority carriers. Similar to the plate in a vacuum tube.

COMMON BASE: Transistor circuit in which base is common to input and output circuits.

COMMON COLLECTOR: Transistor circuit in which collector is common to input and output circuits.

COMMON EMITTER: Transistor circuit in which emitter is common to input and output circuits.

COMMUTATION: The process of changing the alternating current in a generator armature into direct current in the external circuit by a mechanical switch consisting of commutator bars and brushes.

COMMUTATOR: Group of bars providing connections between armature coils and brushes. Mechanical switch to maintain current in one direction in external circuit.

COMPUTER:

ANALOG. A computer which substitutes for any given physical quantity, a mechanical, electrical thermodynamic equivalent quantity that follows in direct proportion, the same laws of behavior as the original quantity. In general, the analog computer gives a continuous solution of the problem. Examples: Slide rule, automobile speedometer.

DIGITAL. A computer switch makes a one-to-one comparison or individual count to calculate. It solves problems in discrete steps, forming a discontinuous solution.

COMPUTER LANGUAGE: Actual set of representations or instructions the computer is fed in order to make it do certain things.

CONDUCTANCE (symbol G): Ability of circuit to conduct current. It is equal to amperes per volt and is measured in siemens (mhos). $G = \dfrac{1}{R}$.

CONDUCTION BAND: Outermost energy level of atom.

CONDUCTIVITY: The ability of a material to conduct an electric current. It is the reciprocal of resistivity.

CONDUCTIVITY, N TYPE: Conduction by electrons in N-type crystal.

CONDUCTIVITY, P TYPE: Conduction by holes in a P-type crystal.

CONDUCTOR: Material which permits free motion of large number of electrons.

CONTINUOUS WAVE (CW): Uninterrupted sinusoidal rf wave radiated into space, with all wave peaks equal in amplitude and evenly spaced along time axis.

CONTROL GRID: Grid in vacuum tube closest to cathode. Grid to which input signal is fed to tube.

COORDINATES: Horizontal and vertical distances to locate point on graph.

COPPER LOSSES: Heat losses in motors, generators and transformers as result of resistance of wire. Sometimes called the I^2R loss.

CORE IRON: Magnetic materials usually in sheet form, used to form laminated cores for electromagnets and transformers.

CORE SATURATION: The tendency of the molecules in an iron core to be oriented in one direction due to direct current.

COULOMB: Quantity representing 6.24×10^{18} electrons.

COUNTER EMF (CEMF): Voltage induced in conductor moving through magnetic field which opposes source voltage.

COUPLING: Percentage of mutual inductance between coils. LINKAGE.

COVALENT BOND: Atoms joined together, sharing each other's electrons to form stable molecule.

CROSSOVER FREQUENCY: Frequency in crossover network at which equal amount of energy is delivered to each of two speakers.

CROSSOVER NETWORK: The network designed to divide audio frequencies into bands for distribution to speakers.

CRYOGENICS: The use of electronic circuits designed to take advantage of increased efficiency at extremely low temperatures.

CRYSTAL DIODE: Diode formed by small semiconductor crystal and cat whisker.

CRYSTAL LATTICE: Structure of material when outer electrons are joined in covalent bond.

CURRENT: Transfer of electrical energy in conductor by means of electrons moving constantly and changing positions in vibrating manner.

CUTOFF BIAS: Value of negative voltage applied to grid of tube which will cut off current flow through tube.

CW: Abbreviation for Continuous Wave.

CYBERNETICS: The study of complex electronic computer systems and their relationship to the human brain.

CYCLE: Set of events occurring in sequence. One complete reversal of an alternating current from positive to negative and back to starting point.

DAMPED WAVE: A wave in which successive oscillation decreases in amplitude.

DAMPING: Gradual decrease in amplitude of oscillations in tuned circuit, due to energy dissipated in resistance.

D'ARSONVAL METER: Stationary-magnet moving coil meter.

DBM: Loss or gain in reference to arbitrary power level of one milliwatt.

DC: Direct current.

DC COMPONENT: The dc value of ac wave which has axis other than zero.

DC GENERATOR: Generator with connections to armature through a commutator. Output is direct current.

DCWV: Abbreviation for Direct Current Working Voltage. It is a specification of a capacitor.

DEBUGGING: Process of correcting mistakes in a computer program.

DECAY: Term used to express gradual decrease in values of current and voltage.

DECAY TIME: The time required for a capacitor to discharge to a specified percentage of its original charge.

DECIBEL: One-tenth of a Bel. A unit used to express the relative increase or decrease in power. Unit used to express gain or loss in a circuit.

DECLINATION: Angle between true north and magnetic north.

DEFLECTION: Deviation from zero of needle in meter. Movement or bending of an electron beam.

DEFLECTION ANGLE: Maximum angle of deflection of electron beam in TV picture tube.

DEGENERATIVE FEEDBACK: Feedback 180 deg. out of phase with input signal so it subtracts from input.

DEMODULATION: Process of removing modulating signal intelligence from carrier wave in radio receiver.

DEPLETION LAYER (in a semiconductor): Region in which mobile carrier charge density is insufficient to neutralize net fixed charge of donors and acceptors. (IRE)

DEPOLARIZER: Chemical agent, rich in oxygen, introduced into cell to minimize polarization.

DETECTION: See DEMODULATION.

DETECTOR:

CRYSTAL. A type of detection which uses the rectification characteristics of a crystal substance such as galena, silicon, germanium, iron pyrite.

DIODE. A detector using a diode tube or semiconductor as the recitifer of the rf signal.

GRID LEAK. A detector in which rectification takes place in the grid circuit of a tube.

PLATE. A vacuum tube detector operated at a bias near cutoff so that detection takes place in the plate circuit of tube.

RADIATION. A device used to detect the presence and level of radiation.

RATIO. A detector for FM signals. It is based upon the ratio of output voltages of two diodes which is detected as the intelligence in the signal.

DEVIATION RATIO: A ratio between the maximum frequency deviation in FM to the highest modulating frequency.

DIAPHRAGM: Thin disc used in an earphone for producing sound.

DIELECTRIC: Insulating material between plates of capacitor.

DIELECTRIC CONSTANT: Numerical figure representing ability of dielectric or insulator to support electric flux. Dry air is assigned the number 1.

DIFFUSION: Movement of carriers across semiconductor junction in absence of external force.

DIGITAL INTEGRATED CIRCUIT: An IC which operates like a switch to give binary output (0 or 1).

DIODE: Two-element tube containing cathode and plate. Also, a solid state device usually designed to permit electron flow in one direction and block flow from the other direction.

DIODE DETECTOR: Detector circuit utilizing unilateral conduction characteristics of diode.

DIRECT CURRENT (dc): Flow of electrons in one direction.

DISTORTION: The deviations in amplitude, frequency and phase between input and output signals of amplifier or system.

AMPLITUDE. Distortion resulting from nonlinear operation of electron tube when peaks of input signals are reduced or cut off by either excessive input signal or incorrect bias.

FREQUENCY. Distortion resulting from signals of some frequencies being amplified more than others or when some frequencies are excluded.

PHASE. Distortion resulting from shift of phase of some signal frequencies.

DOMAIN THEORY: Theory concerning magnetism, assuming that atomic magnets produced by movement of planetary electrons around nucleus have strong tendency to line up together in groups. These groups are called "domains."

DONOR IMPURITY: Impurity added to semiconductor material which causes negative electron carriers.

DOPING: Adding impurities to semiconductor material.

DRY CELL: Nonliquid cell, which is composed of zinc case, carbon positive electrode and paste of ground carbon, manganese dioxide and ammonium chloride as electrolyte.

DYNAMIC CHARACTERISTICS: Characteristics of tube describing the actual control of grid voltage over plate current when tube is operating as an amplifier.

DYNAMIC PLATE RESISTANCE: See AC PLATE RESISTANCE.

DYNAMIC SPEAKER: Speaker which produces sound as result of reaction between fixed magnetic field and fluctuating field of voice coil.

EDISON EFFECT: Effect, first noticed by Thomas Edison, that emitted electrons were attracted to positive plate in vacuum tube.

EDDY CURRENTS: Induced current flowing in rotating core.

EDDY CURRENT LOSS: Heat loss resulting from eddy currents flowing through resistance of core.

EFFECTIVE VALUE: That value of alternating current of sine wave form that has equivalent heating effect of a direct current. ($.707 \times E_{peak}$)

EFFICIENCY: Ratio between output power and input power.

ELECTRODE: Elements in a cell.

ELECTRODYNAMIC SPEAKER: Dynamic speaker that uses electromagnetic fixed field.

ELECTROLYTE: Acid solution in a cell.

ELECTROLYTIC CAPACITOR: Capacitor with positive plate of aluminum and dry paste or liquid forms negative plate. Dielectric is thin coat of oxide on aluminum plate.

ELECTROMAGNET: Coil wound on soft iron core. When current runs through coil, core becomes magnetized.

ELECTROMOTIVE FORCE (EMF): Force that causes free electrons to move in conductor. Unit of measurement is the volt.

ELECTRON: Negatively charged particle.

ELECTRON TUBE: Highly evacuated metal or glass shell which encloses several elements.

ELECTRON VOLT: A measure of energy. It represents the energy acquired by an electron while passing through a potential of one volt.

ELECTROSTATIC FIELD: Space around charged body in which its influence is felt.

ELEMENT: One of the distinct kinds of substances which either singly or in combination with other elements, makes up all matter in the universe.

EMISSION: Escape of electrons from a surface.

EMISSION:

PHOTOELECTRIC. Emission of electrons as result of light striking surface of certain materials.

SECONDARY. Emission caused by impact of other electrons striking surface.

THERMIONIC. Process where heat produces energy for release of electrons from surface of emitter.

EMITTER: Element in a vacuum tube from which electrons are emitted. The CATHODE.

EMITTER:

MAJORITY. In a transistor, the semiconductor section, either P or N type, which emits majority carriers into the interelectrode region.

MINORITY. In a transistor, the semiconductor section, either P or N type, which emits minority carriers into the interelectrode region.

ENERGY: That which is capable of producing work.

ENERGY GAP: The energy range between the valence band and the conduction band in a semiconductor. See FORBIDDEN REGION.

EPITAXY: The physical placement of material on a surface.

EXCITATION: To apply a signal to an amplifier circuit; to apply energy to an antenna system; to apply current to energize the field windings of a generator.

EXECUTE CYCLE: This tells the computer to execute or carry out an instruction.

EXTRINSIC SEMICONDUCTOR: A semiconductor which depends upon impurities for its electrical properties.

FARAD: Unit of measurement of capacitance. A capacitor has a capacitance of one farad when a charge of one coulomb raises its potential one volt. $C = \dfrac{Q}{E}$

FEDERAL COMMUNICATONS COMMISSION (FCC): A federal

board having the power to regulate all electronic communication systems originating in the United States.

FEEDBACK: Transferring voltage from output of circuit back to its input.

FETCH CYCLE: Instruction word that computer memory sends to control unit.

FIELD MAGNETS: Electromagnets which make up field of motor or generator.

FILM RESISTORS: Resistors made from depositing resistive film on a ceramic tube with caps which slip over end of tube to form leads.

FILTER: Circuit used to attenuate specific band or bands of frequencies.

FLIP-FLOP: A multivibrator circuit having two stable (bistable) states and the signal is switched back and forth between them.

FLUX: Greek letter Φ, total number of lines of magnetic force.

FLUX DENSITY (symbol B): Number of lines of flux per cross-sectional area of magnetic circuit.

FORBIDDEN REGION: A region between the valence and conduction band of an atom.

FOSTER-SEELEY: Discriminator circuit used in detection of FM.

FREQUENCY: Number of cycles per second measured in hertz (Hz).

FREQUENCY BANDS: Abbreviations and ranges as follows.

vlf – Very low frequencies 10-30 KHz.
lf – Low frequencies 30-300 KHz.
mf – Medium frequencies 300-3000 KHz.
hf – High frequencies 3-30 MHz.
vhf – Very high frequencies 30-300 MHz.
uhf – Ultra high frequencies 300-3000 MHz.
shf – Super high frequencies 3000-30,000 MHz.
ehf – Extremely high frequencies 30,000-300,000 MHz.

FREQUENCY DEPARTURE: Instantaneous change from center frequency in FM as a result of modulation.

FREQUENCY DEVIATION: Maximum departure from center frequency at the peak of the modulating signal.

FREQUENCY DOUBLER: Amplifer stage in which plate circuit is tuned to twice the frequency of grid tank circuit.

FREQUENCY MODULATION (FM): Modulating transmitter by varying frequency of rf carrier wave at an audio rate.

FREQUENCY RESPONSE: Rating of device indicating its ability to operate over specified range of frequencies.

FREQUENCY SWING: The total frequency swing from maximum to minimum. It is equal to twice the deviation.

FULL-WAVE RECTIFIER: Rectifer circuit which produces a dc pulse output for each half cycle of applied alternating current.

FUNDAMENTAL: A sine wave that has the same frequency as complex periodic wave. Component tone of lowest pitch in complex tone. Reciprocal of period of wave.

FUSE: Safety protective device which opens an electric circuit if overloaded. Current above rating of fuse will melt fusible link.

GAIN: Ratio of output ac voltage to input ac voltage.

GALVANOMETER: Meter which indicates very small amounts of current and voltage.

GAMMA: Greek letter γ, represents the gain of a common-collector connected transistor. It is equal to the ratio of a change in emitter current to a change in base current while the collector voltage is constant.

GATE: A circuit which permits an output only when a predetermined set of input conditions are met.

GAUSS: Measure of flux density in lines per square centimetre.

GENERATOR: Rotating electric machine which converts mechanical energy to electric energy.

GENERATORS:

COMPOUND. Uses both series and shunt winding.

INDEPENDENTLY EXCITED. Field windings are excited by separate dc source.

SERIES. Field windings are in series with armature and load.

SHUNT. Field windings are connected across armature in shunt with load.

GERMANIUM: A rare, grayish-white, metallic chemical element. Symbol Ge; Atomic Wgt., 72.60; Atomic Number 32.

GILBERT: Unit of measurement of magnetomotive force. Represents force required to establish one maxwell in circuit with one Rel of reluctance.

GRID: Grid of fine wire placed between cathode and plate of an electron tube.

GRID BIAS: Voltage between the grid and cathode, usually negative.

GRID CURRENT: Current flowing in grid circuit of electron tube, when grid is driven positive.

GRID DIP METER: A test instrument for measuring resonant frequencies, detecting harmonics and checking relative field strength of signals.

HALF-WAVE RECTIFIER: Rectifier which permits one-half of an alternating current cycle to pass and rejects reverse current of remaining half-cycle. Its output is pulsating dc.

HARDWARE: Material components of a computer system (the machinery).

HARMONIC FREQUENCY: Frequency which is multiple of fundamental frequency. Example: If fundamental frequency is 1000 KHz, then second harmonic is 2 x 1000 KHz or 2000 KHz; third harmonic is 3 x 1000 or 3000 KHz, and so on.

HEATER: Resistance heating element used to heat cathode in vacuum tube.

HEAT SINK: Mass of metal used to carry heat away from component.

HENRY (H): Unit of measurement of inductance. A coil has one henry of inductance if an EMF of one volt is induced when current through inductor is changing at rate of one ampere per second.

HERTZ: The unit of measurement for frequency, named in honor of Heinrich Hertz who discovered radio waves. One hertz equals "one cycle per second."

HETERODYNE: The process of combining two signals of different frequencies to obtain the difference frequency.

HOLE: Positive charge. A space left by removed electron.

HOLE INJECTION: Creation of holes in semiconductor material by removal of electrons by strong electric field around point contact.

HORSEPOWER: 33,000 ft. lb. of work per minute or 550 ft. lb. of work per second equals one horsepower. Also, 746 watts = 1 hp.

HUM: Form of distortion introduced in an amplifier as a result of coupling to stray electromagnetic and electrostatic fields or insufficient filtering.

HYDROMETER: Bulb-type instrument used to measure specific gravity of a liquid.

HYSTERESIS: Property of a magnetic substance that causes magnetization to lag behind force that produces it.

HYSTERESIS LOSS: Energy loss in substance as molecules or domains move through cycle of magnetization. Loss due to molecular friction.

IC: Integrated circuit.

IMPEDANCE MATCHING: Sometimes called Z match; the matching of two different impedances to obtain maximum transfer of power.

IMPEDANCE (Z): Total resistance to flow of an alternating current as a result of resistance and reactance.

IMPURITY: Atoms within a crystalline solid which are foreign to the crystal. See DONOR and ACCEPTOR IMPURITY.

INDUCED CURRENT: Current that flows as result of an induced EMF.

INDUCED EMF: Voltage induced in conductor as it moves through magnetic field.

INDUCTANCE: Inherent property of electric circuit that opposes a change in current. Property of circuit whereby energy may be stored in a magnetic field.

INDUCTION MOTOR: An ac motor operating on principle of rotating magnetic field produced by out-of-phase currents. Rotor has no electrical connections, but receives energy by transformer action from field windings. Motor torque is developed by interaction of rotor current and rotating field.

INDUCTIVE CIRCUIT: Circuit in which an appreciable EMF is induced while current is changing.

INDUCTIVE REACTANCE (X_L): Opposition to an ac current as a result of inductance.

INDUCTOR: A coil or a component with the properties of inductance.

IN PHASE: Two waves of the same frequency are in phase when they pass through their maximum and minimum values at the same instant with the same polarity.

INPUT IMPEDANCE: The impedance of the input terminals of a circuit or device, with the input generator disconnected.

INPUT/OUTPUT UNIT (I/O): This unit feeds data and instructions into and out of the computer memory.

INSTANTANEOUS VALUE: Any value between zero and maximum depending upon instant selected.

INSULATORS: Substances containing very few free electrons and requiring large amounts of energy to break electrons loose from influence of nucleus.

INTEGRATED CIRCUIT (IC): A concentration of transistors, diodes, resistors and capacitors in a microminiature chip.

INTENSITY (H): Magnetizing force per unit length of magnetic circuit.

INTERELECTRODE CAPACITANCE: The capacitance between metal elements in an electron tube.

INTERMEDIATE FREQUENCY (if): Also called "difference signal." In a superheterodyne circuit of a radio receiver, all tuned incoming signals are converted to this single frequency (usually 455 KHz) for "one-dial" tuning.

INTERNAL RESISTANCE: Refers to internal resistance of source of voltage or EMF. A battery or generator has internal resistance which may be represented as a resistor in series with source.

INTERRUPTED CONTINUOUS WAVE (ICW): Continuous wave radiated by keying transmitter into long and short pulses of energy (dashes and dots) conforming to code such as Morse Code.

INTERSTAGE: Existing between stages, such as in interstage transformer between two stages of amplifiers.

INTRINSIC SEMICONDUCTOR: Semiconductor with electrical characteristics similar to a pure crystal.

INVERTER: A circuit that inverts the input signal (example: NOT gate).

ION: An atom which has lost or gained some electrons. It may be positive or negative depending on the net charge.

IONIZATION: An atom is said to be ionized when it has lost or gained one or more electrons.

IONIZATION POTENTIAL: Voltage applied to a gas-filled tube at which ionization occurs.

IONOSPHERE: Atmospheric layer from 40 to 350 miles above the earth, containing a high number of positive and negative ions.

IR DROP: See VOLTAGE DROP.

JOULE: Unit of energy equal to one watt-second.

JUNCTION DIODE: PN junction having unidirectional current characteristics.

JUNCTION TRANSISTOR: Transistor consisting of thin layer of N or P type crystal between P or N type crystals. Designated as NPN or PNP.

KEY: Manually operated switch used to interrupt rf radiation of transmitter.

KEYING: Process of causing CW transmitter to radiate an rf signal when key contacts are closed.

KILO: Prefix meaning one thousand times.

KILOGAUSS: One thousand gauss.

KILOWATT-HOUR (KWh): Means 1000 watts per hours. Common unit of measurement of electrical energy for home and industrial use. Power is priced by the KWh.

KIRCHOFF'S CURRENT LAW: At any junction of conductors in a circuit, algebraic sum of currents is zero.

KIRCHOFF'S LAW OF VOLTAGES: In simple circuit, algebraic sum of voltages around circuit is equal to zero.

LAGGING ANGLE: Angle current lags voltage in inductive circuit.

LAMBDA: Greek letter λ. Symbol for wavelength.

LAMINATIONS: Thin sheets of steel used in cores of transformers, motors and generators.

LARGE SCALE INTEGRATION (LSI): An integrated circuit with more than 100 gates on one chip.

LAWS OF MAGNETISM: Like poles repel; unlike poles attract.

LEAD ACID CELL: Secondary cell which uses lead peroxide and sponge lead for plates, and sulfuric acid and water for electrolyte.

LEADING ANGLE: Angle current leads voltage in capacitive circuit.

LEFT HAND RULE: A method, using your left hand, to determine polarity of an electromagnetic field or direction of electron flow.

LENZ'S LAW: Induced EMF in any circuit is always in such a direction as to oppose effect that produces it.

LIGHT EMITTING DIODE (LED): A PN junction (diode) that gives off light when biased in the forward direction.

LINEAR: In a straight line; a mathematical relationship in which quantities vary in direct proportion.

LINEAR AMPLIFIER: An amplifier whose output is in exact proportion to its input.

LINEAR DETECTOR: Detector using linear portions of characteristic curve on both sides of knee. Output is proportional to input signal.

LINEAR DEVICE: Electronic device or component whose current-voltage relation is a straight line.

LINEAR INTEGRATED CIRCUIT: Amplifier type ICs that control varying voltages.

LINES OF FORCE: Graphic representation of electrostatic and magnetic fields showing direction and intensity.

LOAD: Resistance connected across circuit which determines current flow and energy used.

LOADING A CIRCUIT: Effect of connecting voltmeter across circuit. Meter will draw current and effective resistance of circuit is lowered.

LOCAL OSCILLATOR: Oscillator in superheterodyne receiver, output of which is mixed with incoming signal to produce intermediate frequency.

LODESTONE: Natural magnet called a "leading stone" (lodestone) because it was used to determine direction.

LOGICAL FUNCTIONS: An expression referring to a definite state of condition.

AND FUNCTION. An output is obtained only with a combined group of input signals.

OR FUNCTION. An output is obtained with any one of a group of input signals.

NOT FUNCTION. An output is obtained only when there is no input signal.

MEMORY FUNCTION. An output is continually obtained unless an input signal is applied.

LOUDSPEAKER: Device to convert electrical energy into sound energy.

L PAD: A combination of two variable resistors, one in series and the other across the load, which is used to vary the output of an audio system and match impedances.

L-SECTION FILTER: Filter consisting of capacitor and an inductor connected in an inverted L configuration.

MAGNET: Substance that has the property of magnetism.

MAGNETIC AMPLIFIER: Transformer type device employing a dc control winding. Control current produces more or less magnetic core saturation, thus varying output voltage of amplifier.

MAGNETIC CIRCUIT: Complete path through which magnetic lines of force may be established under influence of magnetizing force.

MAGNETIC FIELD: Imaginary lines along which magnetic force acts. These lines emanate from N pole and enter S pole, forming closed loops.

MAGNETIC FLUX (symbol ϕ phi): Entire quantity of magnetic lines surrounding a magnet.

MAGNETIC LINES OF FORCE: Magnetic line along which compass needle aligns itself.

MAGNETIC MATERIALS: Materials such as iron, steel, nickel and cobalt which are attracted to magnet.

MAGNETIC PICKUP: Phono cartridge which produces an electrical output from armature in magnetic field. Armature is mechanically connected to reproducing stylus.

MAGNETIC SATURATION: This condition exists in magnetic materials when further increase in magnetizing force produces very little increase in flux density. Saturation point has been reached.

MAGNETIZING CURRENT: Current used in transformer to produce transformer core flux.

MAGNETOMOTIVE FORCE (F) (MMF): Force that produces flux in magnetic circuit.

MAGNETOSTRICTION: The effect of a change in dimension of certain elements when placed in a magnetic field.

MAGNET POLES: Points of maximum attraction on a magnet; designated as north and south poles.

MAJORITY CARRIER: Conduction through semiconductor as a result of majority of electrons or holes.

MATTER: Physical substance of common experience. Everything about us in made up of matter.

MAXIMUM POWER TRANSFER: Condition that exists when resistance of load equals internal resistance of source.

MAXIMUM VALUE: Peak value of sine wave either in positive or negative direction.

MAXWELL: One single line of magnetic flux.

MEDIUM SCALE INTEGRATION (MSI): An integrated circuit with approximately 31 to 100 gates.

MEGA: Prefix meaning one million times.

MEMORY UNIT: The computer's storage for data.

MHO: Unit of measurement of conductance (siemen).

MICA CAPACITOR: Capacitor made of metal foil plates separated by sheets of mica.

MICRO: Prefix meaning one millionth of.

MICROAMP (symbol μ A): One millionth of an ampere.

MICROFARAD: (μF): One millionth of a farad.

MICROHENRY (μH): One millionth of a henry.

MICROMHO: One millionth of a mho (siemen).

MICROMICRO: Prefix meaning one millionth of one millionth of.

MICROMICROFARAD ($\mu\mu$ F): One millionth of one millionth of a farad (same as picofarad, pF).

MICROPHONE: Energy converter that changes sound energy into corresponding electrical energy.

MICROPROCESSOR: A computer central arithmetic and logic unit and control unit on one integrated circuit chip.

MICROSECOND: One millionth of a second.

MIL: One thousandth of an inch (.001 in.).

MIL-FOOT: A wire which is one mil in diameter and one foot long.

MILLI: Prefix meaning one thousandth of.

MILLIAMMETER: Meter which measures in milliampere range of currents.

MILLIHENRY (mH): One thousandth of a henry.

MINORITY CARRIER: Conduction through semiconductor opposite to majority carrier. Example: If electrons are majority carrier, then holes are minority carrier.

MINUS (symbol $-$): Negative terminal of junction of circuit.

MISMATCH: Incorrect matching of load to source.

MOBILITY: The velocity of current carriers in a semiconductor. The ratio of the velocity of the carrier to the applied electric field, expressed as cm^2/volt-sec.

MODULATION: Process by which amplitude or frequency of sine wave voltage is made to vary according to variations of another voltage or current called modulation signal.

MODULATION PRODUCT: Sideband frequencies resulting from modulation of a radio wave.

MOLECULE: Smallest division of matter. If further subdivision is made, matter will lose its identity.

MOTOR: Device which converts electrical energy into mechanical energy.

MOTORS, DC:

COMPOUND. Uses both series and parallel field coils.

SERIES. Field coils are connected in series with armature circuit.

SHUNT. Field coils are connected in parallel with armature circuit.

MU: Greek letter (μ) used to represent the amplification factor of a vacuum tube; magnetic permeability; the prefix meaning one millionth of.

MULTIMETER: A combination volt, ampere and ohm meter.

MULTIPLEX: A method employed in radio and telephone communications where several messages are transmitted in one or both directions over a single transmission path.

MULTIPLIER: Resistance connected in series with meter movement to increase its voltage range.

MULTIVIBRATORS:

ASTABLE. A free-running multivibrator.

BISTABLE. A single trigger pulse switches conduction from one tube to the other.

CATHODE COUPLED. Both tubes have a common cathode resistor.

FREE-RUNNING. Frequency of oscillation depending upon value of circuit components. Continuous oscillation.

MONOSTABLE. One trigger pulse is required to complete one cycle of operation.

ONE SHOT. Same as MONOSTABLE.

PLATE COUPLED. The plates of the tubes and grids are connected by RC networks.

MUTUAL INDUCTANCE (M): When two coils are so located that magnetic flux of one coil will cause an EMF in the other, there is mutual inductance.

NATURAL MAGNET: Magnets found in natural state in form of mineral called Magnetite.

NEGATIVE ION: Atom which has gained electrons and is negatively charged.

NEGATIVE RESISTANCE: A condition in a circuit where an increase in voltage produces a decrease in current.

NETWORK: Two or more components connected in either series or parallel.

NEUTRON: Particle which is electrically neutral.

NICKEL CADMIUM CELL: Alkaline cell with paste electrolyte hermetically sealed. Used in aircraft.

NO LOAD VOLTAGE: Terminal voltage of battery or supply when no current is flowing in external circuit.

NONLINEAR DEVICE: Electronic device or component whose current-voltage relation is not a straight line.

NOT CIRUIT: An inverter gate circuit with output always opposite input.

NUCLEONICS: The branch of physics dealing with the science of small particles and the release of energy from the atom.

NUCLEUS: Core of the atom.

NULL INDICATOR: A meter designed to indicate the balance of a circuit; indicator for no current or no voltage in a circuit.

OERSTED: Unit of magnetic intensity equal to one gilbert per centimetre.

OHM (symbol Ω): Unit of measurement of resistance.

OHMMETER: Meter used to measure resistance in ohms.

OHM'S LAW: Mathematical relationship between current, voltage and resistance discovered by George Simon Ohm.

$$I = \frac{E}{R} \qquad E = IR \qquad R = \frac{E}{I}$$

OHMS PER VOLT: Unit of measurement of sensitivity in a meter.

OR CIRCUIT: A circuit with two or more inputs and one input signal must be present to produce an output signal.

OSCILLATOR: An electron tube generator of alternating current voltages.

OSCILLATORS:

ARMSTRONG. An oscillator using tickler coil for feedback.

COLPITTS. An oscillator using split tank capacitor as feedback circuit.

CRYSTAL-CONTROLLED. An oscillator controlled by piezo-electric effect.

ELECTRON COUPLED OSCILLATOR (ECO). Combination oscillator and power amplifier utilizing electron stream as coupling medium between grid and plate tank circuits.

HARTLEY. Oscillator using inductive coupling of tapped tank coil for feedback.

PUSH-PULL. Push-pull circuit utilizing interelectrode capacitance of each tube to feed back energy to grid circuit to sustain oscillations.

RC OSCILLATORS. Oscillators depending upon charge and discharge of capacitor in series with resistance.

OSCILLOSCOPE: Test instrument using cathode ray tube, permitting observation of signal.

OUTPUT: The energy from a circuit or device.

OVERMODULATION: Condition when modulating wave exceeds amplitude of continuous carrier wave, resulting in distortion.

PARALLEL CIRCUIT: Circuit which contains two or more paths for electrons supplied by common voltage source.

PARALLEL RESONANCE: Parallel circuit of an inductor and capacitor at frequency when inductive and capacitive reactances are equal. Current in capacitive branch is 180 deg. out of phase with inductive current and their vector sum is zero.

PARAMETER: A constant whose value varies depending upon application.

PARASITIC OSCILLATION: Oscillations in circuit resulting from circuit components or conditions, occurring at frequencies other than that desired.

PASSIVE DEVICE: Electronic component that does not have the ability to change its output response based on external input signal. Some common passive devices are resistors, capacitors and inductors.

PEAK: Maximum value of sine wave; the highest voltage current or power reached during a particular cycle or operating time.

PEAK INVERSE VOLTAGE (PIV): Value of voltage applied in reverse direction across diode.

PEAK INVERSE VOLTAGE RATING: The inverse voltage a diode will withstand without arcback.

PEAK-TO-PEAK: Measured value of sine wave from peak in positive direction to peak in negative direction.

PEAK VALUE: Maximum value of an alternating current or voltage.

PENTAGRID CONVERTER: Tube with five grids.

PENTAVALENT: Semiconductor impurity having five valence electrons. Donor impurities.

PENTODE: Electron tube with five elements including cathode, plate, control grid, screen grid and suppressor grid.

PERCENTAGE OF MODULATION: Maximum deviation from normal carrier value as result of modulation expressed as a percentage.

PERCENTAGE OF RIPPLE: Ratio of rms value of ripple voltage to average value of output voltage expressed as a percentage.

PERIOD: Time for one complete cycle.

PERMANENT MAGNET (PM): Bars of steel and other substances which have been permanently magnetized.

PERMEABILITY (symbol μ): Relative ability of substance to conduct magnetic lines of force as compared with air.

PHASE: Relationship between two vectors in respect to angular displacement.

PHASE INVERTER: Device or circuit that changes phase of a signal 180 deg.

PHASE SPLITTER: Amplifier which produces two waves that have exactly opposite polarities from single input wave form.

PHOTODIODE: A PN junction diode which conducts upon exposure to light energy.

PHOTOELECTRIC CELL: A cell which produces an electric potential when exposed to light.

PHOTOELECTRIC EFFECT: The property of certain substances to emit electrons when subjected to light.

PHOTOELECTRONS: Electrons emitted as a result of light.

PHOTON: A discrete quantity of electromagnetic energy; a quantum.

PHOTOSENSITIVE: Characteristic of material which emits electrons from its surface when energized by light.

PHOTOTUBE: Vacuum tube employing photo sensitive material as its emitter or cathode.

PHOTOVOLTAIC: The generation of a voltage at the junction of two materials when exposed to light.

PICOFARAD (pF): One millionth of one millionth of a farad.

PIEZOELECTRIC EFFECT: Property of certain crystalline substances of changing shape when an EMF is impressed upon crystal. Action is also reversible.

PI-SECTION FILTER: Filter consisting of two capacitors and an inductor connected in a π configuration.

PITCH: Property of musical tone determined by its frequency.

PLANETARY ELECTRONS: Electrons considered in orbit around the nucleus of an atom.

PLATE: Anode of vacuum tube. Element in tube which attracts electrons.

PLATE MODULATION: Modulation circuit where modulating signal is fed to plate circuit of modulated stage.

PLUS (symbol +): Positive terminal or junction of circuit.

PM SPEAKER: A speaker employing permanent magnet as its field.

PN JUNCTION: The line of separation between N type and P type semiconductor materials.

POINT CONTACT (in a semiconductor): A pressure contact between a semiconductor body and a metallic point.

POINT CONTACT DIODE: Diode consisting of point and a semiconductor crystal.

POLARITY: Property of device or circuit to have poles such as north and south or positive and negative.

POLARIZATION: Defect in cell caused by hydrogen bubbles surrounding positive electrode and effectively insulating it from chemical reaction.

POLES: Number of poles in motor or generator field.

POSITIVE ION: Atom which has lost electrons and is positively charged.

POT: Abbreviation for potentiometer.

POWER: Rate of doing work. In dc circuits, P = I x E.

POWER AMPLIFICATION: Ratio of output power to input grid driving power.

POWER FACTOR: Relationship between true power and apparent power of circuit.

POWER SUPPLY: Electronic circuit designed to provide various ac and dc voltages for equipment operation. Circuit may include transformers, rectifiers, filters and regulators.

POWER TRANSISTOR: Transistors designed to deliver a specified output power level.

PREAMPLIFIER: Sensitive low-level amplifier with sufficient output to drive standard amplifier.

PREFIXES: A word united or joined to the beginning of another word to change its meaning. The approved multiples prefixes are shown in the following chart.

10^{12}	tera-	10^2	hekto-	10^{-3}	milli-
10^9	giga-	10	deka-	10^{-6}	micro-
10^6	mega-	10^{-1}	deci-	10^{-9}	nano-
10^4	myria-	10^{-2}	centi-	10^{-12}	pico-
10^3	kilo-				

PRIMARY CELL: Cell that cannot be recharged.

PRIMARY WINDING: Coil of transformer which receives energy from ac source.

PROBE: A pointed metal end of a test lead, designed to contact specific points in a circuit to be measured.

PROGRAMMED READ-ONLY MEMORY (PROM): An alterable or programmable read-only memory.

PROTON: Positively charged particle.

PULSE: Sudden rise and fall of a voltage or current.

CODE. A train of pulses used to transmit information.

MODULATION. Use of pulse code to modulate a transmitter.

DECAY TIME. The time required for a pulse to drop from 90 percent to 10 percent of its maximum amplitude.

DROOP. A slight decrease in amplitude of flat top square pulse; a form of square wave distortion.

DURATION. The time interval between the first and last instants at which the instantaneous amplitude reaches a stated fraction of the peak pulse amplitude.

RISE TIME. The time required for a pulse to rise from 10 percent to 90 percent of its maximum amplitude.

REPETITION RATE. The frequency of the periodic pulse train.

REPETITION PERIOD. The reciprocal of the repetition frequency.

SPIKE. An unwanted pulse of relatively short duration superimposed on a main pulse.

SYNC. A pulse sent by a TV transmitter to synchronize the scanning of the receiver with the transmitter; a pulse used to maintain predetermined speed and/or phase relations.

PUSH-PULL AMPLIFIER: Two tubes or transistors used to amplify the signal in such a manner that each amplifies one half-cycle of signal. Tubes or transistors operate 180 deg. out of phase.

Q: Letter representation for quantity of electricity (coulomb). Letter symbol for transistor.

Q: Quality, figure of merit; ratio between energy stored in inductor during time magnetic field is being established to losses during same time. $Q = \dfrac{X_L}{R}$

QUANTA: Definte amount of energy required to move an electron to higher energy level.

QUANTUM: Discrete quantities of any physical property such as momentum, energy, mass, etc.

QUIESCENT: At rest. Inactive.

RADIAN: An angle, with its vertex at the center of a circle, will intercept an arc of the circumference equal to the radius of the circle. Circumference of a circle is equal to 2π radians. A radian is approximately 57 deg.

RADIO DETECTOR: Type of FM detector.

RADIO FREQUENCY CHOKE (rfc): Coil which has high impedance to rf currents.

RADIO SPECTRUM: Division of electromagnetic spectrum used for radio.

RADIO WAVE: A complex electrostatic and electromagnetic field radiated from a transmitter antenna.

RANDOM ACCESS MEMORY (RAM): Read/write memories where data can be stored, then retrieved again.

REACTANCE (X): Opposition to alternating current as result of inductance or capacitance.

REACTIVE POWER: Power apparently used by reactive component of circuit.

READ: To transmit data out from the memory to some other peripheral computer device.

READ-ONLY MEMORIES (ROM): A computer memory where the data can be continuously read out.

RECIPROCAL: Reciprocal of number is one divided by the number.

RECTIFIER: Component or device used to convert ac into a pulsating dc.

REED RELAY: Relay made up of two reeds (switch contacts) that are hermetically sealed in a glass capsule filled with an inert gas.

REGENERATIVE FEEDBACK: Feedback in phase with input signal so it adds to input.

REGISTER: Temporary, short term storage circuits in a computer that have the capacity for storing one computer word.

REGULATION: Voltage change that takes place in output of generator or power supply when load is changed.

REGULATION, PERCENTAGE OF: Percentage of change in voltage from no-load to full-load in respect to full-load voltage.

Expressed as: $\dfrac{E_{no\ load} - E_{full\ load}}{E_{full\ load}}$

REJECT CIRCUIT: Parallel tuned circuit at resonance. Rejects signals at resonant frequency.

REL: Unit of measurement of reluctance.

RELAY: Magnetic switch.

RELAXATION OSCILLATOR: Nonsinusoidal oscillator whose frequency depends upon time required to charge or discharge capacitor through resistor.

RELUCTANCE: Resistance to flow of magnetic lines of force.

REPULSION-START MOTOR: A motor which develops starting torque by interaction of rotor currents and single-phase stator field.

RESIDUAL MAGNETISM: Magnetism remaining in material after magnetizing force is removed.

RESISTANCE: Quality of electric circuit that opposes flow of current through it.

RESONANT FREQUENCY: Frequency at which tuned circuit oscillates. (See TUNED CIRCUIT.)

RETENTIVITY: Ability of material to retain magnetism after magnetizing force is removed.

RETRACE: Process of returning scanning beam to starting point after one line is scanned.

RIPPLE VOLTAGE: The ac component of dc output of power supply due to insufficient filtering.

RMS VALUE: Abbreviation for root mean square value.

ROOT MEAN SQUARE VALUE: The same as effective value, .707 x E_{peak}

ROTOR: Rotating part of an ac generator.

ROWLAND'S LAW: Law for magnetic circuits which states that number of lines of magnetic flux is in direct proportion to magnetomotive force and inversely proportional to reluctance of circuit. $\Phi = \dfrac{F}{R}$

SAWTOOTH GENERATOR: Electron tube oscillator producing sawtooth wave.

SAWTOOTH WAVE: Wave shaped like the teeth of a saw.

SCHEMATIC: Diagram of electronic circuit showing electrical connections and identification of various components.

SCREEN GRID: Second grid in electron tube between grid and plate, to reduce interelectrode capacitance.

SECONDARY CELL: Cell that can be recharged by reversing chemical action with electric current.

SECONDARY EMISSION: Emission of electrons as result of electrons striking plate of electron tube.

SECONDARY WINDING: Coil which receives energy from primary winding by mutual induction and delivers energy to load.

SECOND HARMONIC DISTORTION: Distortion of wave by addition of its second harmonic.

SELECTIVITY: Relative ability of receiver to select desired signal while rejecting all others.

SELF-INDUCTANCE: EMF is self-induced when it is induced in conductor carrying current.

SEMICONDUCTOR: Conductor with resistivity somewhere in range between conductors and insulators.

SEMICONDUCTOR, N TYPE: Semiconductor which uses electrons as majority carrier.

SEMICONDUCTOR, P TYPE: Semiconductor which uses holes as majority carrier.

SENSITIVITY: Ability of circuit to respond to small signal voltages.

SENSITIVITY OF METER: Indication of loading effect of meter. Resistance of moving coil and multiplier divided by voltage for full scale deflection. Sensitivity equals one divided by current required for full scale deflection. Ex: A 100 μA meter movement has sensitivity of: $\dfrac{1}{.0001}$ or 10,000 ohms/volt.

SERIES CIRCUIT: Circuit which contains only one possible path for electrons through circuit.

SERIES PARALLEL: Groups of series cells with output terminals connected in parallel.

SERIES RESONANCE: Series circuit of inductor, capacitor and resistor at a frequency when inductive and capacitive reactances are equal and cancelling. Circuit appears as pure resistance and has minimum impedance.

SHADED POLE MOTOR: Motor in which each of its field poles is split to accomodate a short-circuit copper strap called a shading coil. This coil produces a sweeping movement of field across pole face for starting.

SHIELD: Partition or enclosure around components in circuit to minimize effects of stray magnetic and radio frequency fields.

SHIFT REGISTER: A special type of register in which data can be moved (or shifted) to the left or to the right.

SHORT CIRCUIT: Direct connection across source which provides zero resistance path for current.

SHUNT: To connect across or parallel with circuit or component.

SHUNT: Parallel resistor to conduct excess current around meter moving coil. Shunts are used to increase range of meter.

SIDEBANDS: Frequencies above and below carrier frequency as result of modulation.

LOWER. Frequencies equal to difference between carrier and modulating frequencies.

UPPER. Frequencies equal to carrier plus modulating frequencies.

SIDE CARRIER FREQUENCIES: Waves of frequencies equal to sum and difference between carrier wave frequency and modulating wave frequency.

SIEMEN: Unit of measurement of conductance (replaces mhos).

SIGNAL: The intelligence, message or effect to be sent over a communications system; an electrical wave corresponding to intelligence.

SINE WAVE: A wave form of a single frequency alternating current. A wave whose displacement is the sine of an angle proportional to time or distance.

SINGLE ENDED AMPLIFIER: An amplifier whose final power stage is a single vacuum tube or transistor.

SINGLE PHASE MOTOR: Motor which operates on single phase alternating current.

SLIP RINGS: Metal rings connected to rotating armature windings in generator. Brushes sliding on these rings provide connections for external circuit.

SMALL SCALE INTEGRATION (SSI): An integrated circuit with approximately 1 to 30 gates.

SOFTWARE: Stored program of a computer.

SOLENOID: Coil of wire carrying electric current and possessing characteristics of a magnet.

SPACE CHARGE: Cloud of electrons around cathode of an electron tube.

SPACISTOR: A four element semiconductor, similar to a transistor. It uses a space charge region. Its main advantage is its adaptability to ultra high frequencies.

SPECIFIC GRAVITY: Weight of liquid in reference to water which is assigned value of 1.0.

SPLIT PHASE MOTOR: Single-phase induction motor which

develops starting torque by phase displacement between field windings.

SQUARE MIL: The cross-sectional area of a conductor one mil square.

SQUIRREL CAGE ROTOR: Rotor used in an induction motor made of bars placed in slots of rotor core and all joined together at ends.

STABILITY: The ability to stay on a given frequency or in a given state without undesired variation.

STAGE: Section of an electronic circuit, usually containing one electron tube and associated components.

STATIC CHARACTERISTICS: Characteristics of tube taken with constant plate voltage.

STATIC CHARGE: Charge on body either negatively or positively.

STATIC ELECTRICITY: Electricity at rest as opposed to electric current.

STATOR: Stationary coils of an ac generator.

STEADY STATE: Fixed, nonvarying condition.

STORAGE BATTERY: Common name for lead-acid battery used in automotive equipment.

STYLUS: Phonograph needle or jewel, which follows grooves in a record.

SUBHARMONIC: A frequency below harmonic, usually fractional part of fundamental frequency.

SUBSONIC: A frequency below the audio frequency range; infrasonic.

SUPERSONIC: Frequencies above audio frequency range.

SUPERHETERODYNE: Radio receiver in which incoming signal is converted to fixed intermediate frequency before detecting audio signal component.

SUPPRESSOR GRID: Third grid in electron tube, between screen grid and plate, to repel or suppress secondary electrons from plate.

SURFACE ALLOY TRANSISTOR: A silicon junction transistor, in which aluminum electrodes are deposited in shallow pits etched on both sides of a thin silicon crystal, forming P regions.

SURFACE BARRIER TRANSISTOR: A transistor so constructed that the interfaces performing the collection and emission of carriers are located at the surface of the semiconductor crystal.

SWEEP CIRCUIT: Periodic varying voltage applied to deflection circuits of cathode ray tube to move electron beam at linear rate.

SWITCH: Device for directing or controlling current flow in circuit.

SYNCHRONOUS: Having the same period or frequency.

SYNC PULSE: Abbreviation for synchronization pulse, used for triggering an oscillator or circuit.

TANK CIRCUIT: Parallel resonant circuit.

TAP: Connection made to a coil at a point other than its terminals.

TAPE TRANSPORT: The driving mechanism and reels of a tape recorder.

TELEMETRY: The transmission of measurements recorded on instruments by means of radio or telephone to a remote point.

TELEVISION: Method of transmitting and receiving visual scene by radio broadcasting.

TELEVISION CHANNEL: Allocation in frequency spectrum of 6 MHz assigned to each television station for transmission of picture and sound information.

TETRODE: Electron tube with four elements including cathode, plate, control grid and screen grid.

TETRODE TRANSISTOR: A transistor with four elements, usually using either two emitters or two bases.

THERMAL RUNAWAY: In transistor, regenerative increase in collector current and junction temperature.

THERMISTOR: Semiconductor device which changes resistivity with change in temperature.

THETA (θ): Angle of rotation of vector representing selected instants at which sine wave is plotted. Angular displacement between two vectors.

THREE-PHASE ALTERNATING CURRENT: Combination of three alternating currents have their voltages displaced by 120 deg. or one-third cycle.

THYRISTOR: A solid state device that has bistable electrical characteristics similar to those of thyratron tubes. Three thyristor devices are diacs, silicon-controlled rectifiers and triacs.

TICKLER: Coil used to feed back energy from output to input circuit.

TIME CONSTANT (RC): Time period required for the voltage of a capacitor in an RC circuit to increase to 63.2 percent of maximum value or decrease to 36.7 percent of maximum value.

TONE CONTROL: Adjustable filter network to emphasize either high or low frequencies in output of audio amplifier.

TRANSCEIVER: A combined transmitter and receiver.

TRANSCONDUCTANCE (symbol g_m): Grid plate transconductance of vacuum tube expressed as ratio of small change in plate current to small change in grid voltage while plate voltage is held constant. Measured in mhos (siemens).

TRANSDUCER: Device by which one form of energy may be converted to another form, such as electrical, mechanical or acoustical.

TRANSFER CHARACTERISTIC: Relation between input and output characteristics of device.

TRANSFORMER: Device which transfers energy from one circuit to another by electromagnetic induction.

TRANSFORMERS:

ISOLATION. Transformer with one-to-one turns ratio.

STEP-DOWN. Transformer with turns ratio greater than one. The output voltage is less than input voltage.

STEP-UP. Transformer with turns ratio of less than one. Output voltage is greater than input voltage.

TRANSIENT RESPONSE: Response to momentary signal or force.

TRANSISTOR: Name of semiconductor device derived from two words, transfer and resistor.

TRANSISTOR SOCKET: A small device in which the leads of a transistor are placed to provide ease in connection to circuit and to permit replacement of transistor.

TRANSMISSION LINE: Wire or wires used to conduct or guide electrical energy.

TRANSMITTER: Device for converting intelligence into electrical impulses for transmission through lines or through space from radiating antenna.

TRIAC: A full-wave, silicon-controlled switch.

TRIMPOT: The trade name for a precision variable resistor manufactured by Bourns.

TRIODE: Three-element vacuum tube, consisting of cathode, grid and plate.

TRIVALENT: Semiconductor impurity having three valence electrons. Acceptor impurity.

TRUE POWER: Actual power absorbed in circuit.

TRUTH TABLE: In solid state circuitry, a table of input signals to see if an output signal resulted.

TUNE: The process of bringing a circuit into resonance by adjusting one or more variable components.

TUNED AMPLIFIER: Amplifier employing tuned circuit for input and/or output coupling.

TUNED CIRCUIT: Circuit containing capacitance, inductance and resistance in series or parallel, which when energized at specific frequency known as its resonant frequency, an interchange of energy occurs between coil and capacitor.

TURNS RATIO: Ratio of number of turns of primary winding of transformer to number of turns of secondary winding.

TWEETER: A high frequency speaker.

UNIT MEASUREMENT: The standard increments of measurement or quantity used as a means of comparison to other quantities.

UNITY COUPLING: If two coils are positioned so all lines of magnetic flux of one coil will cut across all turns of second coil, it is called UNITY COUPLING.

UNIVERSAL MOTOR: Series ac motor which operates also on dc. Fractional horsepower ac-dc motor.

UNIVERSAL TIME CONSTANT CHART: A graph with curves representing growth and decay of voltages and currents in RC and RL circuits.

VALENCE: The chemical combining ability of an element in reference to hydrogen. The capacity of an atom to combine with other atoms to form molecules.

VALENCE BAND: The energy level of an atom closest to the nucleus.

VAR: The unit of measurement of reactive power. It is derived from Volt-Ampere-Reactive.

VARISTOR: A two-electrode semiconductor device having a voltage-dependent nonlinear resistance.

VECTOR: Straight line drawn to scale showing direction and magnitude of a force.

VECTOR DIAGRAM: Diagram showing direction and magnitude of several forces, such as voltage and current, resistance, reactance and impedance.

VOICE COIL: Small coil attached to speaker cone, to which the signal is applied. Reaction between field of voice coil and fixed magnetic field causes mechanical movement of cone.

VOLT (symbol E in electricity, symbol V in semiconductor circuits): Unit of measurement of electromotive force or potential difference.

VOLTAGE DIVIDER: Tapped resistor or series resistors across source voltage to produce multiple voltages.

VOLTAGE DOUBLER: Rectifier circuit which produces double the input voltage.

VOLTAGE DROP: Voltage measured across resistor. Voltage drop is equal to product of current times resistance in ohms. $E = IR$.

VOLTAGE MULTIPLIER: Rectifier circuits which produce output voltage at multiple greater than input voltage, usually doubling, tripling or quadrupling.

VOLTAIC CELL: Cell produced by suspending two dissimilar elements in acid solution. Potential difference is developed by chemical action.

VOLT-AMPERE: Measurement unit of apparent power.

VOLTMETER: Meter used to measure voltage.

VOM: A common test instrument which combines a voltmeter, ohmmeter and milliammeter in one case.

VTVM: Vacuum tube volt meter.

VU: Number numerically equal to number of decibels above or below reference volume level. Zero Vu represents power level of one milliwatt dissipated in 600 ohm load or voltage of .7746 volts.

WATT: Unit of measurement of power.

WATT-HOUR: Unit of energy measurement, equal to one watt per hour.

WATT-HOUR METER: Meter which indicates instantaneous rate of power consumption of device or circuit.

WATTLESS POWER: Power not consumed in an ac circuit due to reactance.

WATTMETER: Meter used to measure power in watts.

WAVE: A disturbance in a medium which is a function of time and space or both. Energy may be transmitted by waves. Example: Audio and radio waves.

WAVE FORM: The shape of a wave derived from plotting its instantaneous values during a cycle against time.

WAVE LENGTH: Distance between point on loop of wave to corresponding point on adjacent wave.

WHEATSTONE BRIDGE: Bridge circuit used for precision measurement of resistors.

WOOFER: A large speaker designed for reproduction of low frequency sounds. It is used in high fidelity amplifier systems.

WORK: When a force moves through a distance, work is done. It is measured in foot-pounds, Work = Force x Distance.

WORKING VOLTAGE: Maximum voltage that can be steadily applied to a capacitor without danger of arc-over.

WRITE: To transmit data into a memory from some outside computer device.

X-AXIS: The horizontal axis of a graph.

Y-AXIS: The vertical axis of a graph.

Y-AXIS: Axis drawn perpendicular to faces of hexagonal crystal.

Y-CUT CRYSTAL: Crystal cut perpendicular to faces of hexagonal crystal.

Z-AXIS: Optical axis of crystal.

ZENER DIODE: A silicon diode which makes use of the breakdown properties of a PN junction. If a reverse voltage across the diode is progressively increased, a point will be reached when the current will greatly increase beyond its normal cut-off value. This voltage point is called the zener voltage.

ZENER VOLTAGE: The reverse voltage at which the breakdown occurs in a zener diode.

ZERO REFERENCE LEVEL: Power level selected as reference for computing gain of amplifier or system.

INDEX

Software, computer, 325, 331
Solenoid, 67
Sources of electricity, 18
Spacing between plates, capacitors, 133
Speakers, 244, 245
Specific gravity, 25
Specific resistance, 38
Speed, induced voltage, 78, 79
Squirrel cage rotor, 167
Stability factor, transistor circuit, 215, 216
Staircase wave form, 297
Standard abbreviations and symbols, 342
Star connected stator coils, 86, 87
Starters, dc motor, 164, 165
Starting switch, magnetic, 165
Starting windings, 168
Stator windings, 85
Step-down transformer, 120
Step-up transformer, 120
Stereo
  head phones, 246
  receiver, 246
  speakers, 245, 246
Strength of field, induced voltage, 79
Structure of matter, 7, 8
Surge current, diodes, 176
Switches, 57, 58
Switching, 57, 58, 294
  and logic circuits, 294
Symbols, 342
  electronic, 340, 341
  integrated circuits, 313, 314, 315

T

Tank circuit,
  parallel resonance, 153
  oscillator, 251
Theory of energy, 7
Theory of magnetism, 64, 65
Thermocouple, 29
Thermopile, 29
Thin film resistors, 43
Thomson, Sir Joseph John, 8

Three-phase generators, 85
Tickler coil, 252
Time constant, RL, 112, 113
Trailing edge, pulse, 296
Transformers, 108, 118, 178
  coupling, 232, 233
  efficiency and ratings, 123
  losses, 121
  matching, 122, 123
  theory, 119
  voltages, 119, 120
Transistor, 194, 195
  amplifier, 199, 210
  as a switch, 294, 295, 296
  biasing, 196
  circuits, voltage measurements, 203, 204
  current gain, 198
  current paths, 197
  field effect, FET, 221, 222
  potential hills, 196
  detection, 276, 277
  power amplifiers, 228
  symbols, 342
  tester project, 207, 208
  unijunction, 222, 223
Transistorized
  alternator regulation, 87, 88
  ignition systems, 302, 303
  oscillators, 250
  radio, 277
  superheterodyne receiver, 277, 278
  transmitters, 272
Transmitters, transistorized, 272
Trapezoidal wave form 297
Triacs, 190
Triangular wave form, 297
Trigger level, oscilloscope, 103
Trigonometric functions, 346
Trigonometry, 345, 346
Trimmer capacitor, 130
Truth tables, 298, 299, 300, 301, 302
Tuning series RCL circuit, 150
Tunnel diode, 256
  oscillator, 256, 257, 258

Turns ratio, transformer, 120
Turntable, 246
Two-phase generators, 85
Two unequal resistors in parallel, 53, 54

U

Unequal resistors in parallel, 53, 54, 55
Unijunction transistors, 222, 223
Unity coupling, 109
Universal motor, 165
Universal time constant chart, 135

V

Valence, 11
Valence band, 32, 33
Variable capacitor, 130
Variable resistors, 42, 43
Vertical position control, oscilloscope, 102
Volt-Ohm-Ammeter (VOM), 99, 100
Volta, Count Alessandro, 19
Voltage and continuity tester project, 103, 104
Voltage coil, 71
Voltage dividers, 186, 187
Voltage doublers, 181
Voltage drop polarities, 47, 48, 49
Voltage gain, transistor amplifiers, 218
Voltage measurements in transistor circuits, 203, 204
Voltage regulation, 187, 188
Voltage, ripple, 182
Voltmeter, 94, 95, 96
Volts, 45, 46

W

Watt, James, 51
Wave forms, 297, 298
Wavelength, 267, 268
  sine wave, 82
Weber's molecular theory of magnetism, 64, 65
Wheatstone bridge circuits, 56, 57
Work, 18, 19

Z

Zener diode regulation, 188, 189
Zener voltage, 176
Zero beating, 269